COMPUTER ARCHITECTURE AND ORGANIZATION

Pankaj Sharma
(Sr. Asstt. Proff.)
Department of Information Technology
ABES, Ghaziabad

S.K. KATARIA & SONS®
Publisher of Engineering Books
4885/109, Prakash Mahal, Dr. Subhash Bhargav Lane,
Opposite Delhi Medical Association, Daryaganj, New Delhi–110002 (INDIA)
Phone: +91-11-43551243; *Mobile:* +91-9871775858, 7678425209
e-mail: katariabook@gmail.com, katariabooks@yahoo.com
Website: www.skkatariaandsons.com
(e-books, subscriptions and bulk buying also available on our website)

Published by:

S.K. Kataria & Sons®

4885/109, Prakash Mahal, Dr. Subhash Bhargav Lane,
Opposite Delhi Medical Association, Daryaganj, New Delhi-110002 (INDIA)
Phone: +91-11-43551243, *Mobile:* +91-9871775858, 7678425209
e-mail: katariabook@gmail.com, katariabooks@yahoo.com
Website: www.skkatariaandsons.com

First Edition: January, 2011, Reprint: 2026

Price: ₹ 595/-

Laser Typesetting by :
PLUS COMPUTERS, Shahdara, Delhi-110093.

Printed at :
Arora Printing Press, Delhi–110051

Preface to the First Edition.....

Computer Architecture and Organisation is a core subject for undergraduate students and Postgradutate students in EC Branch. This book is primarily intended to serve as a textbook in accordance with the syllabus of Computer Architecture and Organization offered by various universities in India as well as abroad.

In this book, a significant effort has been made to find simple ways to develop theoretical as well as Practical concepts of Web Technology. A very strong emphasis has been given on Practical examples. Neat and Clean diagrams have been used for explanation.

Every effort has been made to make the concepts simple and comprehensive. This book is divided into twelve chapters :

The organisation of these chapters are :

Chapter 1 Introduces the Introduction of Design Methodology of Digital System.

Chapter 2 Describes the Processor Level Design and Processor Basic.

Chapter 3 Describes the Representation of Data and Information.

Chapter 4 Describes the Basic Concept of Instruction Set.

Chapter 5 Describes the Design of Data Path.

Chapter 6 Describes the Design of Control Path.

Chapter 7 Describes the Organization and Types of Memory Chips.

Chapter 8 Describes the Various Input Output Devices and Organization.

Chapter 9 Introduces the Concept of 8085 Microprocessor.

The author hopes that the book will fulfill the need of readers and would welcome any suggestions towards the improvement of the book.

Pankaj Sharma
ABES, Ghaziabad

Acknowledgement.....

I get this opportunity to express my gratitude and thanks to the respected Director (administration) of ABES Engg. College, **Dr. S.K. Kansal** for his valuable technical suggestion and constant encouragement, without which this book would not have to come into existence.

I express kind gratitude of **Prof. R.G. Mendiratta**, Director (Academic) ABES Engg. College, Ghaziabad.

I wish to thank **Prof. C.L. Mittal** (Dean UG) ABES Engg. College, Gzb. I also like to remember the motivation initiated by **Shri M.D Punia** my father-in-Law, (Lead District Manager, PNB, Karnal).

I am grateful to **Prof. V.K. Srivastava,** Director of ABES, IT, Ghaziabad, for helping me through his insightful suggestions and construction inputs.

I wish to thank **Prof. Pradeep Sharma**, Head of Department of Information Technology, Department of ABES Engg. College, Ghaziabad, for their constant co-operation, directly inspiration and encourangement.

I like to remember the motivation initiated by **Prof. D. Pandey** (HOD CSE) and Prof. Sunil Sharma (AHODCSE), ABES, Engg. College. I am also thankful to Prof. K.P. Jayant (HODMCA) for his inspiration.

I thank to my elder brother **Mr. Sanjay Sharma** (KIET), Ghaziabad for inspiration and encourangement.

I thank to **Mr. Aswin Perti,** Lecturer in IT Department for his technical suggestions.

I am grateful to my younger Brother **Sandeep** and daughter **Prisha** for inspiring me for this project.

I deeply express my heartful thanks to the **Mr. Sanjeev Kataria** for **M/s S.K. Kataria & Sons** for publishing this book in such a beautiful getup and well in time.

Lastly I am grateful to **Mr. Manoj Tyagi** and his staff from **Plus Computer's** for thier hard work.

Pankaj Sharma

Dedicated This book to

My Parents, Wife Priyanka, Daughter Prisha, Pihu

and

Readers of this book

Contents.....

Chapter-4

INSTRUCTION SET 111-145

Chapter-5

DATA PATH DESIGN 146-188

Chapter-6

Chapter-7

Chapter-8

Chapter

1

Introduction of Design Methodology

1.0 INTRODUCTION

In computer science and computer engineering, **computer architecture or digital computer organization** is the conceptual design and fundamental operational structure of a computer system. It's a blueprint and functional description of requirements and design implementations for the various parts of a computer, focusing largely on the way by which the Central Processing Unit (CPU) performs internally and accesses addresses in memory. It may also be defined as the science and art of selecting and interconnecting hardware components to create computers that meet functional, performance and cost goals.

Chapter Outline

- ➤ Introduction
- ➤ System Design – System Representation
- ➤ Design Process
- ➤ Gate Level Design
- ➤ Register Level Design
- ➤ PLD
- ➤ Role of Computer Architecture
- ➤ Instruction Set Architecture
- ➤ Computer Organization
- ➤ Design Goals
- ➤ Performance Analysis and Benchmarking
- ➤ Demorgon's Equivalent Symbols
- ➤ Register Level Model of the 74 × 181 ALU
- ➤ Basic Computer of Register Level Design
- ➤ Data Path Unit
- ➤ Using a ROM as PLD

What is a system?

- **From Saltzer & Kaashoek:** A system is a set of interconnected components that has a specified behavior observed at the interface with its environment
- **Components/Environmesnt:** What is and is not part of the system
- **Interface:** Points of interaction between the system and the environment
- **Computer System:** A system intended to store,process, or communicate information

Identifying a system's components

- **Purpose:** What does the system accomplish?

- E.g., your home
- **Granularity:** Which details must be exposed, and which can be hidden?
- E.g., the Internet

Before discussing a system, make sure everyone agrees on what the system actually is

Why build systems?

- **Speed:** SETI, biological simulations, decryption
- **Fault Tolerance:** Power outages, bad components
- **Reliability:** System does not "forget" operations it told you were completed
- **Protection against security threats:** Intentional (Denial of Service Attacks), unintentional (file deletes)
- **Availability:** Anytime, anywhere access
- **Robustness:** Failures to be managed gracefully
- **Maintainability:** Repairs, scheduled maintenance, upgrades, enhancements don't bring down the system
- **Scalability:** A linear increase in the system's load does not produce an exponential decrease in performance

What makes a system hard to build?

- Complexity-lack of understanding of the system's behavior. No absolute measurement so we look for symptoms of complexity:
- Large number of components
- Large number of interactions between components
- Small amount of regularity
- Difficult to describe system methodically
- More than one person required to understand the system

1.1 SYSTEM DESIGN - SYSTEM REPRESENTATION

Computer architecture comprises at least three main subcategories:

- **Instruction set architecture:** Or ISA, is the abstract image of a computing system that is seen by a machine language (or assembly language) programmer, including the instruction set, word size, memory address modes, processor registers, and address and data formats.
- **Microarchitecture:** Also known as computer organization is a lower level, more concrete and detailed, description of the system that involves how the constituent parts of the system are interconnected and how they interoperate in order to implement the ISA.[2] The size of a computer's cache for instance, is an organizational issue that generally has nothing to do with the ISA.
- **System design:** Which includes all of the other hardware components within a computing system such as:
    1. System interconnects such as computer buses and switches
    2. Memory controllers and hierarchies
    3. CPU off-load mechanisms such as direct memory access (DMA)

Once both ISA and microarchitecture have been specified, the actual device needs to be designed into hardware. This design process is called the *implementation*. Implementation is usually not considered architectural definition, but rather hardware design engineering.

Implementation can be further broken down into three (not fully distinct) pieces:

- **Logic Implementation:** Design of blocks defined in the microarchitecture at (primarily) the register-transfer and gate levels.
- **Circuit Implementation:** Transistor-level design of basic elements (gates, multiplexers, latches, etc.) as well as of some larger blocks (ALUs, caches, etc.) that may be implemented at this level, or even (partly) at the physical level, for performance reasons.
- **Physical Implementation:** physical circuits are drawn out, the different circuit components are placed in a chip floorplan or on a board and the wires connecting them are routed.

For CPUs, the entire implementation process is often called CPU design.

More specific usages of the term include more general wider-scale hardware architectures, such as cluster computing and Non-Uniform Memory Access (NUMA) architectures.

1.1.1 Computer Architectures

There are many types of computer architectures:

- Quantum computer vs Chemical computer
- Scalar processor vs Vector processor
- Non-Uniform Memory Access (NUMA) computers
- Register machine vs Stack machine
- Harvard architecture vs von Neumann architecture
- Cellular architecture

The quantum computer architecture holds the most promise to revolutionize computing

Some practitioners of computer architecture at companies such as Intel and AMD use more fine distinctions:

- **Macroarchitecture:** Architectural layers that are more abstract than microarchitecture, e.g. ISA
- **Instruction Set Architecture (ISA):** As defined above
- **Assembly ISA:** A smart assembler may convert an abstract assembly language common to a group of machines into slightly different machine language for different implementations
- **Programmer Visible Macroarchitecture:** higher level language tools such as compilers may define a consistent interface or contract to programmers using them, abstracting differences between underlying ISA, UISA, and microarchitectures. E.g. the C, C++, or Java standards define different programmer visible macroarchitecturer–although in practice the C microarchitecture for a particular computer includes
- **Microcode Instruction Set Architecture (MISA):** a family of machines with different hardware level microarchitectures may share a common microcode architecture, and hence a UISA.

Pin Architecture–the set of functions that a microprocessor is expected to provide, from the point of view of a hardware platform. E.g. the x86 A20M, FERR/IGNNE or FLUSH pins, and the messages that the processor is expected to emit after completing a cache invalidation so that external caches can be invalidated. Pin architecture functions are more flexible than ISA functions-external hardware can adapt to changing encodings, or changing from a pin to a message-but the functions are expected to be provided in successive implementations even if the manner of encoding them changes.

1.1.2 The Role Of Computer Architecture

The coordination of abstract levels of a processor under changing forces, involving design, measurement and evaluation. It also includes the overall fundamental working principle of the internal logical structure of a computer system.

It can also be defined as the design of the task-performing part of computers, *i.e.,* how various gates and transistors are interconnected and are caused to function per the instructions given by an assembly language programmer.

1.1.3 Instruction Set Architecture

1. The ISA is the interface between the software and hardware.
2. It is the set of instructions that bridges the gap between high level languages and the hardware.
3. For a processor to understand a command, it should be in binary and not in High Level Language. The ISA encodes these values.
4. The ISA also defines the items in the computer that are available to a programmer. For example, it defines data types, registers, addressing modes, memory organization, etc.
5. Register are high speed storage for numbers that can be accessed by a processor. Data as well as instructions can be in a register.

Addressing modes are the ways in which the instructions locate their operands.

Memory organization defines how instructions interact with the memory.

1.1.4 Computer Organization

Computer organization helps optimize performance-based products. For example, software engineers need to know the processing ability of processors. They may need to optimize software in order to gain the most performance at the least expense. This can require quite detailed analysis of the computer organization. For example, in a multimedia decoder, the designers might need to arrange for most data to be processed in the fastest data path.

Computer organization also helps plan the selection of a processor for a particular project. Multimedia projects may need very rapid data access, while supervisory software may need fast interrupts.

Sometimes certain tasks need additional components as well. For example, a computer capable of virtualization needs virtual memory hardware so that the memory of different simulated computers can be kept separated.

The computer organization and features also affect the power consumption and the cost of the processor.

1.1.5 Design goals

The exact form of a computer system depends on the constraints and goals for which it was optimized. Computer architectures usually trade off standards, cost, memory capacity, latency and throughput. Sometimes other considerations, such as features, size, weight, reliability, expandability and power consumption are factors as well.

The most common scheme carefully chooses the bottleneck that most reduces the computer's speed. Ideally, the cost is allocated proportionally to assure that the data rate is nearly the same for all parts of the computer, with the most costly part being the slowest. This is how skillful commercial integrators optimize personal computers.

1.1.6 Performance

Computer performance is often described in terms of clock speed (usually in MHz or GHz). This refers to the cycles per second of the main clock of the CPU. However, this metric is somewhat misleading, as a machine with a higher clock rate may not necessarily have higher performance. As a result manufacturers have moved away from clock speed as a measure of performance.

Computer performance can also be measured with the amount of cache a processor has. If the speed, MHz or GHz, were to be a car then the cache is like the gas tank. No matter how fast the car goes, it will still need to get gas. The higher the speed, and the greater the cache, the faster a processor runs.[dubious – *discuss*]

Modern CPUs can execute multiple instructions per clock cycle, which dramatically speeds up a program. Other factors influence speed, such as the mix of functional units, bus speeds, available memory, and the type and order of instructions in the programs being run.

There are two main types of speed, latency and throughput. Latency is the time between the start of a process and its completion. Throughput is the amount of work done per unit time. Interrupt latency is the guaranteed maximum response time of the system to an electronic event (*e.g.* when the disk drive finishes moving some data). Performance is affected by a very wide range of design choices–for example, pipelining a processor usually makes latency worse (slower) but makes throughput better. Computers that control machinery usually need low interrupt latencies. These computers operate in a real-time environment and fail if an operation is not completed in a specified amount of time. For example, computer-controlled anti-lock brakes must begin braking almost immediately after they have been instructed to brake.

The performance of a computer can be measured using other metrics, depending upon its application domain. A system may be CPU bound (as in numerical calculation), I/O bound (as in a webserving application) or memory bound (as in video editing). Power consumption has become important in servers and portable devices like laptops.

1.2 DESIGN PROCESS

CPU design focuses on these areas:

1. Datapaths (such as ALUs and pipelines)
2. Control unit: logic which controls the datapaths
3. Memory components such as register files, caches
4. Clock circuitry such as clock drivers, PLLs, clock distribution networks
5. Pad transceiver circuitry
6. Logic gate cell library which is used to implement the logic

CPUs designed for high-performance markets might require custom designs for each of these items to achieve frequency, power-dissipation, and chip-area goals.

CPUs designed for lower performance markets might lessen the implementation burden by:

- Acquiring some of these items by purchasing them as intellectual property
- Use control logic implementation techniques (logic synthesis using CAD tools) to implement the other components - datapaths, register files, clocks

Common logic styles used in CPU design include:

- Unstructured random logic

- Finite-state machines
- Microprogramming (common from 1965 to 1985)
- Programmable logic array (common in the 1980s, no longer common)

Device types used to implement the logic include:

- Transistor-transistor logic Small Scale Integration logic chips - no longer used for CPUs
- Programmable Array Logic and Programmable logic devices - no longer used for CPUs
- Emitter-coupled logic (ECL) gate arrays - no longer common
- CMOS gate arrays - no longer used for CPUs
- CMOS ASICs - what's commonly used today, they're so common that the term ASIC is not used for CPUs
- Field-programmable gate arrays (FPGA) - common for soft microprocessors, and more or less required for reconfigurable computing

A CPU design project generally has these major tasks:

- Programmer-visible instruction set architecture, which can be implemented by a variety of microarchitectures
- Architectural study and performance modeling in ANSI C/C++ or SystemC
- High-level synthesis (HLS) or RTL (*e.g.* logic) implementation
- RTL Verification
- Circuit design of speed critical components (caches, registers, ALUs)
- Logic synthesis or logic-gate-level design
- Timing analysis to confirm that all logic and circuits will run at the specified operating frequency
- Physical design including floorplanning, place and route of logic gates
- Checking that RTL, gate-level, transistor-level and physical-level representations are equivalent
- Checks for signal integrity, chip manufacturability

As with most complex electronic designs, the logic verification effort (proving that the design does not have bugs) now dominates the project schedule of a CPU.

Key CPU architectural innovations include index register, cache, virtual memory, instruction pipelining, superscalar, CISC, RISC, virtual machine, emulators, microprogram, and stack.

1.2.1 Goals

The first CPUs were designed to do mathematical calculations faster and more reliably than human computers[1].

Each successive generation of CPU might be designed to achieve some of these goals:

- higher performance levels of a single program or thread
- higher throughput levels of multiple programs/threads
- less power consumption for the same performance level
- lower cost for the same performance level
- greater connectivity to build larger, more parallel systems
- more specialization to aid in specific targeted markets

Re-designing a CPU core to a smaller die-area helps achieve several of these goals.

- Shrinking everything (a "photomask shrink"), resulting in the same number of transistors on a smaller die, improves performance (smaller transistors switch faster), reduces power (smaller wires have less parasitic capacitance) and reduces cost (more CPUs fit on the same wafer of silicon).

- Releasing a CPU on the same size die, but with a smaller CPU core, keeps the cost about the same but allows higher levels of integration within one VLSI chip (additional cache, multiple CPUs, or other components), improving performance and reducing overall system cost.

1.2.2 Performance Analysis and Benchmarking

Because there are too many programs to test a CPU's speed on all of them, benchmarks were developed. The most famous benchmarks are the SPECint and SPECfp benchmarks developed by Standard Performance Evaluation Corporation and the ConsumerMark benchmark developed by the Embedded Microprocessor Benchmark Consortium EEMBC.

Some important measurements include:

- Instructions per second – Most consumers pick a computer architecture (normally Intel IA32 architecture) to be able to run a large base of pre-existing pre-compiled software. Being relatively uninformed on computer benchmarks, some of them pick a particular CPU based on operating frequency (see Megahertz Myth).
- FLOPS – The number of floating point operations per second is often important in selecting computers for scientific computations.
- Performance per watt – System designers building parallel computers, such as Google, pick CPUs based on their speed per watt of power, because the cost of powering the CPU outweighs the cost of the CPU itself. [1][2]
- Some system designers building parallel computers pick CPUs based on the speed per dollar.
- System designers building real-time computing systems want to guarantee worst-case response. That is easier to do when the CPU has low interrupt latency and when it has deterministic response. (DSP)
- Computer programmers who program directly in assembly language want a CPU to support a full featured instruction set.
- Low power – For systems with limited power sources (*e.g.* solar, batteries, human power).
- Small size or low weight – For portable embedded systems, systems for spacecraft.
- Environmental impact – Minimizing environmental impact of computers during manufacturing and recycling as well during use. Reducing waste, reducing hazardous materials. (see Green computing).

Some of these measures conflict. In particular, many design techniques that make a CPU run faster make the "performance per watt", "performance per dollar", and "deterministic response" much worse, and vice-versa.

1.3 GATE LEVEL DESIGN

A **logic gate** performs a logical operation on one or more logic inputs and produces a single logic output. The logic is called Boolean logic and is most commonly found in digital circuits. Logic gates are primarily implemented electronically using diodes or transistors, but can also be constructed using electromagnetic relays (relay logic), fluidic logic, pneumatic logic, optics, molecules, or even mechanical elements.

1.3.1 Truth table

A truth table is a table that describes the behavior of a logic gate or any combination of logic gates. It lists the value of the output for every possible combination of the inputs and can be used to simplify the number of logic gates and level of nesting in an electronic circuit. In general the truth table does not lead to an efficient implementation; a minimization procedure, using Karnaugh maps, the Quine–McCluskey algorithm or an heuristic algorithm is required for reducing the circuit complexity.

1.3.2 Background

The simplest form of electronic logic is diode logic. This allows AND and OR gates to be built, but not inverters, and so is an incomplete form of logic. Further, without some kind of amplification it is not possible to have such basic logic operations cascaded as required for more complex logic functions. To build a functionally complete logic system, relays, valves (vacuum tubes), or transistors can be used. The simplest family of logic gates using bipolar transistors is called resistor-transistor logic (RTL). Unlike diode logic gates, RTL gates can be cascaded indefinitely to produce more complex logic functions. These gates were used in early integrated circuits. For higher speed, the resistors used in RTL were replaced by diodes, leading to diode-transistor logic (DTL). Transistor-transistor logic (TTL) then supplanted DTL with the observation that one transistor could do the job of two diodes even more quickly, using only half the space. In virtually every type of contemporary chip implementation of digital systems, the bipolar transistors have been replaced by complementary field-effect transistors (MOSFETs) to reduce size and power consumption still further, thereby resulting in complementary metal–oxide–semiconductor (CMOS) logic.

For small-scale logic, designers now use prefabricated logic gates from families of devices such as the TTL 7400 series by Texas Instruments and the CMOS 4000 series by RCA, and their more recent descendants. Increasingly, these fixed-function logic gates are being replaced by programmable logic devices, which allow designers to pack a large number of mixed logic gates into a single integrated circuit. The field-programmable nature of programmable logic devices such as FPGAs has removed the 'hard' property of hardware; it is now possible to change the logic design of a hardware system by reprogramming some of its components, thus allowing the features or function of a hardware implementation of a logic system to be changed.

Electronic logic gates differ significantly from their relay-and-switch equivalents. They are much faster, consume much less power, and are much smaller (all by a factor of a million or more in most cases). Also, there is a fundamental structural difference. The switch circuit creates a continuous metallic path for current to flow (in either direction) between its input and its output. The semiconductor logic gate, on the other hand, acts as a high-gain voltage amplifier, which sinks a tiny current at its input and produces a low-impedance voltage at its output. It is not possible for current to flow between the output and the input of a semiconductor logic gate.

Another important advantage of standardized integrated circuit logic families, such as the 7400 and 4000 families, is that they can be cascaded. This means that the output of one gate can be wired to the inputs of one or several other gates, and so on. Systems with varying degrees of complexity can be built without great concern of the designer for the internal workings of the gates, provided the limitations of each integrated circuit are considered.

The output of one gate can only drive a finite number of inputs to other gates, a number called the 'fanout limit'. Also, there is always a delay, called the 'propagation delay', from a change in input of a gate to the corresponding change in its output. When gates are cascaded, the total propagation delay is approximately the sum of the individual delays, an effect which can become a problem in high-speed circuits. Additional delay can be caused when a large number of inputs are connected to an output, due to the distributed capacitance of all the inputs and wiring and the finite amount of current that each output can provide.

1.3.3 Logic gates

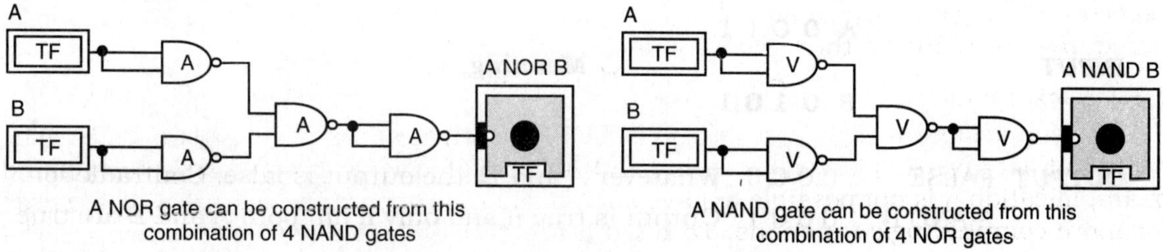

Fig. 1.1.

All other types of Boolean logic gates (*i.e.*, AND, OR, NOT, XOR, XNOR) can be created from a suitable network of NAND gates. Similarly, all gates can be created from a network of NOR gates. Historically, NAND gates were easier to construct from MOS technology and thus NAND gates served as the first pillar of Boolean logic in electronic computation.

For an input of 2 variables, there are 16 possible boolean algebraic functions. These 16 functions are enumerated below, together with their outputs for each combination of inputs variables.

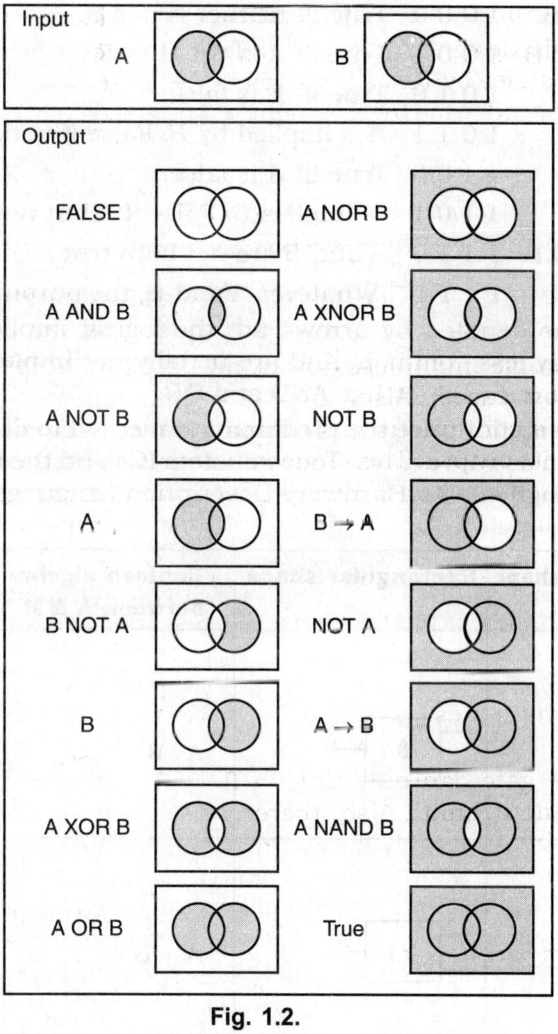

Fig. 1.2.

Venn Diagrams for Logic Gates

		A 0 0 1 1	
INPUT			**Meaning**
		B 0 1 0 1	
OUTPUT	FALSE	0 0 0 0	Whatever *A* and *B*, the output is false. Contradiction.
	A AND B	0 0 0 1	Output is true if and only if (iff) both *A* and *B* are true.
	A ↛ B	0 0 1 0	*A* doesn't imply *B*. True iff *A* but not *B*.
	A	0 0 1 1	True whenever *A* is true.
	A ↚ B	0 1 0 0	*A* is not implied by *B*. True iff not *A* but *B*.
	B	0 1 0 1	True whenever *B* is true.
	A XOR B	0 1 1 0	True iff *A* is not equal to *B*.
	A OR B	0 1 1 1	True iff *A* is true, or *B* is true, or both.
	A NOR B	1 0 0 0	True iff neither *A* nor *B*.
	A XNOR B	1 0 0 1	True iff *A* is equal to *B*.
	NOT B	1 0 1 0	True iff *B* is false.
	A ← B	1 0 1 1	*A* is implied by *B*. False if not *A* but *B*, otherwise true.
	NOT A	1 1 0 0	True iff *A* is false.
	A → B	1 1 0 1	*A* implies *B*. False if *A* but not *B*, otherwise true.
	A NAND B	1 1 1 0	*A* and *B* are not both true.
	TRUE	1 1 1 1	Whatever *A* and *B*, the output is true. Tautology.

The four functions denoted by arrows are the logical implication functions. These functions are generally less common, and are usually not implemented directly as logic gates, but rather built out of gates like AND and OR.

In the 1980s, schematics were the predominant method to design both circuit boards and custom ICs known as gate arrays. Today custom ICs and the field-programmable gate array are typically designed with Hardware Description Languages (HDL) such as Verilog or VHDL.

Type	**Distinctive shape**	**Rectangular shape**	**Boolean algebra between A & B**	**Truth table**
AND	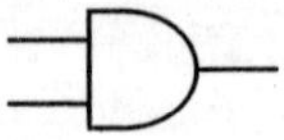	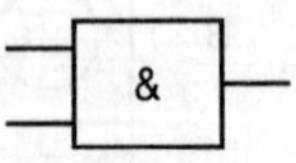	A . B	Input: A B / Output: A AND B 0 0 0 0 1 0 1 0 0 1 1 1
OR	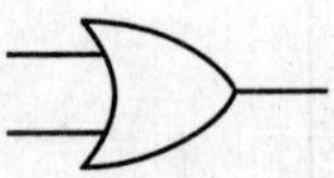	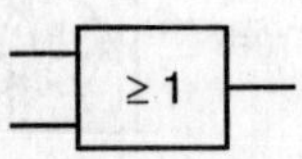	A + B	Input: A B / Output: A OR B 0 0 0 0 1 1 1 0 1 1 1 1

<table>
<tr><td>NOT</td><td>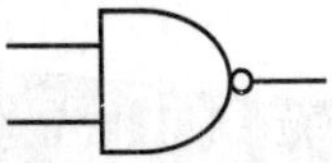</td><td>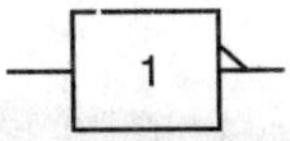</td><td>$\overline{A}$</td><td></td><td><table><tr><th>Input</th><th>Output</th></tr><tr><td>A</td><td>NOT A</td></tr><tr><td>0</td><td>1</td></tr><tr><td>1</td><td>0</td></tr></table></td></tr>
</table>

In electronics a NOT gate is more commonly called an inverter. The circle on the symbol is called a *bubble*, and is used in logic diagrams to indicate a logical inversion between the external logic state and the internal logic state (1 to 0 or vice-versa). On a circuit diagram it must be accompanied by a statement asserting that the *positive logic convention* or *negative logic convention* is being used (high voltage level = 1 or high voltage level = 0), respectively). The *wedge* is used in circuit diagrams to directly indicate an active-low (high voltage level = 0) input or output without requiring a uniform convention throughout the circuit diagram. This is called *Direct Polarity Indication*. See IEEE Std 91/91A and IEC 60617-12. Both the *bubble* and the *wedge* can be used on distinctive-shape and rectangular-shape symbols on circuit diagrams, depending on the logic convention used. On pure logic diagrams, only the *bubble* is meaningful.

<table>
<tr><th>Type</th><th>Distinctive shape</th><th>Rectangular shape</th><th>Boolean algebra between A & B</th><th>Truth table</th></tr>
<tr><td>NAND</td><td>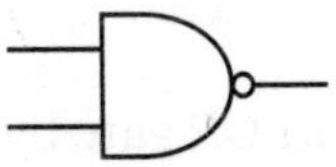</td><td></td><td>$\overline{A \cdot B}$</td><td><table><tr><th colspan="2">Input</th><th>Output</th></tr><tr><td>A</td><td>B</td><td>A AND B</td></tr><tr><td>0</td><td>0</td><td>1</td></tr><tr><td>0</td><td>1</td><td>1</td></tr><tr><td>1</td><td>0</td><td>1</td></tr><tr><td>1</td><td>1</td><td>0</td></tr></table></td></tr>
<tr><td>NOR</td><td>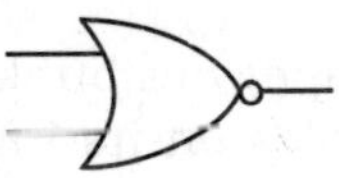</td><td>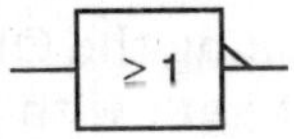</td><td>$\overline{A + B}$</td><td><table><tr><th colspan="2">Input</th><th>Output</th></tr><tr><td>A</td><td>B</td><td>A NOR B</td></tr><tr><td>0</td><td>0</td><td>1</td></tr><tr><td>0</td><td>1</td><td>0</td></tr><tr><td>1</td><td>0</td><td>0</td></tr><tr><td>1</td><td>1</td><td>0</td></tr></table></td></tr>
<tr><td>XOR</td><td>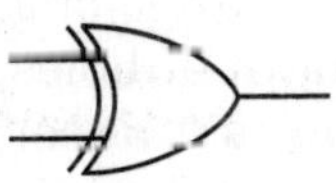</td><td>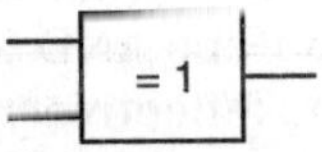</td><td>$A \oplus B$</td><td><table><tr><th colspan="2">Input</th><th>Output</th></tr><tr><td>A</td><td>B</td><td>A XOR B</td></tr><tr><td>0</td><td>0</td><td>0</td></tr><tr><td>0</td><td>1</td><td>1</td></tr><tr><td>1</td><td>0</td><td>1</td></tr><tr><td>1</td><td>1</td><td>0</td></tr></table></td></tr>
<tr><td>XNOR</td><td>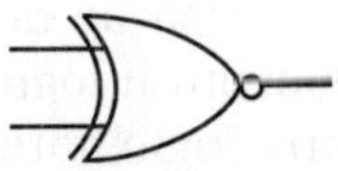</td><td>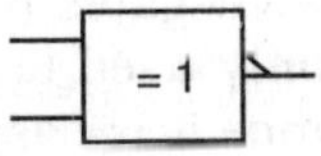</td><td>$\overline{A \oplus B}$ or $A \odot B$</td><td><table><tr><th colspan="2">Input</th><th>Output</th></tr><tr><td>A</td><td>B</td><td>A XNOR B</td></tr><tr><td>0</td><td>0</td><td>1</td></tr><tr><td>0</td><td>1</td><td>0</td></tr><tr><td>1</td><td>0</td><td>0</td></tr><tr><td>1</td><td>1</td><td>1</td></tr></table></td></tr>
</table>

Charles Sanders Peirce (1880) showed that NAND gates alone (or alternatively NOR gates alone) can be used to reproduce the functions of all the other logic gates, but his work on it was unpublished until 1935. The first published proof was by Henry M. Sheffer in 1913.

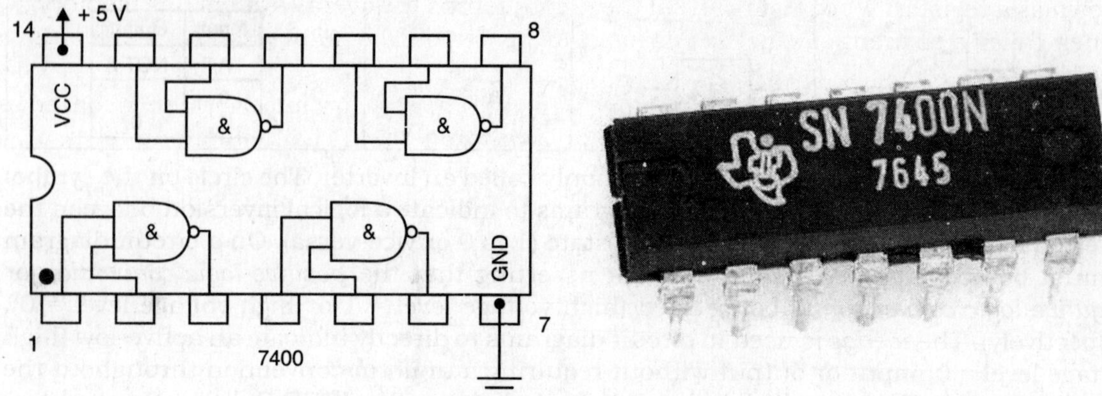

Fig. 1.3.

The 7400 chip, containing four NANDs. The two additional pins supply power (+5 V) and connect the ground.

Two more gates are the exclusive-OR or XOR function and its inverse, exclusive-NOR or XNOR. The two input Exclusive-OR is true only when the two input values are *different*, false if they are equal, regardless of the value. If there are more than two inputs, the gate generates a true at its output if the number of trues at its input is *odd* ([2]). In practice, these gates are built from combinations of simpler logic gates.

1.3.4 De Morgan equivalent symbols

By use of De Morgan's theorem, an *AND* gate can be turned into an *OR* gate by inverting the sense of the logic at its inputs and outputs. This leads to a separate set of symbols with inverted inputs and the opposite core symbol. These symbols can make circuit diagrams for circuits using active low signals much clearer and help to show accidental connection of an active high output to an active low input or vice-versa.

Symbolically, a NAND gate can also be shown using the OR shape with bubbles on its inputs, and a NOR gate can be shown as an AND gate with bubbles on its inputs. The bubble signifies a logic inversion. This reflects the equivalency due to De Morgans law, but it also allows a diagram to be read more easily, or a circuit to be mapped onto available physical gates in packages easily, since any circuit node that has bubbles at both ends can be replaced by a simple bubble-less connection and a suitable change of gate. If the NAND is drawn as OR with input bubbles, and a NOR as AND with input bubbles, this gate substitution occurs automatically in the diagram (effectively, bubbles "cancel"). This is commonly seen in real logic diagrams – thus the reader must not get into the habit of associating the shapes exclusively as OR or AND shapes, but also take into account the bubbles at both inputs and outputs in order to determine the "true" logic function indicated.

All logic relations can be realized by using NAND gates (this can also be done using NOR gates). De Morgan's theorem is most commonly used to transform all logic gates to NAND gates or NOR gates. This is done mainly since it is easy to buy logic gates in bulk and because many electronics labs stock only NAND and NOR gates.

1.3.5 Data storage

Logic gates can also be used to store data. A storage element can be constructed by connecting several gates in a "latch" circuit. More complicated designs that use clock signals and that change only on a rising or falling edge of the clock are called edge-triggered "flip-flops". The combination of multiple flip-flops in parallel, to store a multiple-bit value, is known as a register. When using any of these gate setups the overall system has memory; it is then called a sequential logic system since its output can be influenced by its previous state(s).

These logic circuits are known as computer memory. They vary in performance, based on factors of speed, complexity, and reliability of storage, and many different types of designs are used based on the application.

1.3.6 Three-state logic gates

A tristate buffer can be thought of as a switch. If *B* is on, the switch is closed. If B is off, the switch is open.

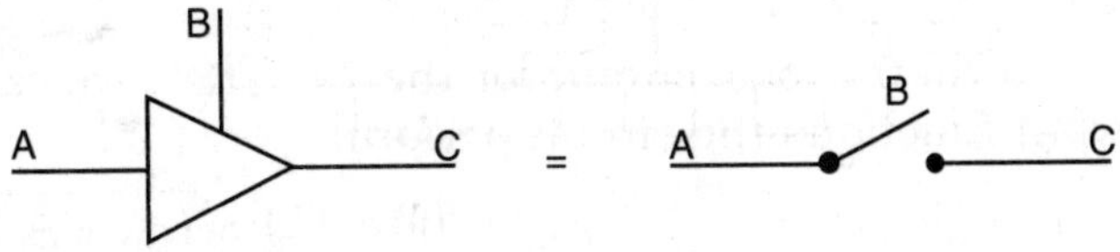

Three-state, or 3-state, logic gates are a type of logic gates that have three states of the output: high (H), low (L) and high-impedance (Z). The high-impedance state plays no role in the logic, which remains strictly binary. These devices are used on buses also known as the data buses of the CPU to allow multiple chips to send data. A group of three-states driving a line with a suitable control circuit is basically equivalent to a multiplexer, which may be physically distributed over separate devices or plug-in cards.

1.4 REGISTER LEVEL DESIGN

A digital system can be treated at different level of abstraction or complexity.

Logic Circuit for the 74X181 ALU

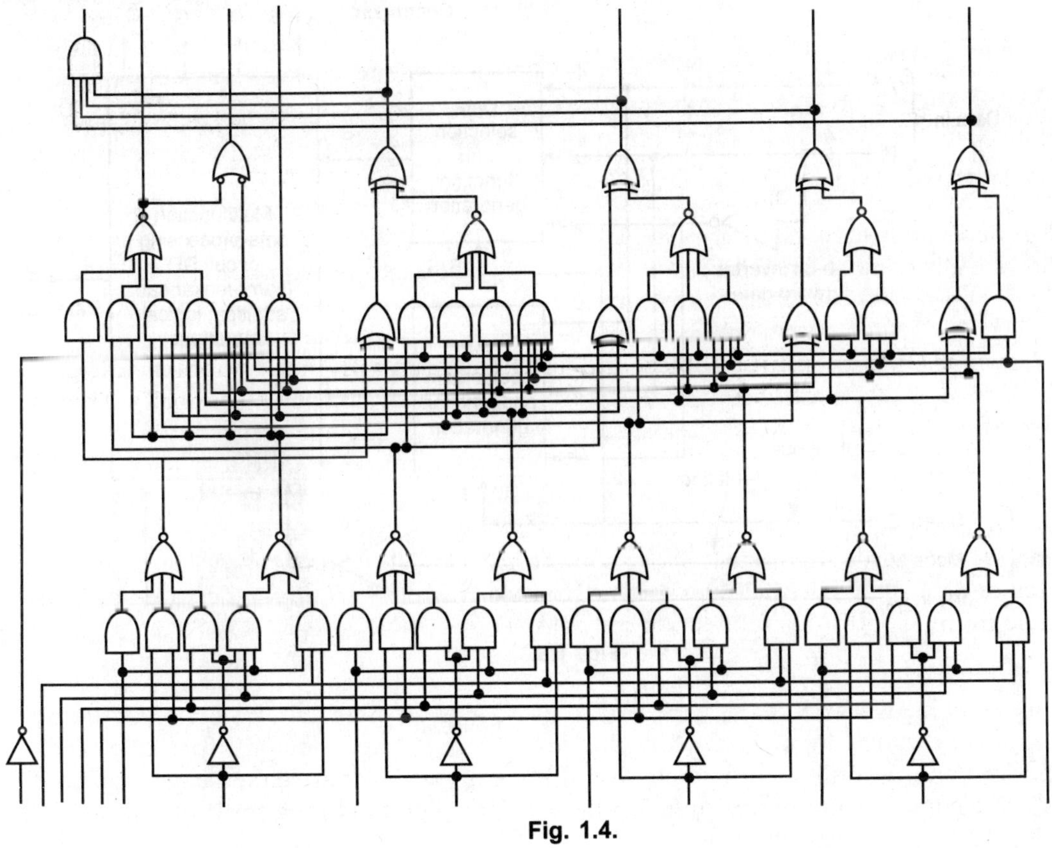

Fig. 1.4.

So far, we have seen it at the gate level and the transistor level. At a higher level than the gate level, we have the register level. The central component at this level is an n-bit storage device, or *register*.

	Central Component	
Transistor Level	Transistor Level	
Gate Level	Logic Gate	Higher level of abstraction ↓
Register Level	Register	

Form the logic circuit for the 74 × 181, we can wirte which gives little information on what functions the circuit perform.

$$F_0 = \overline{(M + C_n)\ \varpi\ (a_0 + B_0S_0 + \bar{B}_0S_1)\ \varpi\ (A_0\bar{B}_0S_2 + A_0B_0S_3)}$$

Using a truth table to describe the functions of the circuit is not a good idea either as the circuit has 14 inputs, which means the truth table will have 2^{14} = 16384 rows!

The 74 × 181 is actually a 4-bit ALU which performs 32 different arithmetic and logical operations on two 4-bit inputs $A = A_3\ A_2\ A_1\ A_0$ and $B = B_3\ B_2\ B_1\ B_0$ producing the result $F = F_3\ F_2\ F_1\ F_0$.

The operation to be performed is determined by the control signals M and $S = S_3\ S_2\ S_1\ S_0$. In particular, the molde control signal M is set to 1 for logical operations, and 0 for arithmetic operations. The select control signal S is use to select the required operations.

Register-Level Model of the 74X181 ALU

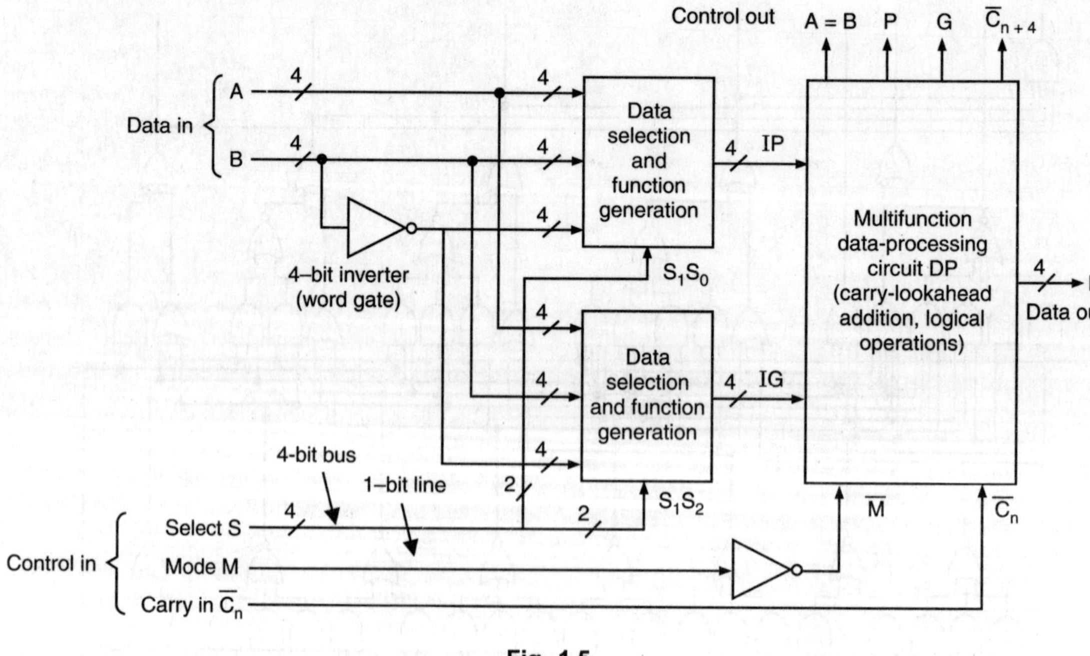

Fig. 1.5.

74X181 Arithmetic and Logical operations

SELECTIONS	ACTIVE-HIGH DATA		
	M = H	M = L ; ARITHMETIC OPERATIONS	
S3 S2 S1 S0	LOGIC FUNCTIONS	$\bar{C}_n = H$ (no carry)	$\bar{C}_n = L$ (with carry)
L L L L	$F = \bar{A}$	$F = A$	$F = A$ PUS 1
L L L H	$F = \overline{A + B}$	$F = A + B$	$F = (A + B)$ PLUS 1
L L H L	$F = \bar{A}B$	$F = A + \bar{B}$	$F = (A + \bar{B})$ PLUS 1
L L H H	$F = 0$	$F =$ MINUS 1 (2's COMPL)	$Z =$ ZERO
L H L L	$F = \overline{AB}$	$F = A$ PLUS $A\bar{B}$	$F = A$ PLUS $A\bar{B}$ PLUS 1
L H L H	$F = \bar{B}$	$F = (A + B)$ PLUS $A\bar{B}$	$F = (A + B)$ PLUS $A\bar{B}$ PLUS 1
L H H L	$F = A \oplus B$	$F = A$ MINUS B MINUS 1	$F = A$ MINUS B
L H H H	$F = A\bar{B}$	$F = A\bar{B}$ MINUS 1	$F = A\bar{B}$
H L L L	$F = \bar{A} + B$	$F = A$ PLUS AB	$F = A$ PLUS AB PLUS 1
H L L H	$F = \overline{A \oplus B}$	$F = A$ PLUS B	$F = A$ PLUS B PLUS 1
H L H L	$F = B$	$F = (A + \bar{B})$ PLUS AB	$F = (A + \bar{B})$ PLUS AB PLUS 1
H L H H	$F = AB$	$F = AB$ MINUS 1	$F = AB$
H H L L	$F = 1$	$F = A$ PLUS $A1$	$F = A$ PLUS A PLUS 1
H H L H	$F = A + \bar{B}$	$F = (A + B)$ PLUS A	$F = (A + B)$ PLUS A PLUS 1
H H H L	$F = A + B$	$F = (A + \bar{B})$ PLUS A	$F = (A + \bar{B})$ PLUS A PLUS 1
H H H H	$F = A$	$F = A$ MINUS 1	$F = A$

Register-level design is geared toward the processing of words as the basic units of information or signals. In contrast, in gate-level design individual bits are treated as basic signals. The word length is register-level design varies, but it is always a multiple of 4. Another important aspect of register-level design is the data-flow representation which views the function of a digital circuit as processing data words that flow between computational and storage devices. The computational devices are typically multifunction combinatorial circuits (*e.g.* the ALU), and the storage devices are registers.

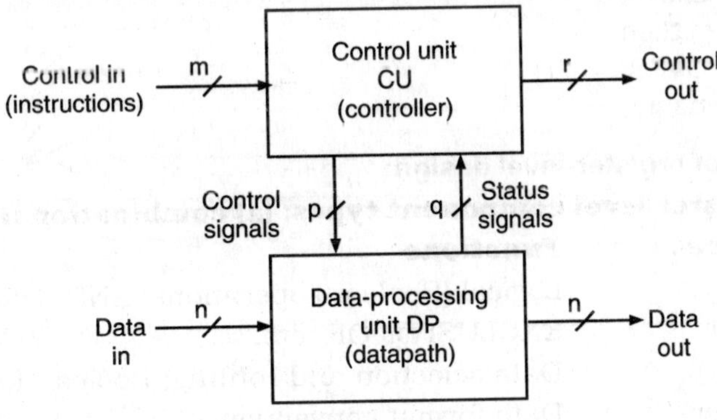

Fig. 1.6.

Complex digital cirucit can be divided into a data-processing (DP) part and a control part (CU) at the register level.

To manage these devices and control the data flow, control signals and circuits that generate these signals are required.

Example of a small processor

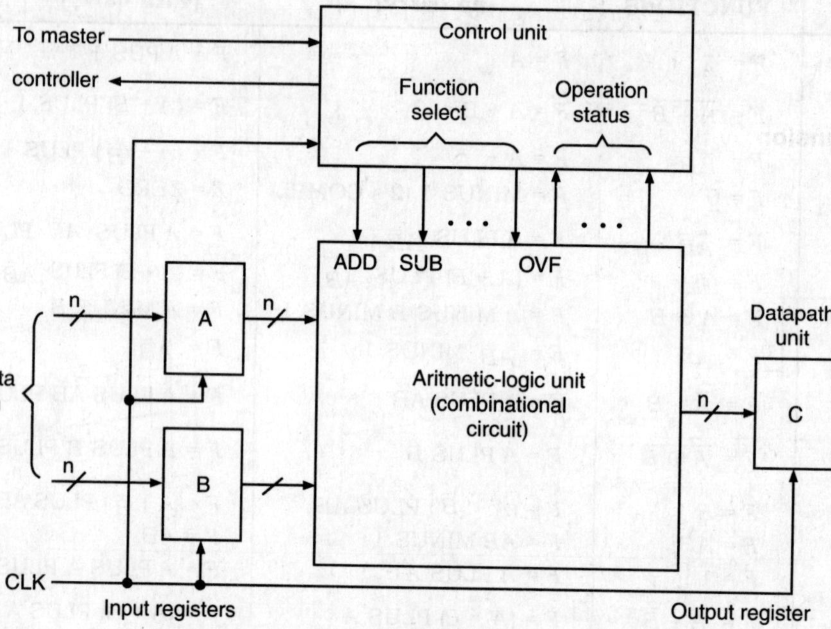

Fig. 1.7.

The behavior of a register-level circuit can be described using behavioral style VHDL code.

For example, the operations of ALU of the small processor can be described using the following VHDL code:

```
            if ADD  =  'a' then
                   C < =  A + B;
        elsif SUB  =  '1' then
                   C < =  A – B;
             end if;
 if overflow occurs then
           OVF < =  1;
          end if;
```

Basic components of register-level design:

Basic register-level component types: (a) combinational; (b) sequential.

Component types	***Functions***
Word gate	Logical (Boolean) operations: AND, OR, NOT, NAND, EXCLUSIVE-OR, etc.
Multiplexer	Data selection and routing; Boolean functions generation
Decoder/encoder	Data format conversion

Adder	Binary addition and sutrraction
Comparator	Data comparsion
ALU	Arithmetic and logical operations

(a)

Componets type	***Functions***
(Parallel) register	Data storage
Shift register	Data storage and format conversion
Counter	Counter state generations; arithmetic operations; timing

(b)

Component Expansion by Bit Slicing

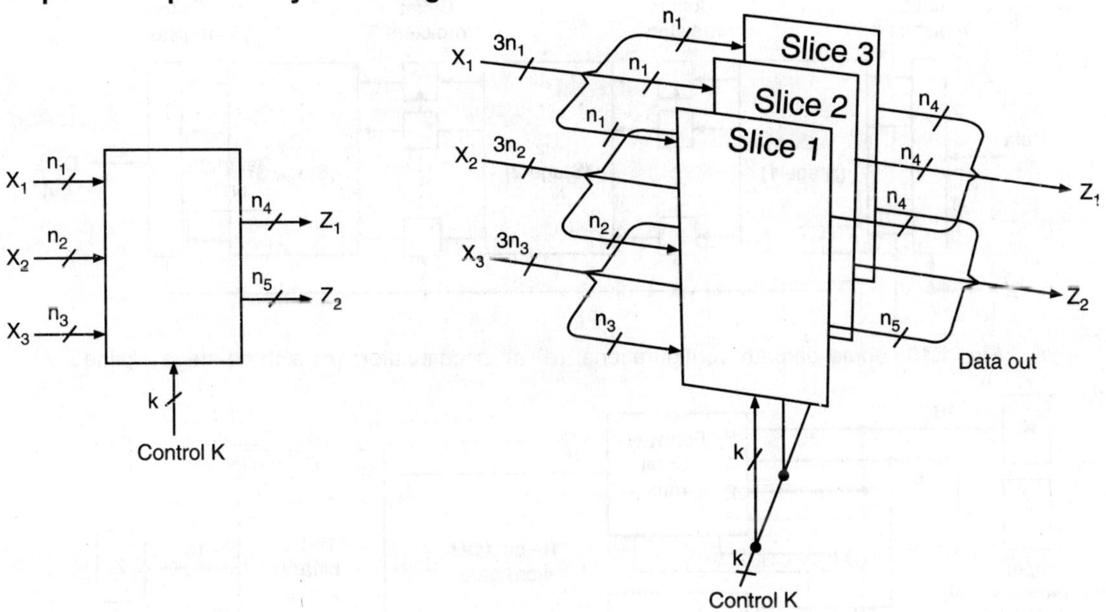

Fig. 1.8.

Shift Register Expansion using Bit Slicing

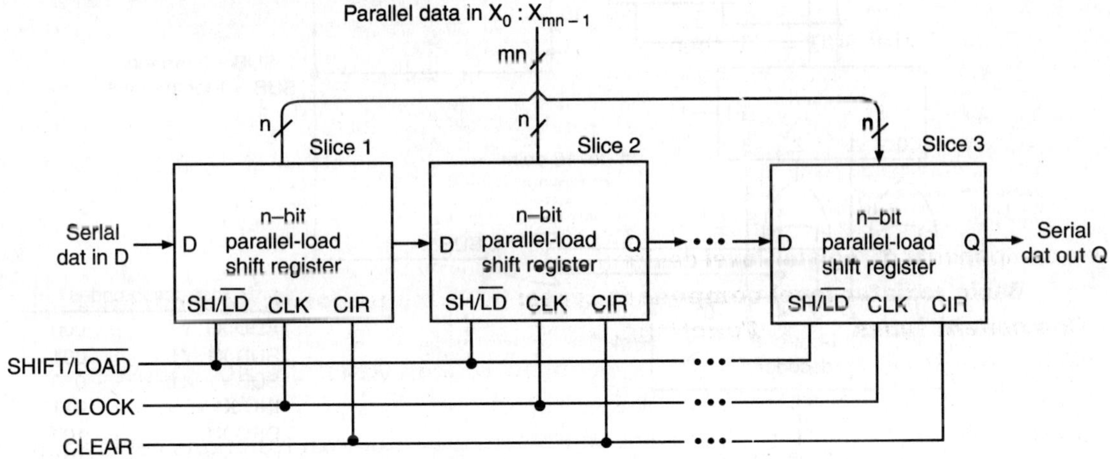

Fig. 1.9.

Datapath Unit (DP)

DP's are used to move and process data (in words).

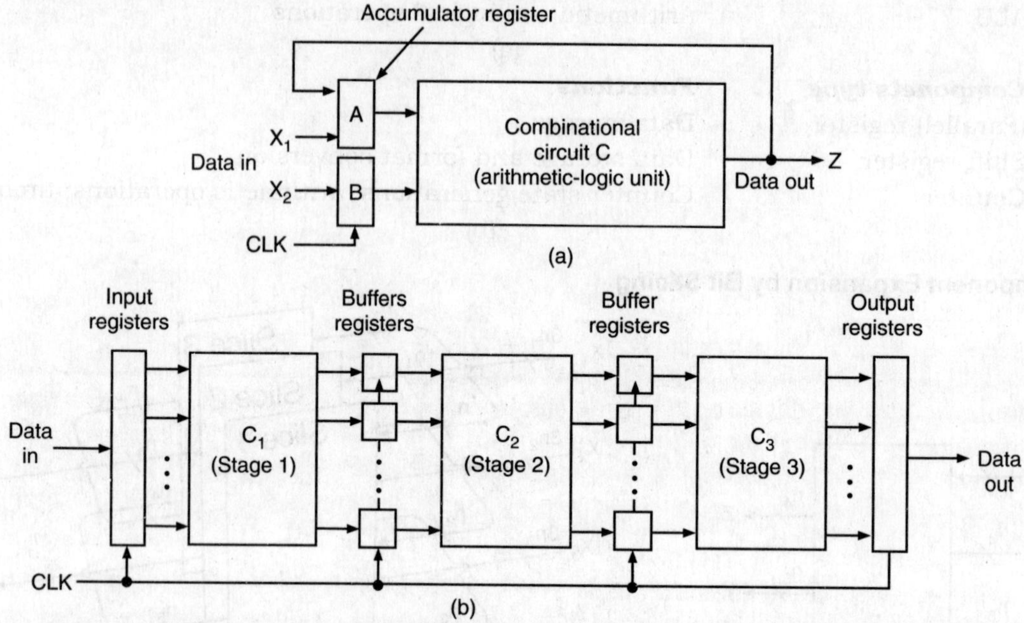

Fig. 1.10. Some datpath configurations: (a) an accumulator; (b) a three-stage pipline

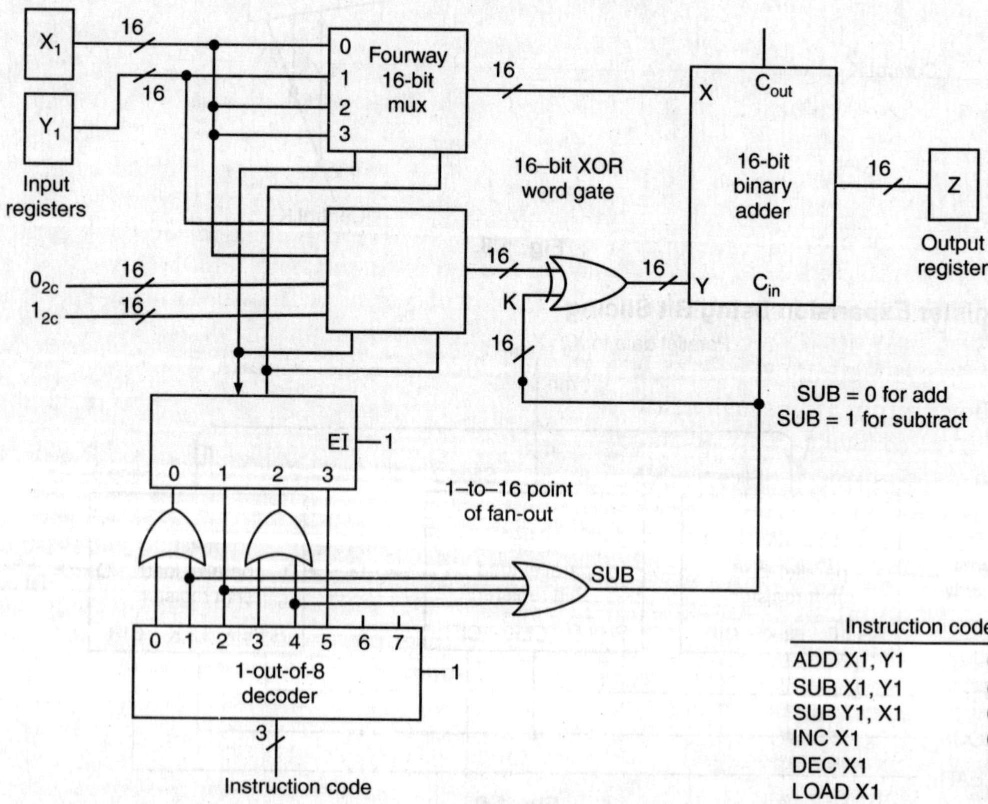

Instruction codes	
ADD X1, Y1	000
SUB X1, Y1	001
SUB Y1, X1	010
INC X1	011
DEC X1	100
LOAD X1	101

Fig. 1.11.

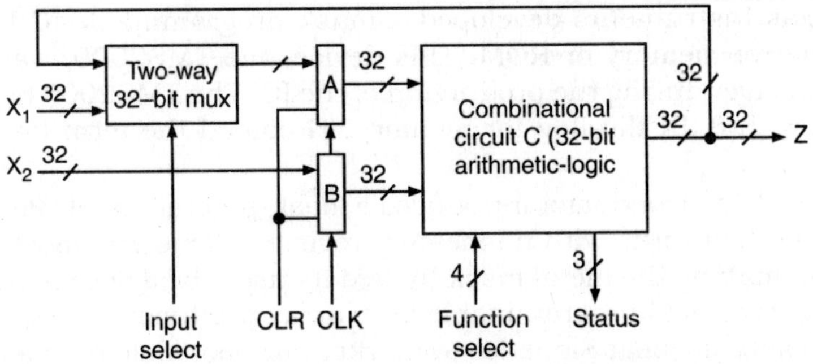

Fig. 1.12. A more detailed logic diagram for the accumulator-based datapath

1.5 PLD

A programmable logic device or PLD is an electronic component used to build reconfigurable digital circuits. Unlike a logic gate, which has a fixed function, a PLD has an undefined function at the time of manufacture. Before the PLD can be used in a circuit it must be programmed, that is, reconfigured.

1.5.1. Using a ROM as a PLD

Before PLDs were invented, read-only memory (ROM) chips were used to create arbitrary combinational logic functions of a number of inputs. Consider a ROM with m inputs (the address lines) and n outputs (the data lines). When used as a memory, the ROM contains 2^m words of n bits each. Now imagine that the inputs are driven not by an m-bit address, but by m independent logic signals. Theoretically, there are 2^m possible Boolean functions of these m signals, but the structure of the ROM allows just 2^n of these functions to be produced at the output pins. The ROM therefore becomes equivalent to n separate logic circuits, each of which generates a chosen function of the m inputs.

The advantage of using a ROM in this way is that any conceivable function of the m inputs can be made to appear at any of the n outputs, making this the most general-purpose combinational logic device available. Also, PROMs (programmable ROMs), EPROMs (ultraviolet-erasable PROMs) and EEPROMs (electrically erasable PROMs) are available that can be programmed using a standard PROM programmer without requiring specialized hardware or software. However, there are several disadvantages:

- they are usually much slower than dedicated logic circuits,
- they cannot necessarily provide safe "covers" for asynchronous logic transitions so the PROM's outputs may glitch as the inputs switch,
- they consume more power,
- they are often more expensive than programmable logic, especially if high speed is required.

Since most ROMs do not have input or output registers, they cannot be used stand-alone for sequential logic. An external TTL register was often used for sequential designs such as state machines. Common EPROMs, for example the 2716, are still sometimes used in this way by hobby circuit designers, who often have some lying around. This use is sometimes called a 'poor man's PAL'.

1.5.2. Early Programmable Logic

In 1969, Motorola offered the XC157, a mask-programmed gate array with 12 gates and 30 uncommitted input/output pins.

In 1970, Texas Instruments developed a mask-programmable IC based on the IBM read-only associative memory or ROM. This device, the TMS2000, was programmed by altering the metal layer during the production of the IC. The TMS2000 had up to 17 inputs and 18 outputs with 8 JK flip-flop for memory. TI coined the term Programmable Logic Array for this device.

In 1973 National Semiconductor introduced a mask-programmable PLA device (DM7575) with 14 inputs and 8 outputs with no memory registers. This was more popular than the TI part but cost of making the metal mask limited its use. The device is significant because it was the basis for the field programmable logic array produced by Signetics in 1975, the 82S100. (Intersil actually beat Signetics to market but poor yield doomed their part.)[2][3]

In 1971, General Electric Company (GE) was developing a programmable logic device based on the new PROM technology. This experimental device improved on IBM's ROAM by allowing multilevel logic. Intel had just introduced the floating-gate UV erasable PROM so the researcher at GE incorporated that technology. The GE device was the first erasable PLD ever developed, predating the Altera EPLD by over a decade. GE obtained several early patents on programmable logic devices.

In 1974 GE entered into an agreement with Monolithic Memories to develop a mask-programmable logic device incorporating the GE innovations. The device was named the 'Programmable Associative Logic Array' or PALA. The MMI 5760 was completed in 1976 and could implement multilevel or sequential circuits of over 100 gates. The device was supported by a GE design environment where Boolean equations would be converted to mask patterns for configuring the device. The part was never brought to market.

1.5.3. PAL

MMI introduced a breakthrough device in 1978, the Programmable Array Logic or PAL. The architecture was simpler than that of Signetics FPLA because it omitted the programmable OR array. This made the parts faster, smaller and cheaper. They were available in 20 pin 300 mil DIP packages while the FPLAs came in 28 pin 600 mil packages. The PAL Handbook demystified the design process. The PALASM design software (PAL Assembler) converted the engineers' Boolean equations into the fuse pattern required to program the part. The PAL devices were soon second-sourced by National Semiconductor, Texas Instruments and AMD.

1.5.4. GALs

An innovation of the PAL was the generic array logic device, or GAL, invented by Lattice Semiconductor in 1985. This device has the same logical properties as the PAL but can be erased and reprogrammed. The GAL is very useful in the prototyping stage of a design, when any bugs in the logic can be corrected by reprogramming. GALs are programmed and reprogrammed using a PAL programmer, or by using the in-circuit programming technique on supporting chips.

Lattice GALs combine CMOS and electrically erasable (E^2) floating gate technology for a high-speed, low-power logic device.

A similar device called a PEEL (programmable electrically erasable logic) was introduced by the International CMOS Technology (ICT) corporation.

1.5.5. CPLDs

PALs and GALs are available only in small sizes, equivalent to a few hundred logic gates. For bigger logic circuits, complex PLDs or CPLDs can be used. These contain the equivalent of several PALs linked by programmable interconnections, all in one integrated circuit. CPLDs can replace thousands, or even hundreds of thousands, of logic gates.

Some CPLDs are programmed using a PAL programmer, but this method becomes inconvenient for devices with hundreds of pins. A second method of programming is to solder the device to its printed circuit board, then feed it with a serial data stream from a personal computer. The CPLD contains a circuit that decodes the data stream and configures the CPLD to perform its specified logic function.

Each manufacturer has a proprietary name for this programming system. For example, Lattice Semiconductor calls it "in-system programming". However, these proprietary systems are beginning to give way to a standard from the Joint Test Action Group, JTAG.

1.5.6. FPGAs

While PALs were busy developing into GALs and CPLDs (all discussed above), a separate stream of development was happening. This type of device is based on gate array technology and is called the field-programmable gate array (FPGA). Early examples of FPGAs are the 82s100 array, and 82S105 sequencer, by Signetics, introduced in the late 1970s. The 82S100 was an array of AND terms. The 82S105 also had flip-flop functions.

FPGAs use a grid of logic gates, similar to that of an ordinary gate array, but the programming is done by the customer, not by the manufacturer. The term "field-programmable" means the array is done outside the factory, or "in the field."

FPGAs are usually programmed after being soldered down to the circuit board, in a manner similar to that of larger CPLDs. In most larger FPGAs the configuration is volatile, and must be re-loaded into the device whenever power is applied or different functionality is required. Configuration is typically stored in a configuration PROM or EEPROM. EEPROM versions may be in-system programmable (typically via JTAG).

The difference between FPGAs and CPLDs is that FPGAs are internally based on Look-up tables (LUTs) whereas CPLDs form the logic functions with sea-of-gates (e.g. sum of products). CPLDs are meant for simpler designs while FPGAs are meant for more complex designs. In general, CPLDs are a good choice for wide combinational logic applications while FPGAs are more suitable for large state machines (i.e. microprocessors).

1.5.7. PLD Programming Languages

Many PAL programming devices accept input in a standard file format, commonly referred to as 'JEDEC files'.They are analogous to software compilers. The languages used as source code for logic compilers are called hardware description languages, or HDLs.

PALASM and ABEL are frequently used for low-complexity devices, while Verilog and VHDL are popular higher-level description languages for more complex devices. The more limited ABEL is often used for historical reasons, but for new designs VHDL is more popular, even for low-complexity designs.

SOLVED PROBLEMS

Q.1. Define the following terms in brief :

(*a*) Register transfer language

(*b*) Micro-operation

(*c*) Registers

(*d*) Register Transfer

(*e*) Control function.

Solution: (*a*) Register transfer language : The symbolic notation used to describe the micro-operation transfers among registers is called a register transfer language. The term register transfer implies the availability of hardware logic circuits that can perform a stated micro-operation and transfer the result of the operation to the same or another registers the word "language" is borrowed from programmes who apply this term to programming language.

(*b*) Micro-operation : The operations executed on data stored in registers are called micro-operations. A micro operation is an elementary operation performed on the information stored in one or more registers. The result of operation may replace the previous binary information of a register or may be transferred to another register *e.g.*, shift, count, clear and load.

(*c*) Registers : The register that gold the address of the memory units is called a memory address register and is designated by the name MAR other designations for registers are:

PC (Program counter)

IR (Instruction register)

R_1 (Processor Register)

(*d*) Register Transfer : Information transferred from one register to another is designated in symbolic form by means of a replacement operator. The statement $R_2 \leftarrow R_1$, denotes a transfer of the content of transfer R_1 into register R_2. It designates a replacement the contents of R_2 by the content of R_1. By definition, the content of source register R_1 does not change after the transfer.

(*e*) Control Function : The control function is a Boolean variable that is equal to 1 or 0. The control function is included in the statement as follows :

P, $R_2 \leftarrow R_1$

The control condition is terminated with a colon, it symbolizes the requirement that the transfer operation be executed by hardware only of P = 1.

Q.2 Describe three state bus buffer. Draw the diagram of three state buffer.

Solution: A bus system can be constructed with three state gates instead of multiplexers. A three state gate is a digital circuit that exhibits three states. Two of these states are signals equivalent to logic 1 and 0 as in a conventional gate the third state is a high independence state behaves like an open circuit, which means that the output is disconnected and does not have a logic significance. Three state gates may perform any conventional logic such as AND or NAND. However, the one most commonly used in the design of a bus system is the buffer gate.

The graphic symbol of a three state buffer gate is shown below :

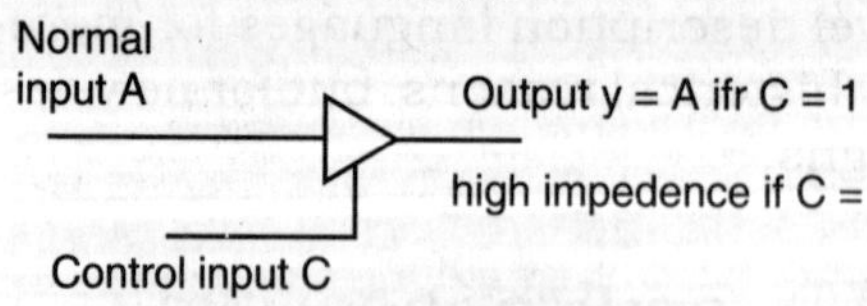

Fig. 1.13.

To construct a common bus for four registers of *n*-bit using three state buffers, we need *n* circuit with four buffers. Each group of four buffers receives one significant bit from the four registers. Each common output produces one of the lines for the common bus for a total of *n* lines only one decoder is necessary to select between the four registers.

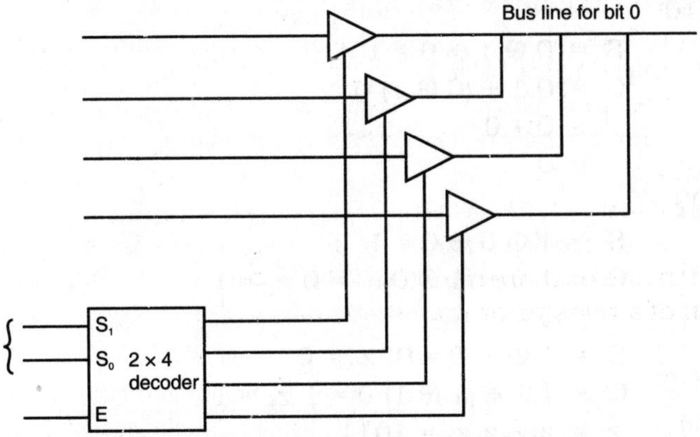

Fig. 1.14. Bus line with three state buffers

Q.3. Consider the following circuit A = $a_2a_1a_0$ and B = $b_2b_1b_0$ are three bit binary numbers input to the circuit. The output is Z is z_3z_2,z_0. R_0 R_1 and are registers with loading clock shown. The registers are loaded with their input data with the falling edge of the clock pulse. The bits of the input numbers A, B and full address are shown in the circuit.

(a) Find the output 2 for the inputs

A = 110

B = 101

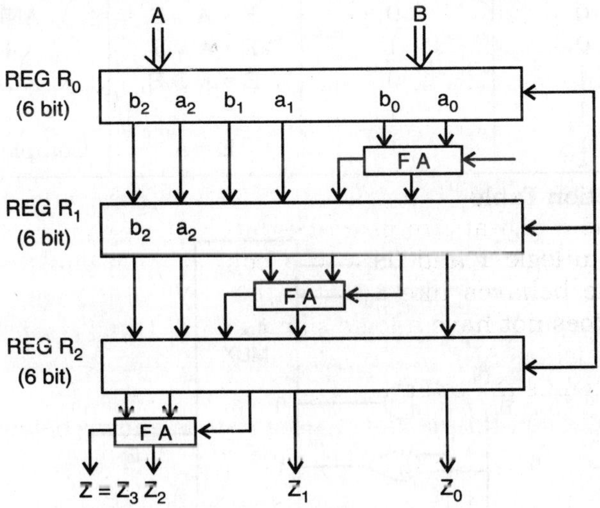

Fig. 1.15.

Solution: Given value

$$A = 110 = a_2a_1a_0$$
$$B = 101 = b_2b_1b_0$$
$$a_0b_0 = 01$$
$$a_1b_1 = 10$$
$$a_2b_2 = 11$$

For bit $a_0b_0 = 01$

and third bit = 01

∴ First bit is 010

$$S = 0 \oplus 1 \oplus 0 = 1$$
$$C = 0.1 + (0 \oplus 1)\,0$$
$$= 0 + 0$$
$$= 0$$

Second bit is 100

$$S = 1 \oplus 0 \oplus 0 = 1$$
$$C = 1.0 + (1 \oplus 0).0 = 0 \quad z_1 = 1$$

Third bit is 110

$$S = 1 \oplus 1 \oplus = 0 \quad z_2 = 0$$
$$C = 1.1 + (1 \oplus 1)\,0 = 1 \quad z_3 = 1$$
$$z = z_3 z_2 z_1 z_0 = 1011$$

Q.4. Describe the hardware implementation of logic-micro-operation. Draw the diagram of the stage of logic circuit used with AND, Or, XOR and NOT GATE with the help of function table.

Solution: The hardware implementation of logic micro-operations requires that logic gates be inserted for each bit or pair of bits in the register to perform the required logic function. Although there are 16 logic micro-operations, most computers use only four AND, OR XOR and complement.

S_1	S_0	Output	Operation
0	0	$E = A \wedge B$	AND
0	1	$E = A \vee B$	OR
1	0	$E = A \oplus B$	XOR
1	1		
1	1	$E = A$	Complement

Function Table

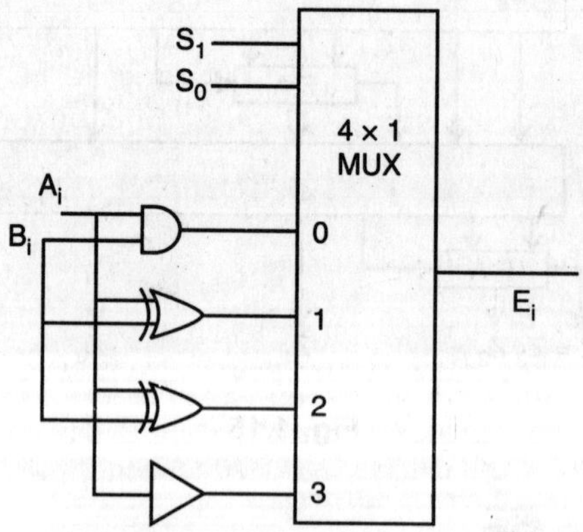

Logic diagram of one store of logic circuit

Fig. 1.16.

One stage of circuit that generates the four basic logic micro-operation. It consists of four gates and multiplexer. Each of the four logic operations is generated through a gate that perform the required logic the outputs of gates are applied to the data inputs of the multiplexer. The two selection inputs S_1 and S_0. Choose one of the data inputs of the multiplexer and direct its value to the output. For a logic circuit with *n* bits the diagram mist be repeated *n* times for $i = 0, 1, 2, 3, n$ the selection variables arc applied to all stages.

Q.5. Define the term and describe its operation :

(*a*) Selective set
(*b*) Selective complement
(*c*) Selective clear
(*d*) Mask operation
(*e*) Insert operation
(*f*) Clear operation

Solution: (*a*) Selective set : The selective set operation sets to 1. The bits in the register A where there are corresponding is in register B. It does not affect that have O's in the B.

e.g.,
1010 A before
1100 B (logical operand)
1110 A (after)

The two left most bits of B are 1's B are 1's, so the corresponding bits of A are set to 1. The two bits of A with corresponding O's in B remain unchanged. Therefore the OR micro-operation can be used to selectively get bits of a register.

(*b*) Selective complement : The selective complement operation complements bits in A where there are corresponding 1's in B. It does not affect bit positions that have O's in B.

e.g.,
1010 A before
1100 B (logic operand)

0110

Therefore exclusive–Or operation can be used to selectively complement bits of register.

(*c*) Selective clear : Selective clear operation clears to O. The bits in A only where there are corresponding 1's in B.

e.g.,
1010 A before
1100 B (logic operand)

0010 A after

The boolean operation performed on the individual bits in AB^1. The corresponding logic operation

$$A \leftarrow A \cap B^1$$

(*d*) Mask operation : The mask operation is similar to the selective clear operation except that the bits of A are cleared only above. There are corresponding O's in B. The mask operation is an AND micro-operation as seen from the following numcrical example

1010 A before
1100 B (mask)
1000 A after masking

(*e*) Insert operation : The insert operation inserts a new value into a group of bits this is done by first masking the bits and then OR in them the required value.

e.g.,
0110 1010 A before
0000 1111 B (mask)

0000 1010 After masking

0000	1010	A before
1001	0000	B (insert)
1001	1010	A after insertion

***(f)* Clear operation :** Clear operation compares the words in A and B and produces an all O's result if the two numbers are equal. This operation is achieved by an exclusive–OR micro operation

e.g.,

1010 A
1010 B
0000

$$A \leftarrow A \oplus B$$

Q.6. Show the bock diagram of the hardware that implements the following register transfer statements

$$yT_2 : R_2 \leftarrow R_1, R_1 \leftarrow R_2$$

Solution:

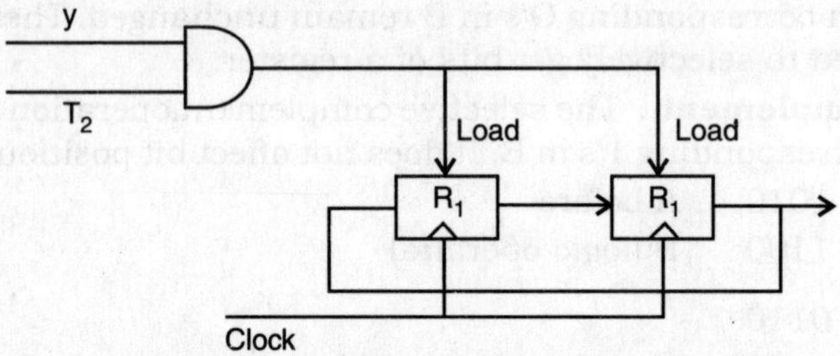

Fig. 1.17.

Q.7. The outputs of four registers R_0, R_1, R_2 and R_3 are connected through 4–to–1 line multiplexers to the inputs of a fifth register, R_5. Each register is eight bits long. The required transfers are dictated by four timing variables to through T_3 as follows :

$$T_0 : R_5 \leftarrow R_0$$
$$T_1 : R_5 \leftarrow R_1$$
$$T_2 : R_5 \leftarrow R_2$$
$$T_3 : R_5 \leftarrow R_3$$

The timing variables are mutually exclusive, which means that only one variable is equal to 1 at any given time, while the other three are equal to 0. Draw a block diagram showing the hardware implementation of register transfers, include the connection S necessary from the four timing variables to the selection inputs of the multiplexers and to load input of register R_5.

Solution:

T_0	T_1	T_2	T_3	S_1	S_0	R_3 load
0	0	0	0	×	×	0
1	0	0	0	0	0	1
0	1	0	0	0	1	1
0	0	1	0	1	0	1
0	0	0	1	1	1	1

$$S_1 = T_2 + T_3$$
$$S_0 = T_1 + T_3$$
$$\text{load} = T_0 + T_1 + T_2 + T_3$$

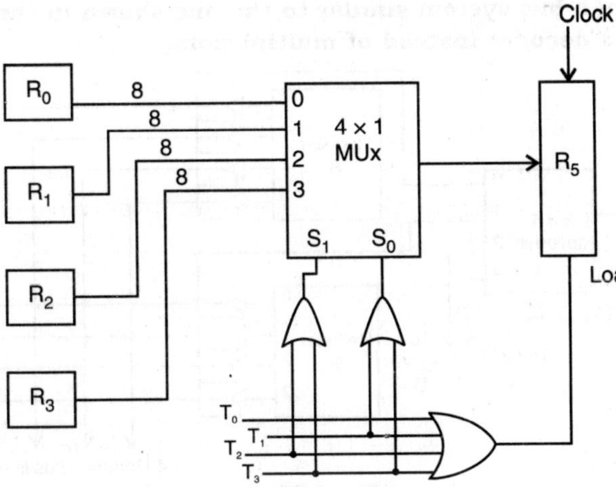

Fig. 1.18.

Q.8. Represent the following conditional, control statement by two register transfer statements with control functions.

If (P = 1) then

($R_1 \leftarrow R_2$)

else if (Q = 1) then

($R_1 \leftarrow R_3$)

Solution: $P : R_1 \leftarrow R_2$

$P^1Q : R_1 \leftarrow R_3$

Q.9. What has to the bus system of figure given below to be able to transfer information from any register to any other register ? Specifically. Show the connections that must be included to provide a path from the outputs of register C to the inputs of register A.

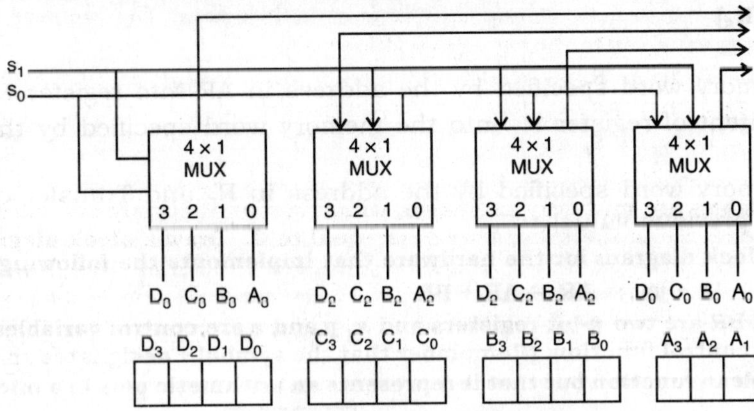

Fig. 1.19. Bus system for four register

Solution: Connect the 4–line common bus to the fair inputs of each register. Provide a "load" control input in each register provide a clock input for each register to transfer from register C to register A.

Apply $S_1S_0 = 10$ (to select C for the bus)

Enable the load input of A and apply a clock pulse.

Q.10. Draw a diagram of a bus system similar to the one shown in the figure below but use three-state buffers and a decoder instead of multiplexers.

Solution:

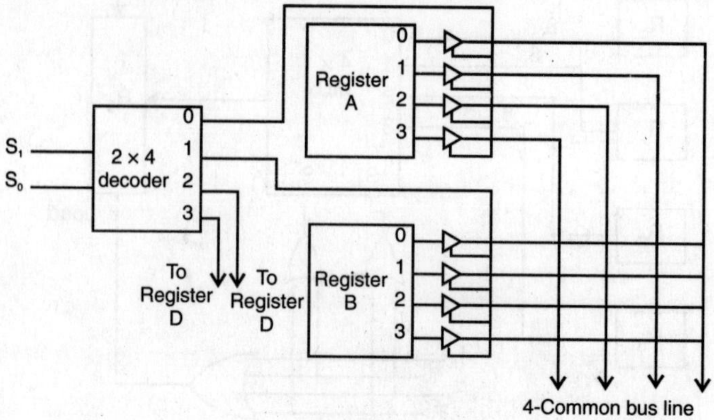

Fig. 1.20.

Q.11. A digital computer has a common bus system for 16 registers of 32 bits each. The bus is constructed with multiplexers.

(a) How many selection inputs are there in each multiplexers?

(b) What size of multiplexers are needed?

(c) How many multiplexers are there in the bus?

Solution: (*a*) 4 selection lines to select one of 16 register

(*b*) 16 × 1 multiplexers are needed.

(*c*) 32 multiplexers, one for each bit of the registers.

Q.12. The following transfer statements specify a memory. Explain the memory operation in each case.

(a) $R_2 \leftarrow M[AR]$

(b) $M[AR] \leftarrow R_3$

(c) $R_5 \leftarrow M[R_5]$

Solution:

(*a*) Read memory word specified by the address in AR into register R_2

(*b*) Write content of register R_3 into the memory word specified by the address in AR

(*c*) Read memory word specified by the address in R_5 and Transfer content to R_5 (destroys previous value)

Q.13. Draw the block diagram for the hardware that implements the following statements :

$$x + yz : AR \leftarrow AR + BR$$

Where AR and BR are two *n*-bit registers and *x*, *y* and *z* are control variables. Include the logic gates for the control function. (Remember that the symbol + designates an OR operation in a control or Boolean function but that it represents an arithmetic plus in a micro-operation)

Solution:

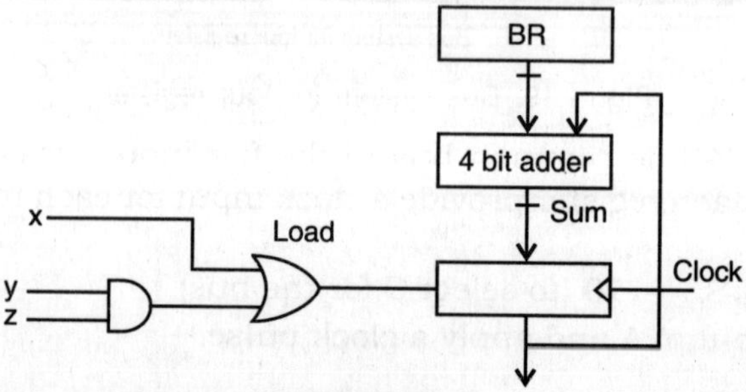

Fig. 1.21.

Show the hardware that implements the following statement. Include the logic gates for the control function and a block diagram for the binary counter with a count enable input.

$$xyT_0 + T_1 + y^1T_2 \ : \ AR \leftarrow AR + 1$$

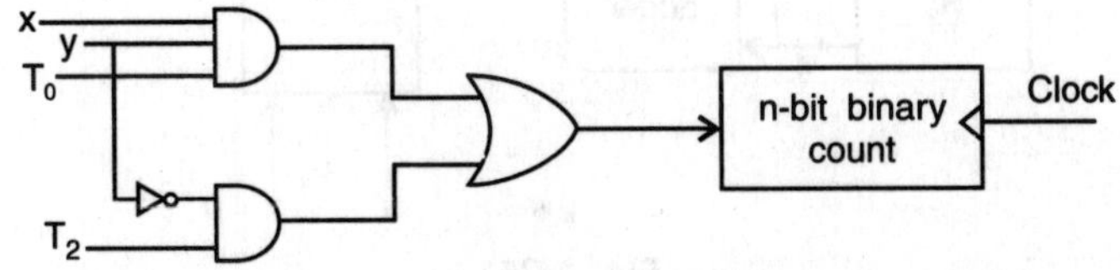

Fig. 1.22.

Q.14. Consider the following register transfer statements for two 4–bit register R_1 and R_2

$$xT \ : \ R_1 \leftarrow R_2$$
$$x^1T \ : \ R_1 \leftarrow R_2$$

Every time that variable $T = 1$, either the content of R_2 is added to the content of R_1 if $x = 1$, or the content of R_2 is transferred to R_1 if $x = 0$. Draw a diagram showing the hardware implementation of the two statements. Use block diagrams for the two. 4-bit registers, 4-bit adder and a quotable 2-to-1 line multiplexer that selects the inputs to R_1. In the diagram, show how the control variables x and T select the inputs of the multiplexers and the load input of register R_1.

Solution:

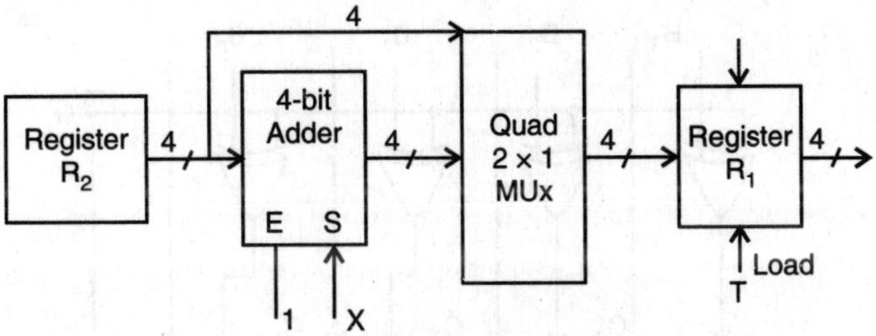

Fig. 1.23.

Q.15. Using a 4-bit counter with parallel load and a 4-bit adder as in figure given below, draw a block diagram that shows how to implement the following statements

$$x \ : \ R_1 \leftarrow R_1 + R_2 \qquad \textbf{Add } R_2 \textbf{ to } R_1$$
$$x^1y \ : \ R_1 \leftarrow R_1 + 1 \qquad \textbf{Increment } R_1$$

Where R_1 is a counter with parallel load and R_2 is a 4-bit register.

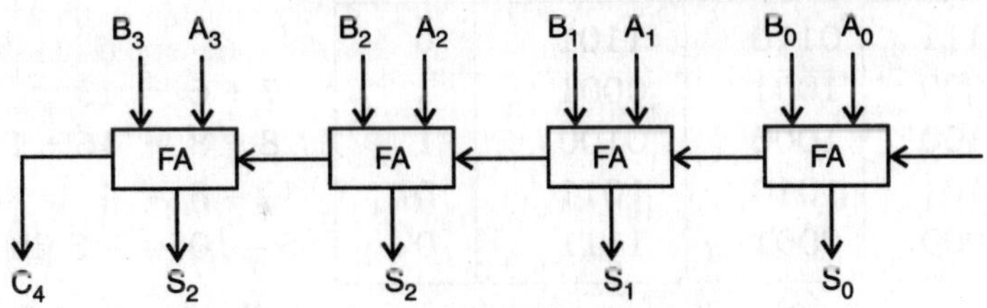

Fig. 1.24.

Solution:

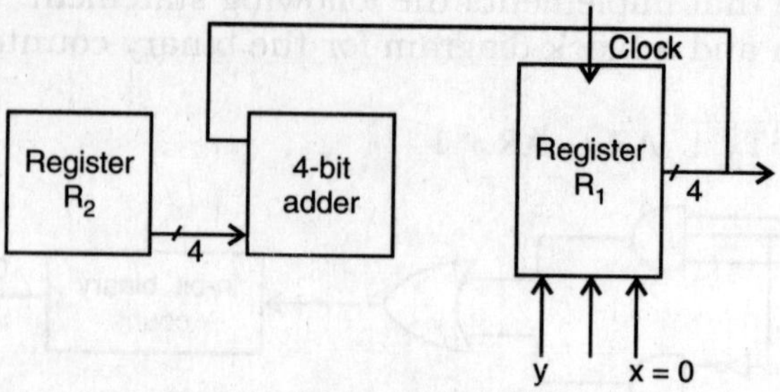

Fig. 1.25.

Q.16. The adder-subtracter circuit of figure given below has the following values for input mode M and data inputs A and B. In each case, determine the values of the outputs : S_3, S_2, S_2, S_1, S_0 and C4

	M	*A*	*B*
(a)	0	0111	0110
(b)	0	1000	1001
(c)	1	1100	1000
(d)	1	0101	1010
(e)	1	0000	0001

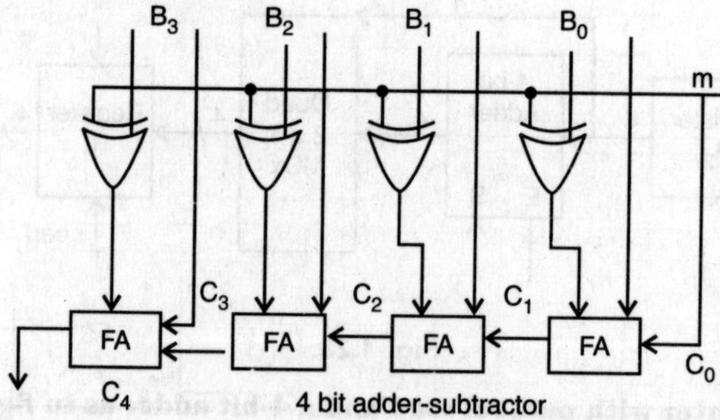

Fig. 1.26.

Solution:

M	*A*	*B*	*Sum*	C_4	
0	0111	0110	1101	0	
0	1000	1001	0001	1	7 + 6 = 13
1	1100	1000	0100	1	8 + 8 = 16 + 1
1	0101	1010	1011	0	12 – 8 = 4
1	0000	0001	1111	0	5 – 10 = – 5 (in 2's comp. form)

0 – 1 = – 1 (in was comp. form)

Q.17. Design a 4-bit combinational circuit decrements using four full adder circuits.

Solution: A 's complement of 1 = P + 1111

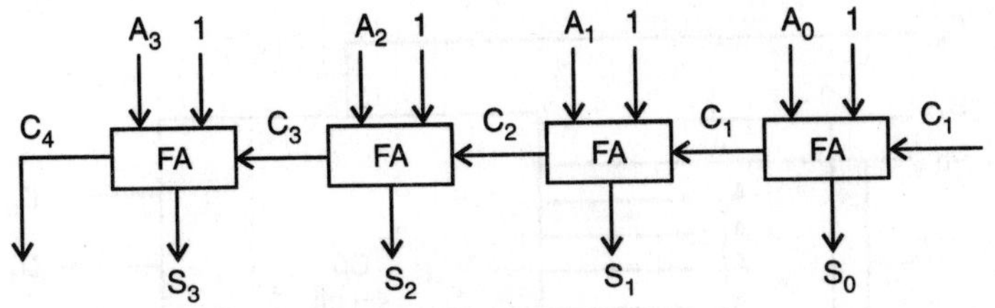

Fig. 1.27.

Q.18. Assume that the 4-bit arithmetic circuit of figure given below is enclosed in on IC package. Show the connections among two such ICs form an 8-bit arithmetic circuit.

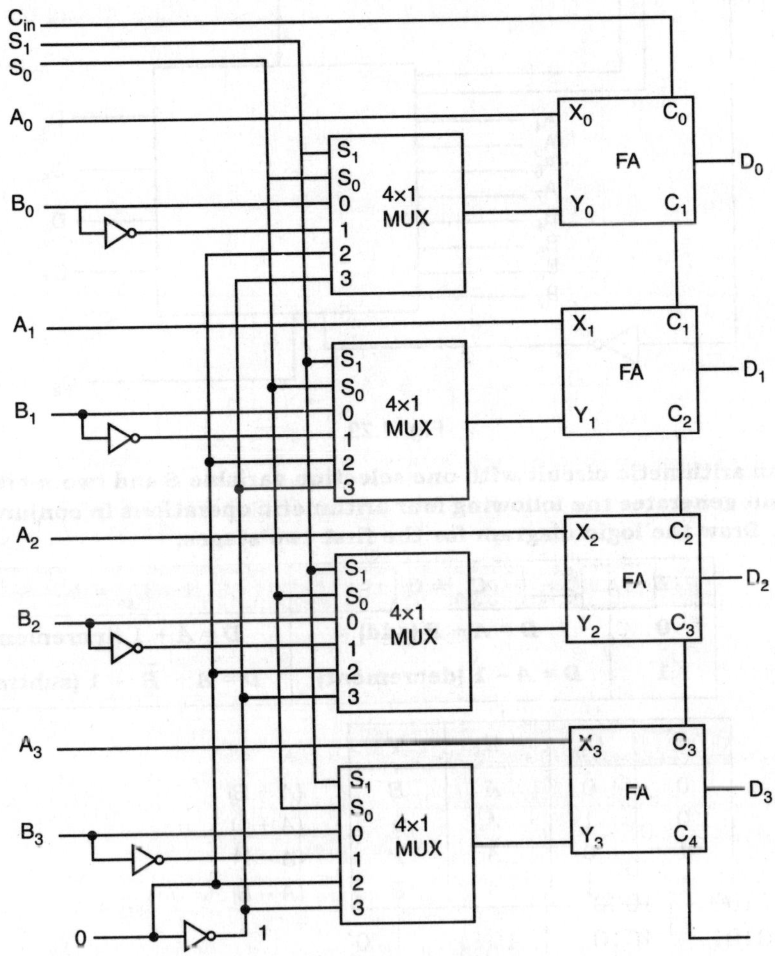

Fig. 1.28. 4-bit arithmetic circuit

Solution:

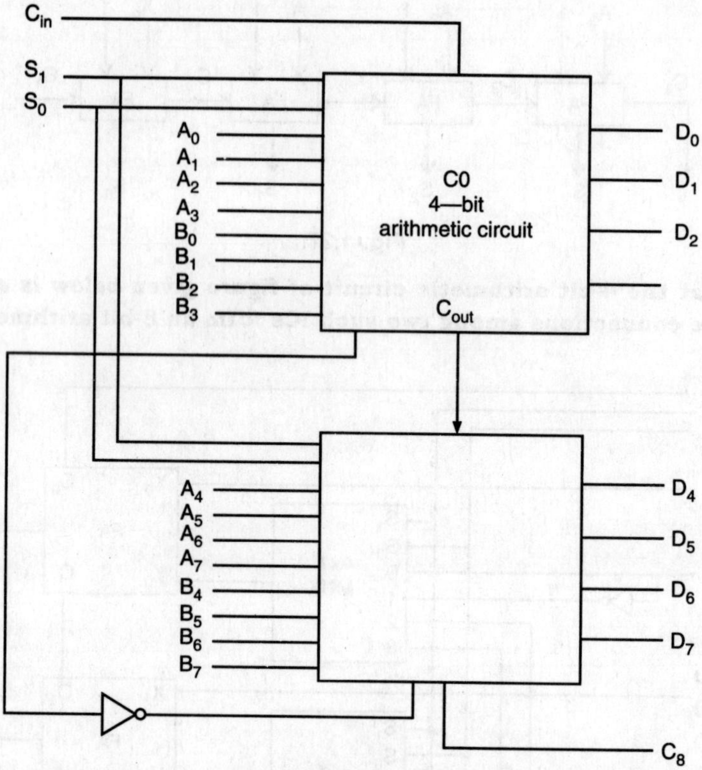

Fig. 1.29.

Q. 19. Design an arithmetic circuit with one selection variable *S* and two *n*-bit data inputs *A* and *B*. The circuit generates the following four arithmetic operations in conjunction with the input carry C_{in}. Draw the logic diagram for the first two stages.

S	$C_{in} = 0$	$C_{in} = 1$
0	$D = A + B$ (add)	$D = A + 1$ (increment)
1	$D = A - 1$ (decrement)	$D = A + \overline{B} + 1$ (subtract)

Solution:

S	C_{in}	*X*	*Y*	
0	0	*A*	*B*	$(A + B)$
0	1	*A*	0	$(A + 1)$
1	0	*A*	1	$(A - 1)$
1	1	*A*	*B*	$(A - B)$

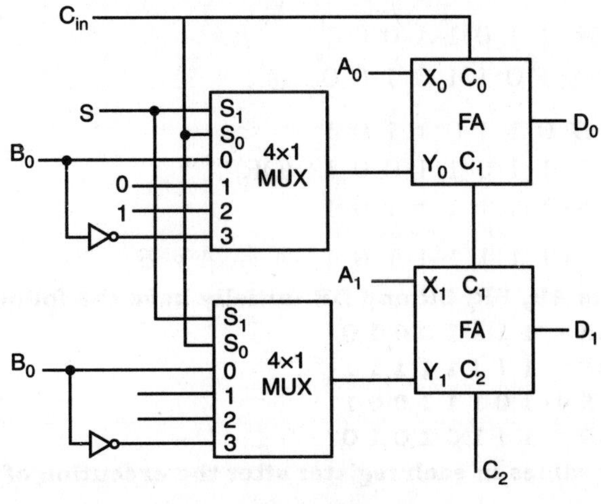

Fig. 1.30.

Q.20. Derive a combinational circuit that selects and generates any of the 16 logic functions listed in Table given below.

Truth Table for 16 Functions of Two Variables

x	y	F_0	F_1	F_2	F_3	F_4	F_5	F_6	F_7	F_8	F_9	F_{10}	F_{11}	F_{12}	F_{13}	F_{14}	F_{15}
0	0	0	0	0	0	0	0	0	0	1	1	1	1	1	1	1	1
0	1	0	0	0	0	1	1	1	1	0	0	0	0	1	1	1	1
1	0	0	0	1	1	0	0	1	1	0	0	1	1	0	0	1	1
1	1	0	1	0	1	0	1	0	1	0	1	0	1	0	1	0	1

Solution:

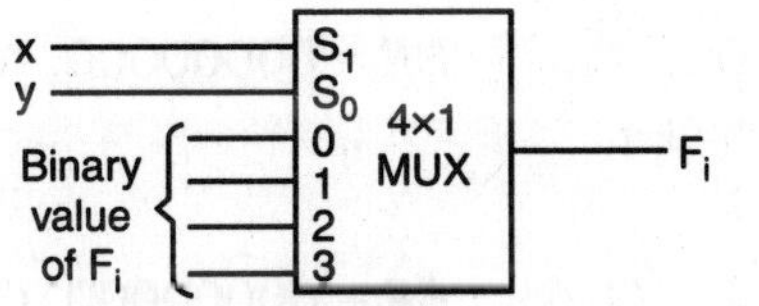

Fig. 1.31.

Q.21. Design a digital circuit that performs the four logic operations of exclusive OR, exclusive NOR, NOR and NAND. Use two selection variables. Show the logic diagram of one typical stage.

Solution:

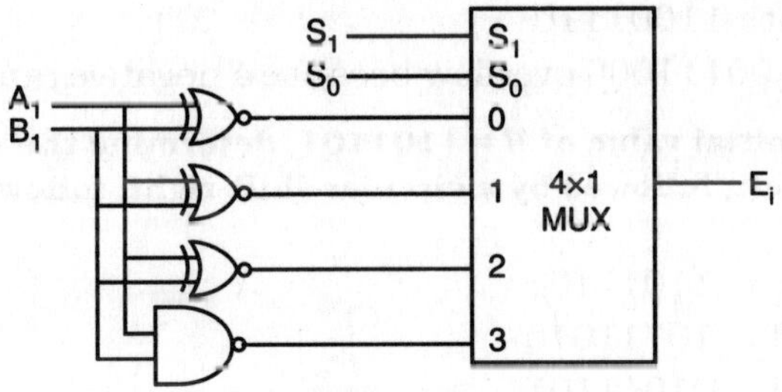

Fig. 1.32.

Q.22. Register *A* holds the 8-bits binary 11011001. Determine the *B* operand and the logic micro-operation to be performed in order to change the value in *A* to :

(*a*) 01101101

(*b*) 11111101

Solution:

(*a*) A = 1 1 0 1 1 0 0 1

B = 1 0 1 1 0 1 0 0 ⊕

$A \leftarrow A \oplus B$ 0 1 1 0 1 1 0 1

(*b*) A = 1 1 0 1 1 0 0 1 (OR)

B = 1 1 1 1 1 1 0 1

1 1 1 1 1 1 0 1 $A \leftarrow A \vee B$

Q.23. The 8-bit registers AR, BR, CR and DR initially have the following values :

AR **= 1 1 1 1 0 0 1 0**

BR **= 1 1 1 1 1 1 1 1**

CR **= 1 0 1 1 1 0 0 1**

DR **= 1 1 1 0 1 0 1 0**

Determine the 8-bit values in each register after the execution of the following sequence of micro-operations.

$AR \leftarrow AR + BR$	**Add** ***BR*** **to** ***AR***
$CR \leftarrow CR \wedge DR$, $BR \leftarrow +1$	**AND** ***DR*** **to** ***CR*****, increment** ***BR***
$AR \leftarrow AR - CR$	**Subtract** ***CR*** **from** ***AR***

Solution:

(*a*) AR = 1 1 1 1 0 0 1 0

BR = 1 1 1 1 1 1 1 1 (+)

AR = 1 1 1 1 0 0 0 1 BR = 11111111, CR = 10111001, DR = 11101010

(*b*) CR = 1 0 1 1 1 0 0 1 BR = 11111111

DR = 1 1 1 0 1 0 1 0 (AND)

CR = 1 0 1 0 1 0 0 0 BR = 00000000, AR = 11110001, DR = 11101010

(*c*) AR = 1 1 1 1 0 0 0 1 (–)

CR = 1 0 1 0 1 0 0 0

AR = 0 1 0 0 1 0 0 1 BR = 00000000, CR = 10101000, DR = 11101010

Q.24. An 8-bit register contains the binary value 10011100. What is the register value after an arithmetic shift right ? Starting from the initial number 10011100, determine the register value after an arithmetic shift left, and state whether there is an overflow.

Solution: R = 10011100

Arithmetic shift right : 11001110

Arithmetic shift left : 00111000 overflow because a negative number changed to positive.

Q.25. Starting from an initial value of ***R*** **= 1101101, determine the sequence of binary values in R after a logical shift-left, followed by a circular shift-right, followed by a logical shift-right and a circular shift-left.**

Solution: R = 11011101

Logical shift left : 10111010

Circular shift right : 01011101

Logical shift right : 00101110

Circular shift left : 01110100

Q.26. What is the value of output ***H*** **in figure given below if input is .....**

I_R = 1 and I_L = 0 ?

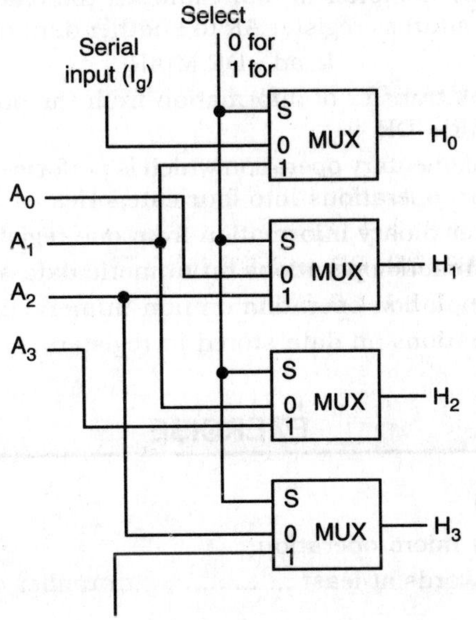

Fig. 1.33. 4-bit combinational circuit shifter

Function Table

Select	*Output*			
S	H_0	H_1	H_2	H_3
0	I_R	A_0	A_1	A_2
1	A_1	A_2	A_3	I_L

Solution: S = 1 Shift left

A_0	A_1	A_2	A_3	I_L	
1	0	0	1	0	
H = 0	0	1	0		Shif left

SUMMARY

1. Computer registers are designated by capital letters.
2. Symbols are a handy tool to represent the micro-operation sequences in registers and the control functions that initiate them, in a lucid & concise form. Such a symbolic notation is referred to as register transfer language.
3. A bus structure consists of a set of common lines, one for each bit of a register, through which binary information is transferred one at a time. Control signals determine which register is selected by the bus during each particular register transfer.
4. A micro-operation is an elementary operation performed with the data stored in registers.
5. These micro-operations perform some basic arithmetic operations on the numeric data stored in the registers.
6. A read operation implies transfer of information to the outside environment from a memory word, whereas storage of information into the memory is defined as write operation. Symbolizing a memory word by the letter M, it is selected by the memory address during the transfer

which is a specification for transfer operations. The address is specified by enclosing it in square brackets following the letter M. For example, the read operation for the transfer of a memory unit M from an address register AR to another data register DR can be illustrated as:

Read : DR M[AR]

7. A write operation implies transfer of information from the outside world to a memory word. For example, Write: M[AR]<-DR.
8. A micro-operation is an elementary operation which is performed on the data stored in registers. We can classify the micro-operations into four categories:
9. Register transfer: transfer binary information from one register to another.
10. Arithmetic: perform arithmetic operations on numeric data stored in registers.
11. Logic: perform bit manipulation operation on non-numeric data stored in registers.
12. Shift: perform shift operations on data stored in registers.

EXERCISE

(I) Fill in the Blanks (Solved)

1. .... is shift left micro-operation.
2. To address 35 memory words at least .... number of address bits are required.

(II) True or False (Solved)

1. Memory read micro-operation takes 10 clock cycles.
2. A microprogram is a set of micro-operations.

(III) Fill in the Blanks (Unsolved)

1. The channel through which CPU and memory accomplish using ....
2. Mirco-operation for adding content of memory location 5 to the content of register R1 is ....

(IV) True or False (Unsolved)

1. Accumulator is a decoder.
2. Control unit is a part of ALU.

(V) Detailed Questions

1. Starting from an initial value of R=11011101, determine the sequence of binary values in R after a logical shift-left, followed by a circular shift-right followed by a logical shift right and a circular shift left.
2. What is wrong with the following register transfer statements?
   (*i*) $xT: AR \leftarrow AR, AR \leftarrow 0$ (*ii*) $yT: R1 \leftarrow R2, R1 \leftarrow R3$
   (*iii*) $zT: PC \leftarrow AR, PC \leftarrow PC + 1$.

Answers

(I) Fill in the Blanks (Solved)

1. Shl 2. 6

(II) True or False (Solved)

1. False 2. True

Chapter

2

Processor Level Design and Processor Basic

Chapter Outline

- ➤ Introduction
- ➤ Processor Level Components
- ➤ CPU Organization-Fundamentals
- ➤ Instruction Unit
- ➤ Data Processing Unit
- ➤ Program Unit
- ➤ Memory and I/O Interface Unit
- ➤ Register Organization
- ➤ Register Stack
- ➤ Memory Stack
- ➤ Infix and Post Fix Notation
- ➤ Instruction Formats
- ➤ Addressing Modes

2.0 INTRODUCTION

At the next higher level, we have the architecture level. This level deals with memories and processing units that store and process blocks of data words known as program instructions and program data.

The interconnection of the digital system to the outside world is also

dealt with at this level through input/output circuits.

2.1 PROCESSOR LEVEL COMPONENTS

The major components at this level are: CPU, memory and I/O circuits.

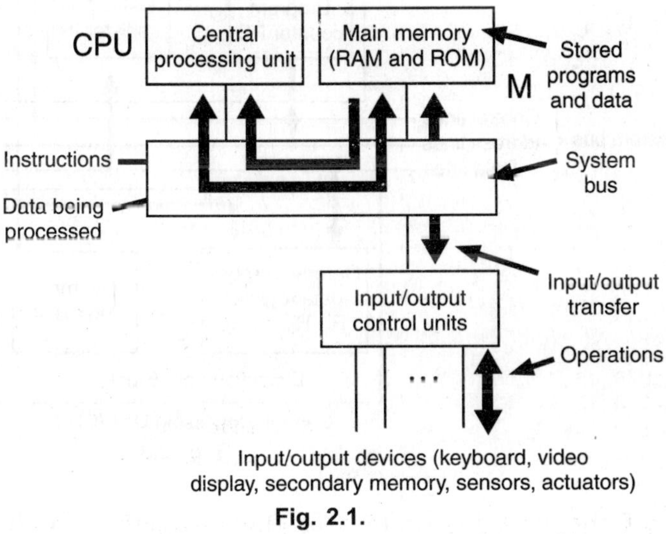

Fig. 2.1.

The operations of these components are determined by executing a series of control words, or program instructions.

The primary function of the CPU is to fetch instructions from the main memory and execute them.

The CPU can be dived into 2 parts: the control part and the data part. The control part is called the *instructions unit (I-unit).* It contains an Instruction Register (IR), a Program Counter (PC), an instruction decoder.

The data part is called the *excution unit (E-mail).* It contains at least a datpath and a set of registers for temporary data storage. Some E-units also contains a fast local memory called the *cache memory.*

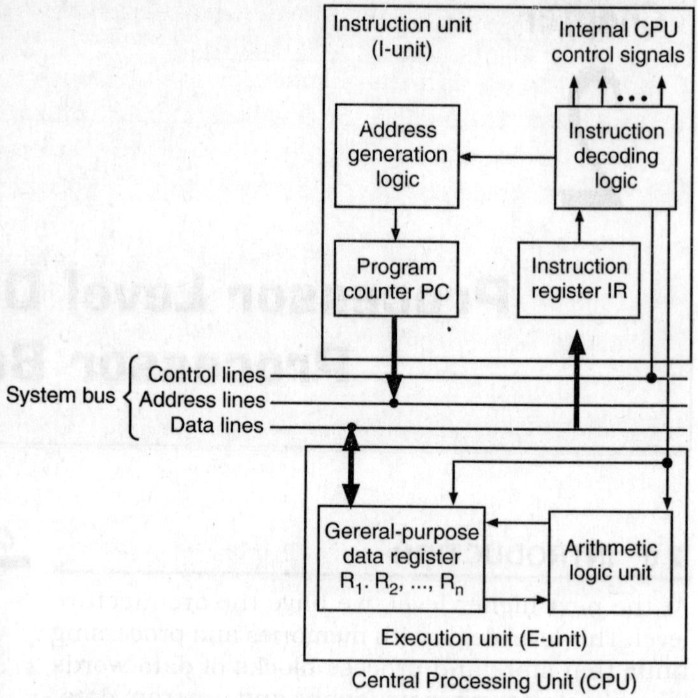

Fig. 2.2.

The Insturction Unit (I-unit)

The I-unit is responsible for generating the control signals to ensure the proper operations and correct data flow are performed on the input data.

Its components are:

- Instruction register
- Program counter
- Instruction decoder
- Address generator

In order to perform computing operations, a CPU executes instructions from a program written in its machine language. The machine language of a CPU is simply a collection of instructions (called the instruction set) that define all the possible operations of the CPU.

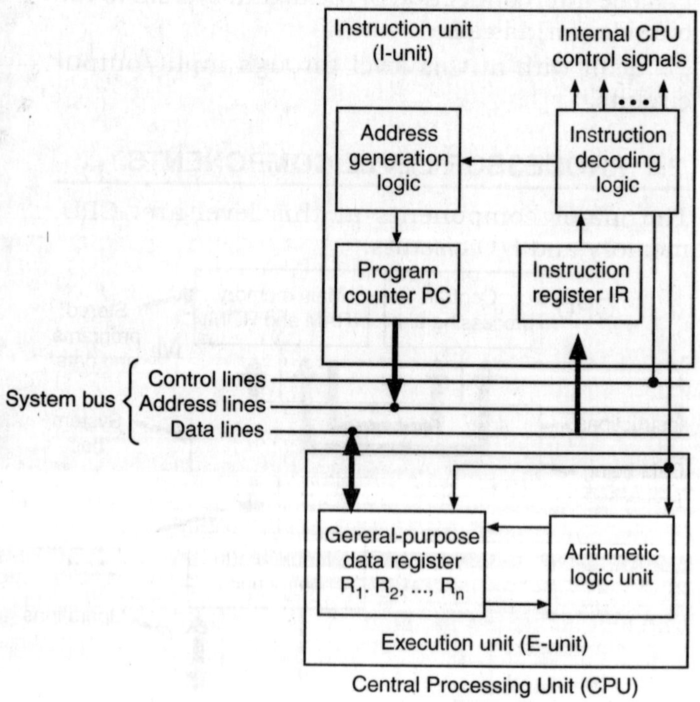

Fig. 2.3.

A typical sequence of operations performed by the I-Unit is:

1. Load the location (address) of the first instruction of the program into the Program Counter (PC).

2. Load the instruction into the instruction register (IR).
3. Initiate the instruction decoder to decode the instruction so that the appropriate control and address signals for the datapath unit are generated.
4. Increment the value of the Program Counter (PC) by 1.
5. Repeated steps 2 to 4 until the end of the program is reached.

The datapath consists of at least 2 main components: the register file, and the ALU.

- The ALU performs logical and arithmetic operations on the input data.
- The register file provides a set of fast local storage locations.

The interconnections of these units determine the allowable data flow paths. The properties of these units and how they are connected together define the architecture of the CPU.

Some datapaths also contain a local memory called the cache memory. It allows the CPU to access the main memory (M) more efficiently.

Instruction Set

A machine language instruction is a binary-encoded word. In other words, it is just a binary code. The code is derived from the property of the datapath and the data flow paths.

A machine language instruction usually consists of two parts: the opcode and the operand. Some instructions do not have operands.

For example:

The processor in Lab 7 uses 8-bit machine language instructions. For instructions with operands, the most significant 4-bits used as opcode and the remaining 4-bits as operand. For instructions without operands, the most significant 4-bits and the least significant bit form a 5-bit opcode. The instruction set of a typical commercial CPU is quite large, but can be categorized into the following groups:

1. Data-transfer instructions (load, store and I/O)
2. Data-processing instructions (arithmetic and logic operations)
3. Program-control instructions (conditional and unconditional branching)

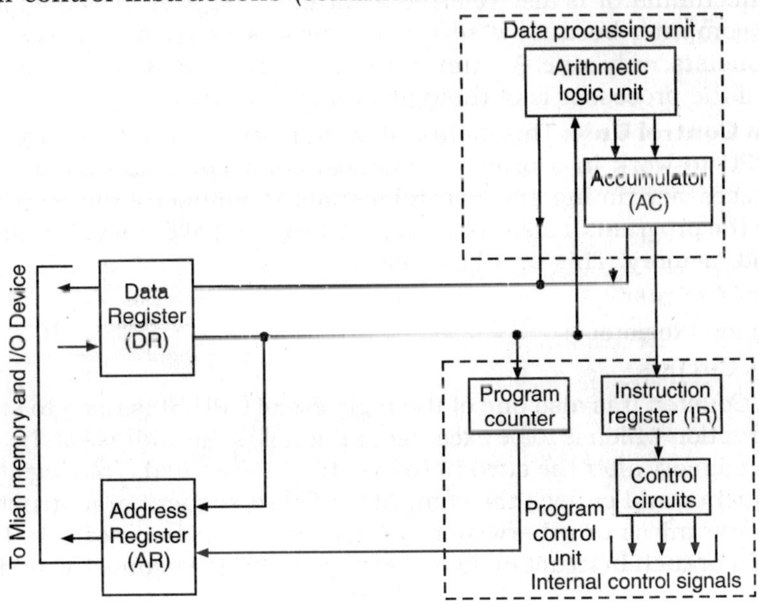

Fig. 2.4.

The figure below shows the simple accumulator based CPU given by Neumann which is the basis for almost all CPUs. It consists minimum number of registers and necessary circuits. In this CPU one register called the accumulator is used to store an input data or an output result in the execution of most instructions. Hence it is known as accumulator based CPU.

The accumulator is the main register of the arithmetic logic unit. The data register acts as buffer between the CPU and main memory. The other major registers are the program counter PC which stores the address of the next instruction which is to be executed. Instruction register IR which holds the opcode of the current instruction which is to be executed and the memory address register AR which holds the address of the current instruction.

The accumulator based CPU consists the following units:

1. Data Processing Unit
2. Program Control Unit
3. Memory and I/O interface Unit

1. Data Processing Unit: This is the unit where the data is processed to get the output i.e., some results. This unit can be subdivided into:

(*i*) Arithmetic Logic Unit ALU
(*ii*) Accumulator AC

(*i*) *Arithmetic Logic Unit*: The arithmetic logic unit performs all arithmetic and logical operations. The necessary data is supplied by the accumulator and other registers. It performs the required operation on the operands contained in the instruction whenever it is activated by control unit. The results of all the arithmetic and logical operations are stored back in accumulator.

(*ii*) *Accumulator*: The accumulator is the main operand register of the1 arithmetic logic unit. The data is first loaded into the accumulator before it is transferred to the arithmetic logic unit for any operation. In performing these operations generally one of the operand is held in the accumulator. The result after performing the operation is stored again in accumulator. Accumulator is also referred as primary register. It is the most complicated register in the computer because it serves as a processor register. A simple accumulator based CPU consists only one Accumulator. It is also possible to have two or more accumulators if the processor is of the type multiprocessor.

2. Program Control Unit: This is the subsystem of CPU which helps in making various parts of the CPU to work in a proper sequence, *i.e.* it controls various parts of CPU. It provides the necessary timing and control signals to all operations depending upon the instructions in the program. It also controls the movement of data within the CPU registers, I/O devices and memory. This unit basically consists of:

(i) Program Counter
(ii) Instruction Register
(iii) Control Circuits.

(i) *Program Counter*: It is also one of the registers of CPU. It is used to store the address of the next instruction which is to be executed, *i.e.* it holds the address of the next instruction to be read from memory after the current instruction is executed. This register goes through a counting sequence and causes the computer to read sequential instructions previously stored in memory unless or otherwise a branch instruction comes into the picture. The address part of a branch instruction is transferred to PC to become the address of the next instruction.

(*ii*) *Instruction Register*: This register which holds the opcode of the current instruction is one of the high speed registers under the program control unit of the CPU. Generally this register is not directly available to the programmer.

The operation code (Opcode) of an instruction is a group of bits that define such operations as add, subtract, multiply, division, shift, and compliment, etc.

(*iii*) *Control Circuits*: This is the most important part of the program control unit. It is responsible for all-most every operation that takes place in CPU. These circuits provide the necessary control signals for fetching instructions decoding the opcodes. In most cases the output of the instruction register is given to these circuits. The internal control signals are the output of these circuits.

3. Memory and I/O interface Unit: To this unit I/O devices and memory are connected. This unit consists of

(*i*) Data register and

(*ii*) Address registers.

(*i*) *Data Registers*: This register DR acts as a buffer between the CPU and main memory. It is used as input register to store an operand during the execution of an instruction.

(*ii*) *Address Register*: This contains the address of the present instruction which is obtained from the program control unit. In most of the cases the address is contained in the memory hence it is also known as Memory Address Register MAR.

In addition to this, the accumulator based CPU consists special registers such as stack pointer and flag registers. The stack pointer is used to hold the address of the most recently entered item into the stack. It always shows the top of the stack. Some times temporarily it holds the contents of program counter. The operations performed with in the stack are very fast.

The flag registers indicates the status. Depending upon the result the flags are set. Most commonly used flags in the flag register are zero flag and carry flag.

2.2 CPU ORGANIZATION- FUNDAMENTALS

2.2.1 Processor Organization

The part of the computer that performs the bulk of data-processing operations is called CPU. The CPU is made up of three major parts, as shown in Fig. 2.5. The register set stores intermediate data used during the execution of the instructions. The ALU performs the required micro-operations for executing the instructions. The control unit supervises the transfer of information among the registers and instructs the ALU as to which operation to perform.

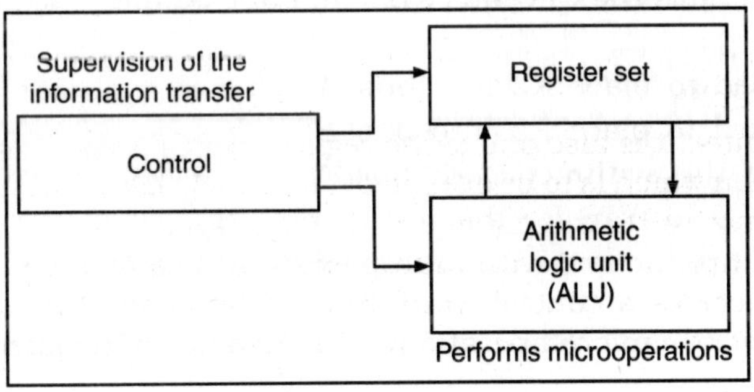

Fig. 2.5. Major components of CPU

Instruction set provides the specification for the design of the CPU. The design of CPU involves choosing the hardware for implementin, the machine instructions. In the following two sections we present the two main CPU organizations.

A programmer using assembly language has to be aware of the register set, the memory structure, the type of data supported by the instructions, and the function of each

2.2.2 General Register Organization

Storing pointers, counters, return addresses, temporary results, and partial products into the memory is not efficient: referring to memory locations is time consuming, it is more efficient to use registers. If CPU has a large number of registers, a common bus is used to connect the registers. Arithmetic Logic Unit (ALU) is used to perform various micro-operations. A bus organization of seven CPU registers is shown in Fig. 2.6.

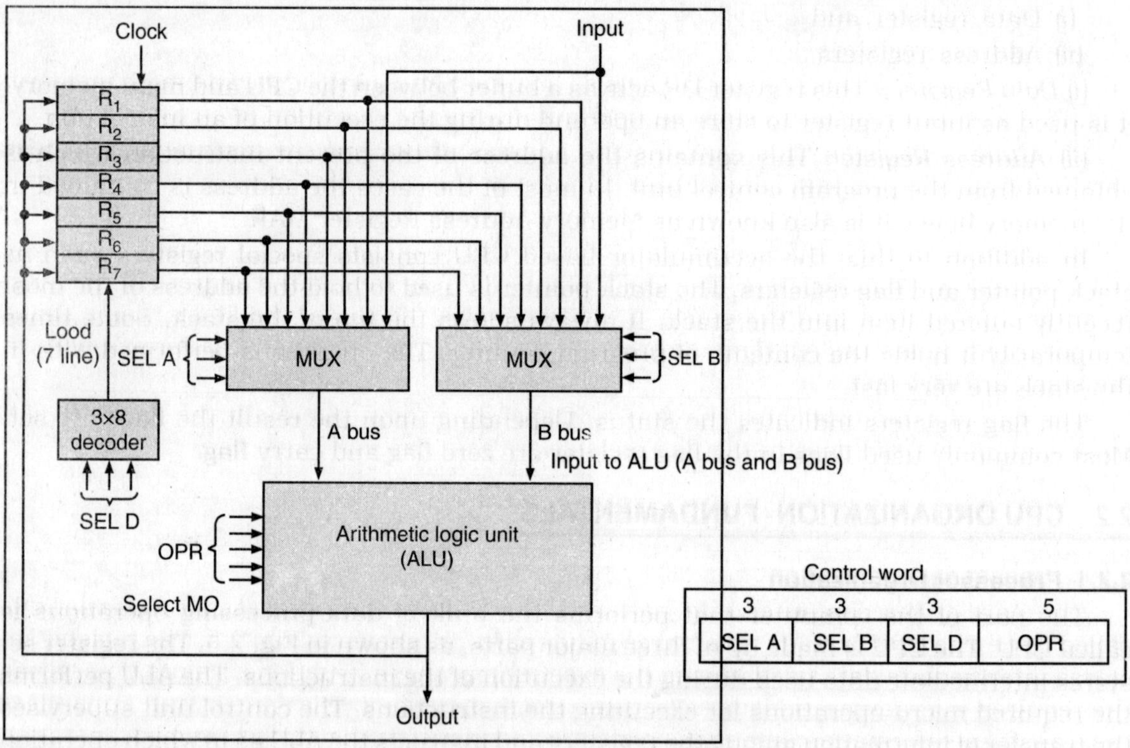

Fig. 2.6. Register set with common ALU

Control unit operates the CPU bus system. For example:

*R*1 *R*2 + *R*3

1. SELA is used to place R2 into bus *A*.
2. SELB is used to place R3 into bus *B*.
3. OPR selects the arithmetic addition.
4. SELD is used to transfer the result into *R*1.

The buses are implemented with multiplexers of 3-state gates. The state of 14 binary selection inputs specifies a control word. The 14-bit control word (when applied to the selection inputs specify a micro-operation). The encoding of register selections in specified in Table 2.1.

Table 2.1. Encoding of register selection fields

Binary Code	SELA	SELB	SELD
000	Input	Input	None
001	R1	R1	R1
010	R2	R2	R2
011	R3	R3	R3
100	R4	R4	R4
101	R5	R5	R5
110	R6	R6	R6
111	R7	R7	R7

Input = external input

If SELD = 000 then the content of the output in bus is available in the external output

ALU provides arithmetic and logic operations. The function table of the ALU (presented here) is listed in Table 2.2.

Table 2.2. Encoding of ALU operations

OPR Select	Operations	Symbol
00000	Transfer *A*	TSFA
00001	Increment *A*	INCA
00010	Add *A* + *B*	ADD
00101	Subtract A – B	SUB
00110	Decrement A	DECA
01000	AND A and B	AND
01010	OR A and B	OR
01100	XOR A and B	XOR
01110	Complement A	COMA
10000	Shift right A	SHRA
11000	Shift left A	SHLA

Examples

There are some examples of micro-operations and corresponding control words:

Table 2.3. Examples of micro-operations for the CPU

Symbolic Designation					
Microoperation	**SEL A**	**SEL B**	**SEL D**	**OPR**	**Control Word**
R1 ← R2 ← R3	R2	R3	R1	SUB	010 011 001 00101
R4 ← R4 V R5	R4	R5	R4	OR	100 101 100 01010
R6 ← R6 + 1	R6	—	R6	INCA	110 000 110 00001
R7 ← R1	R1	—	R7	TSFA	001 000 111 00000
Output ← R2	R2	—	None	TSFA	010 000 000 00000
Output ← Input	Input	—	None	TSFA	000 000 000 00000
R4 ← sh1 R4	R4	—	R4	SHLA	100 000 100 11000
R5 ← 0	R5	R5	R5	XOR	101 101 101 01100

(XOR (x,x) = 0

000 (not used)

The most efficient way to generate control words with a large number of bits is to store them in a memory unit, which is referred to as control memory (control store). By reading consecutive control words from memory, it is possible to initiate the desired sequence of microoperations for the CPU, microprogrammed control.

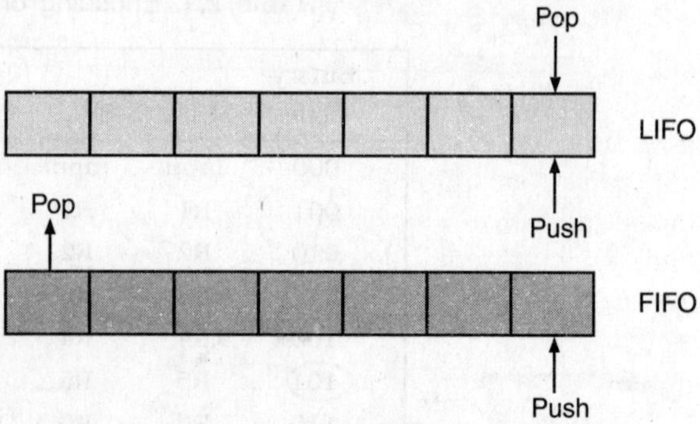

In computer science, a stack is a data structure that works on the principle of Last In First Out (LIFO). This means that the last item put on the stack is the first item that can be taken off, like a physical stack of plates. A stack-based computer system is one that is based on the use of stacks, rather than being register based.

2.2.3 Stack Organization

1. Register Stack

Stack can reside in a portion of a large memory unit or it can be organized as a collection of a finite number of (fast) registers. Fig. 2.7. organization of 64-word register stack.

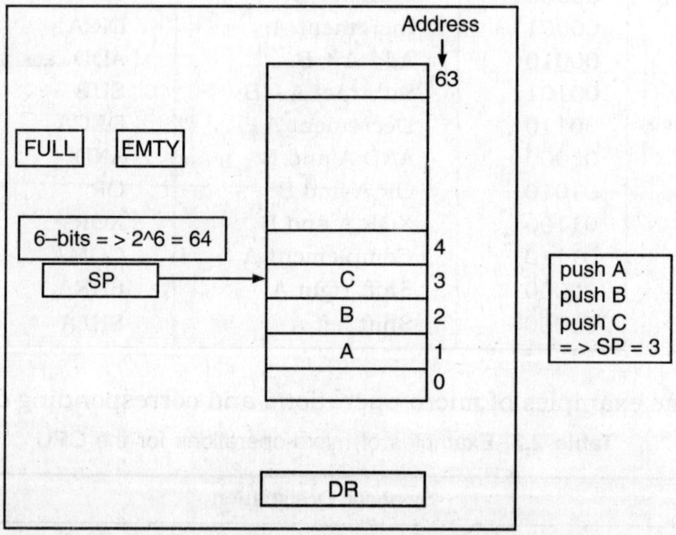

Fig. 2.7. Block diagram of a 64-word stack

1. Push (performed if stack is not full, *i.e.* if FULL = 0):

SP ← SP + 1	Increment stack pointer
M[SP] ← DR	Write item on top of the stack
If (SP = 0) then (FULL ← 1)	Check if stack is full
EMTY ← 0	Mark the stack not empty

2. Pop (performed if stack is not empty, *i.e.* if EMTY = 0):

DR ← M[SP]	Read item form the top of stack
SP ← SP – 1	Decrement stack pointer
If (SP = 0) then (EMPTY ← 1)	Check if stack is empty
FULL ← 0	Mark the stack not full

2. Memory Stack

Stack can also be implemented with RAM attached to a CPU: a portion of memory is assigned to a stack operation, a processor register is used as a stack pointer. Fig. 2.8 shows how a portion of memory partitioned into three segments: program, data, and stack. Most computer do not provide hardware for checking stack overflow or underflow if registers are used to store the upper limit (*e.g.* 3000) and the lower limit (*e.g.* 4000), then after push SP can be compared against the upper limit register, and after pop against the lower limit register. The advantage of the memory stack is that CPU can refer it without having to specify an address: the address in always in SP and automatically updated during a push or pop instruction.

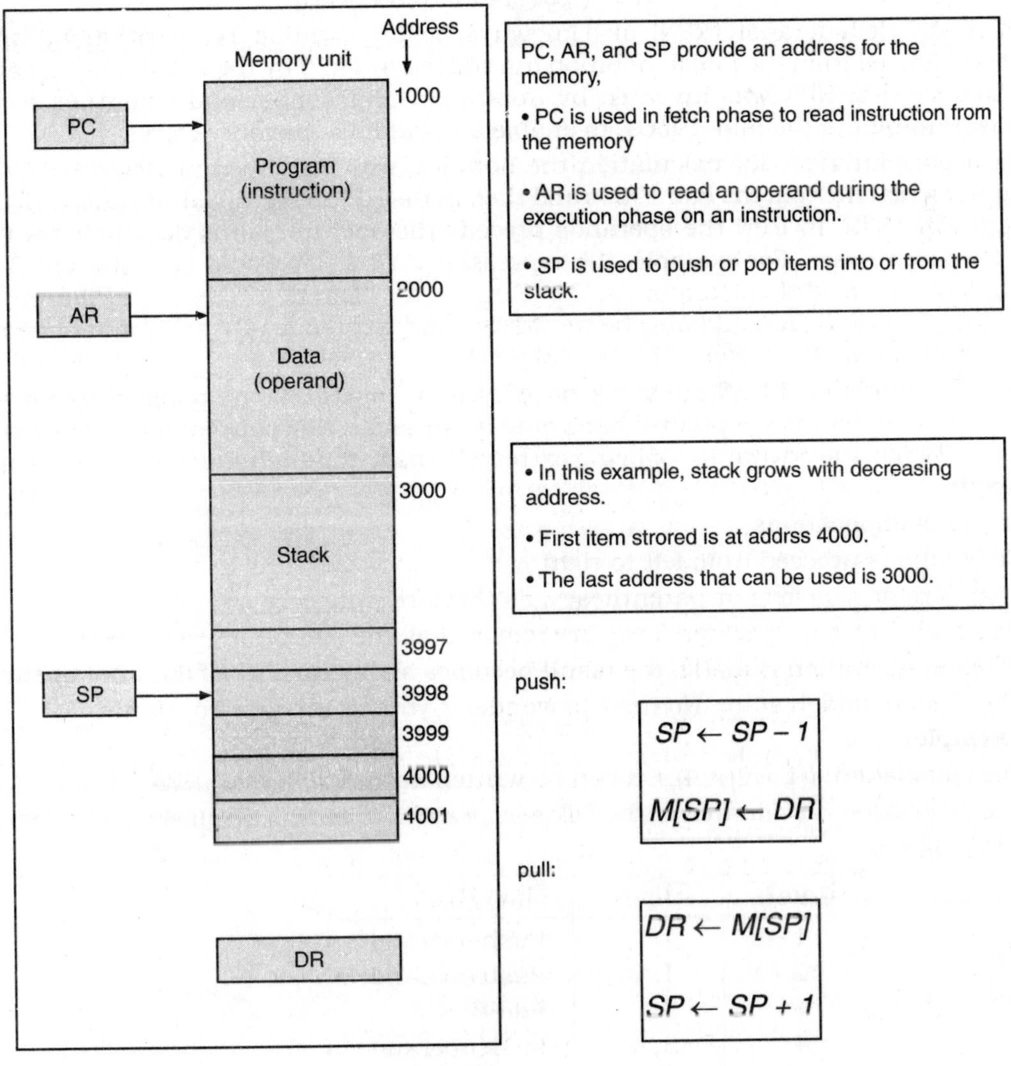

Fig. 2.8. Computer memory with program, data and stack segments

3. Reverse Polish Notation

A stack is effective for evaluating arithmetic expressions. Arithmetic operations are usually written in infix notation: each operator resides between the operands, *e.g.*:

(A * B) + (C * D),

where * denotes multiplication: A * B and C * D has to be computed and stored. after the two products, sum (A * B) + (C * D) is computed, there is no straight forward way to determine the next operation that is performed.

Arithmetic expressions can be presented in prefix notation (also referred to as Polish notation by Polish mathematician Lukasiewicz): operators are placed before the operands. The postfix notation (reverse Polish notation (RPN)) places the operator after the operand. *e.g.*:

A + B , infix notation

+AB , prefix notation

AB+ , postfix notation (RPN)

Reverse Polish notation (RPN), also known as postfix notation, is an arithmetic formula notation, derived from the Polish notation introduced in 1920 by the Polish mathematician Jan Lukasiewicz. RPN was invented by Australian philosopher and computer scientist Charles Hamblin in the mid-1950s, to enable zero-address memory stores.

As a user interface for calculation the notation was first used in Hewlett-Packard's desktop calculators from the late 1960s and then in the HP-35 handheld scientific calculator launched in 1972. In RPN the operands precede the operator, thus dispensing with the need for parentheses. For example, the expression 3 * (4 + 7) would be written as 3 4 7 + *, and done on an RPN calculator as "3", "Enter", "4", "Enter", "7", "+", "*". (Alternatively, and more-compactly, it could also be re-ordered and written as 4 7 + 3 *, and done on an RPN calculator as "4", "Enter", "7", "+", "3", "*".)

Implementations of RPN are stack-based; that is, operands are popped from a stack, and calculation results are pushed back onto it. Although this concept may seem obscure at first, RPN has the advantage of being extremely easy, and therefore fast, for a computer to analyze.

Practical implications

Calculations proceed from left to right

There are no brackets or parentheses, as they are unnecessary.

Operands precede operator. They are removed as the operation is evaluated.

When an operation is made, the result becomes an operand itself (for later operators).

There is no hidden state. No need to wonder if you hit an operator or not.

Example:

The calculation: ((1 + 2) * 4) + 3 can be written down like this in RPN: 1 2 + 4 * 3 +

The expression is evaluated in the following way (the Stack is displayed after Operation has taken place):

Input	Stack	Operation
1	1	Push operand
2	1, 2	Push operand
+	3	Addition
4	3, 4	Push operand
*	12	Multiplication
3	12, 3	Push operand
+	15	Addition

The final result, 15, lies on the top of the stack at the end of the calculation.

Converting from infix notation

Like the evaluation of RPN, conversion from infix notation to RPN is stack-based. Infix expressions are the form of math most people are used to, for instance 3+4 or 3+4*(2-1). For the conversion there are 2 text variables (strings), the input and the output. There is also a stack holding operators not yet added to the output stack. To convert, the program reads each letter in order and does something based on that letter.

A simple conversion

Input: 3+4

#Add 3 to the output queue (whenever a number is read it is #added to the output)

#Add 4 to the output queue

#Push + (or its ID) onto the operator stack

#After reading expression pop the operators off the stack and #add them to the output.

#In this case there is only one, "+".

#Output 3 4 +

This already shows a couple of rules:

- All numbers are added to the output when they are read.
- At the end of reading the expression, pop all operators off the stack and onto the output.

The algorithm in detail

- While there are tokens to be read:

  Read a token.

  - If the token is a number, then add it to the output queue.
  - If the token is an operator, o1, then:

1. while there is an operator, o2, at the top of the stack, and either o1 is left-associative and its precedence is less than or equal to that of o2, or o1 is right-associative and its precedence is less than that of o2, pop o2 off the stack, onto the output queue;
2. push o1 onto the operator stack.

  - If the token is a left parenthesis, then push it onto the stack.
  - If the token is a right parenthesis, then pop operators off the stack, onto the output queue, until the token at the top of the stack is a left parenthesis, at which point it is popped off the stack but not added to the output queue. If the stack runs out without finding a left parenthesis, then there are mismatched parentheses.
  - When there are no more tokens to read, pop all the tokens, if any, off the stack, add each to the output as it is popped out and exit. (These must only be operators; if a left parenthesis is popped, then there are mismatched parentheses.)

Example:

Consider the arithmetic expression

(3*4)+(5*6)

In RPN, it is expressed as

34*56*+

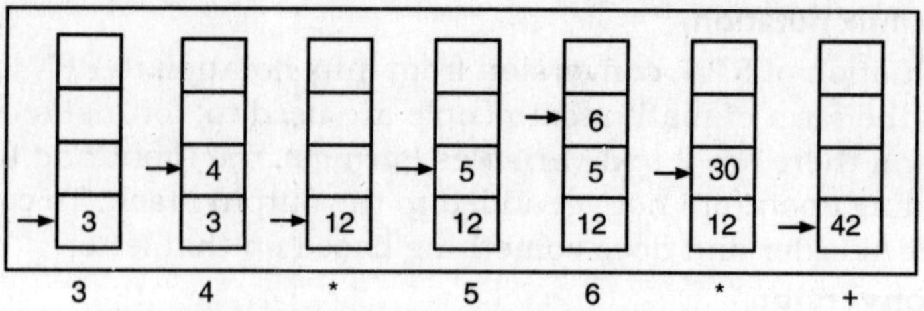

Fig. 2.9. Stack operations to evaluate 3*4+5*6

Example:

Consider the expression

$$X = (A + B)*(C + D)$$

Convert the expression to the reverse polish notation and show the stack operations for evaluating the result.

The conversion : $(A + B)*(C + D)$ à A B + C D + *

Instructions to be executed in stack:

```
PUSH  A
PUSH  B
ADD
PUSH  C
PUSH  D
ADD
MUL
POP   X
```

The stack operations will be as follows:

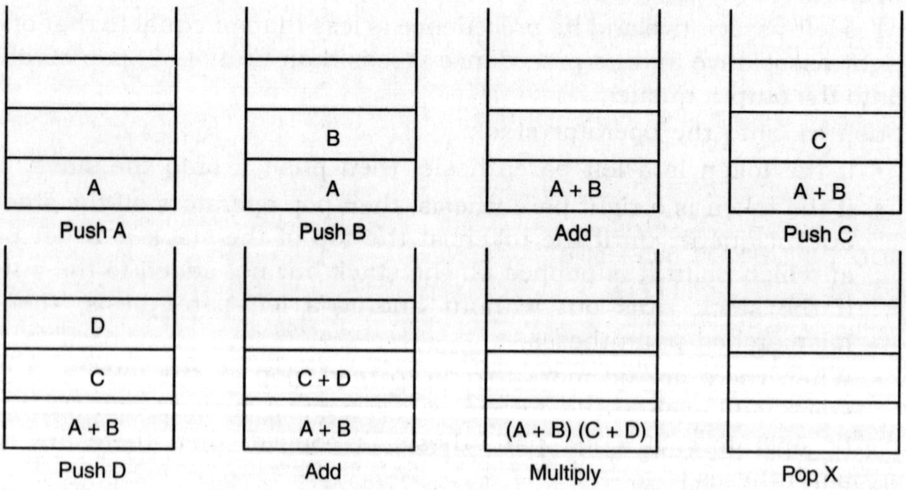

Fig. 2.10.

2.2.4 Additional Features

2.2.4.1 Instruction Format

The format of an instruction is usually depicted in a rectangular box symbolizing the bits of the instruction as they appear in memory words or in a control register. The bits of the instruction are divided into groups called fields. The most common fields found in instruction formats are:

1. Operation code specifying the operation: add, subtract, complement, etc.
2. Address field designating a memory address or a register
3. Mode field for specifying the way for determining the effective address of an operand.

The number of address fields in the instruction format of a computer depends on the internal organization of its registers. E.g.: MIPS (a RISC microprocessor architecture developed by MIPS Computer Systems Inc.).

Computers may have instructions of several different lengths containing varying number of addresses. The number of address fields in the instruction format of a computer depends on the internal organization of its register. Most computers fall into one of three types of CPU organizations:

1. Single accumulator organization (uses one address field).
2. General register organization (uses two or three address fields).
3. Stack organization (uses zero address field).

Example:

To illustrate the influence of the number of addresses on computer programs, we will evaluate the arithmetic statement using zero, one, two, or three address instructions.

$$X = (A + B)*(C + D)$$

1. Using three address instructions

```
ADD    R1, A, B     R1 ← M[A] + M[B]
ADD    R2, C, D     R2 ← M[C] + M[D]
MUL    X, R1, R2    X ← R1*R
```

2. Using two address instructions

```
MOV  R1, A      R1 M[A]
ADD  R1, B      R1 R1 + M[B]
MOV  R2, C      R2 M[C]

ADD  R2, D      R2 R2 + M[D]
MUL  R1, R2     R1 R1* R2
MOV  X, R1      X R1
```

3. Using one address instructions

```
LOAD    A   AC M[A]
ADD     B   AC AC + M[B]
STORE   T   M[T] AC
LOAD    C   AC M[C]
ADD     D   AC AC + M[D]
MUL     T   AC AC*M[T]
STORE   X   M[X] AC
```

4. Using zero address instructions (RPN: AB+CD+*)

```
PUSH    A   TOS A
PUSH    B   TOS B
ADD         TOS (A + B)
PUSH    C   TOS C
PUSH    D   TOS D
ADD         TOS (C + D)
MUL         TOS (C + D) * (A + B
POP     X   M[X] TOS
```

2.2.4.2 Addressing Modes

The addressing mode specifies a rule for interpreting or modifying the address field of the instruction before the operand is actually referenced. Computers use addressing mode techniques for the purpose of accommodating one or both of the following provisions:

1. To provide programming versatility for the user: pointers to memory, counters for loop control, indexing data, etc.
2. To reduce the number of bits in the addressing field of the instruction.

The decoding phase of an instruction cycle determines the addressing mode(s) and the locations (registers and/or memory locations) of operands. Depending on the CPU, an instruction can have more than one address field, and each address field may be associated with its own particular addressing mode.

Classifying Instructions According to Number of Operands

Possible to have anywhere from 0 to 4 or more operands. Fewer operands make for shorter but less flexible instructions since some of the operands are implicit.

Zero Address - any stack architecture - *e.g.,* IJVM Mic -1

One Address - accumulator is implicit operand - *e.g.,* PDP-8, IAS

Two Address - one operand is both source and destination - *e.g.,* Intel 80 × 86

Three Address - two source operands and one destination operand

VAX has 3 address instructions (a good example of a CISC architecture)

ARC has 3 register instructions (a good example of RISC architecture)

Four Address - 4th operand is either address of next instruction (obsolete technique) or for some reason the instruction requires more than three operands (*e.g.,* integer division produces quotient and remainder).

The "Standard" Addressing Modes - A Functional View

Level refers to the number of times memory is referenced to obtain the operand. "Half-levels" (non-canonical) refer to register look up.

EAddr denotes "effective address"; C( ) denotes "contents of" or "contents at".

Binding time refers to the latest time that the effective address can be fixed.

The value of the Operand Field of an instruction can be considered as either an Address or a Displacement value

- **Immediate Mode** (level 0) - operand is part of instruction; *i.e.,* operand contained in Operand Field - binding time is at assembly time: ARC, IJVM, and Intel 80 × 86

  Register Addressing (level 1/2?) - operand contained in register - binding time is assembly time: ARC, Intel 80 × 86

  Direct (level 1) - effective address is part of instruction or directly obtained (calculated) from instruction - binding time is assembly tiime: PDP-8, ARC, IJVM, Intel 80x86

  Indirect (level 2) - sometimes called "deferred"; operand field of instruction contains address of the address of the operand. This is like a "pointer" which can be modified -binding time is run time.

  Register Indirect Addressing - effective address of operand contained in a register EAddr = C(Reg) ARC, Intel 80 × 86.

  Memory Indirect Addressing - effective address of operand contained in a word in memoty Operand Field contains an Address so EAddr = C(Address) ,PDP-8 .

- **Indexed or Base (or Index)) + Offset** (Indirect) or Register + Offset (Indirect): Level 1 1/2- effective address obtained by adding value of operand field or offset to contents of register - EAddr = C(Reg) + Offset - binding time is run time Sometimes called displacement addressing

- **Array Type:** Fixed Base (address) + Variable Register Offset Operand Field contains a base Address so EAddr = base Address + C(Reg) used to implement arrays

  Record Type: Fixed Register + Variable Offset (address) Operand Field contains a Displacement so EAddr = C(Reg) + Displacement used to implement records and stack frames (local variables for subroutines)

  Program Counter Based: PC + offset : relative addressing Operand Field contains a Displacement so EAddr = C(PC) + Displacement Near and Short Jump instructions on the Intel 80 × 86 use this

- **Base + Index** (Indirect) or **Double Indexed** (Indirect): Level 1 1/2 - effective address is obtained by adding the contents of two registers - EAddr = C(Reg1)+C(Reg2) - binding time is run time: ARC, Intel 80 × 86:

  Base + Index + Offset (Indirect) - Level 1 1/2 - effective address obtained by adding contents ot two registers plus an offset - EAddr = C(Reg1) + C(Reg2) + offset - binding time is run time: Intel 80 × 86.

  Indexed With Scaling : Base + Register Offset * Scaling Factor - binding time is run time variant of Indexing useful for array calculations where size of component is multiple bytes long

  Operand Field contains base Address then EAddr = base Address + C(Reg) * Scaling Factor

  Auto-indexing : Register Indirect or Indexing with auto-increment /decrement of register

  Auto-Increment: EAddr = C(Reg); Reg = C(Reg) + 1. Note indirect-increment order!

  Auto-Decrement: Reg = C(Reg) - 1; EAddr = C(Reg). Note decrement-indirect order!

- PDP-8 auto-indexing mode increments then uses indirection; IReg = C(IReg) + 1; EAddr = C(IReg)
- Stack Addressing : push & pop - a variant of register indirect with auto-increment/ decrement using the SP register implicitly

  Push: EAddr = C(SP). SP = C(SP) -1 ; C(C(SP)) = Source auto-decrement!

- Pop: EAddr = C(SP); Destination = C(C(SP)); SP = C(SP) + 1; auto-increment

  Some "Not-So-Standard" Addressing Modes

  Auto-increment/Auto-decrement Indirect

  EAddr = C(C(Reg)); Reg = C(Reg) + 1; PDP-11 Mode 3 Addressing

  The PDP-11 used the PC as the register to implement direct addressing this way!

  Reg = C(Reg) - 1; EAddr = C(C(Reg)); PDP-11 Mode 5 Addressing

- Index 80 × 86 Supports this using Base Registers (BX, BP) and Index Registers (SI, DI)

  Indexing Indirect : Indexing coupled with indirection (rare?)

- postindexing: indexing after indirection: EAddr = C(Address) + C(Reg) preindexing: indirection after indexing: EAddr = C(Address + C(Reg))

  Multiple Indirection : Requires indirect bit as part of address field (PDP-10 could do this)

In this chapter, we will cover the following modes:

1. **Implied mode:** The operands are implicitly specified by the instruction, e.g.: "complement accumulator".

2. **Immediate mode:** The operand is specified in the instruction. Can be used, *e.g.* to initialize register to constant value (immediate operand).
3. **Register mode:** Operands are in registers that reside within the CPU. The particular register is selected with the register field of the instruction.
4. **Register indirect mode:** The content of a register specifies the address of the operand in memory.
5. **Autoincrement or autodecrement mode:** Similar to register indirect mode but the content of the register is automatically incremented/decremented after data access.
6. **Direct address mode:** The effective address is equal to the address part of the instruction. The operand resides in this address.
7. **Indirect address mode:** The address field gives the address where the effective address is stored in memory.
8. **Relative address mode:** The content of the program counter (PC) is added to the address part of the instruction in order to obtain the effective address, *i.e.* the effective address is relative to the address of the next instruction. This address mode can be used in branch type instructions when the branch address is in the area surrounding the instruction word.
9. **Indexed addressing mode:** Content of an index register (special CPU register) is added to the address part of the instruction to obtain the effective address. The index register can be incremented for accessing consecutive operands.
10. **Base register addressing mode:** The content of a base register in added to the address part of the instruction to obtain the effective address. The address part of the instruction gives the displacement relative to the base address.

Numerical Example

For each possible mode, calculate the effective address and the operand that must be loaded into AC. Use the instruction defined in Fig. 2.11. Table 2.4 show the tabular list of the numerical example.

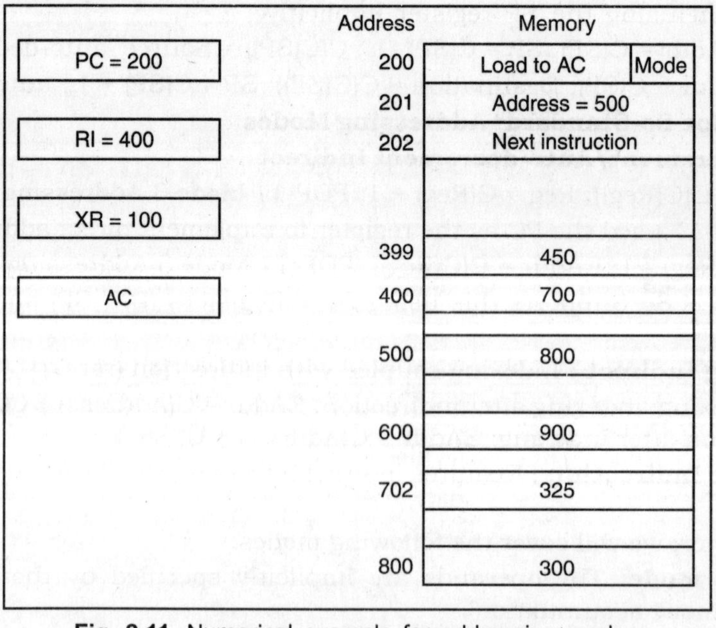

Fig. 2.11. Numerical example for addressing modes

Table 2.4. Tabular list of the numerical example

TABLE 2.4. Tabular List of Numerical Example

Addressing mode	Effective Address	Content of AC
Direct address	500	800
Immediate operand	201	500
Indirect address	800	300
Relative address	702	325
Indexed address	600	900
Register	—	400
Register indirect	400	700
Autoincrement	400	700
Autodecrement	399	450

Decrements R1 prior the execution → Autodecrement

SOLVED PROBLEMS

Q. 1. What is the stack organization ? Describe all functionality regarding to stack ?

Sol. Stack is very useful in the CPU of most computers. Stack is also called last in first out (LIFO). A stack is a storage device that stores information in such a manner that the item last is the first item retrieved. The stack in digital computers is essentially a memory unit with an address register that can count only after an initial value is loaded into it.

Various Function of Stack :

(*a*) **Stack Pointer :** The register that holds the address for the stack is called a stack pointer (SP) because its value always points at the top item in the stack.

(*b*) **PUSH :** The operation of insertion is called push, because it can be thought of as the result of pushing a new item on the top.

(*c*) **POP :** The operation of election is called POP., because it can be though of as the result of removing one item so that stack pops up.

Q.2. Define the following

(*a*) Register stack; (*b*) Memory stack; (*c*) Stack limits

Solution: (*a*) **Register Stack :** A stack can be placed in a portion of large memory or it can be organized as a collection of a finite number of memory words or registers. The organization of 64-word register is stack shown below :

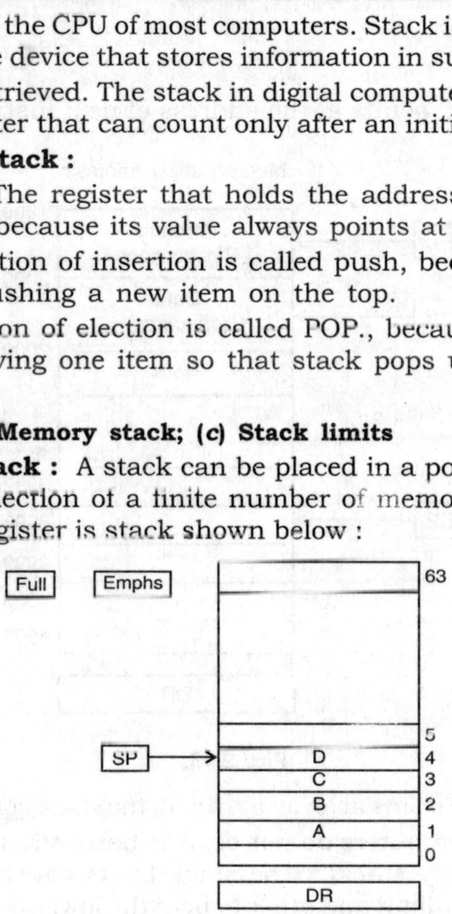

Fig. 2.12. Block diagram of 64-word stack

The stack point register SP contains a binary numbers whose value is equal to the address of the word is currently on top of the stack. So, in above block diagram, the content of SP is now 4. To remove the top item, the stack is popped by reading the memory word at address 3 and determination the content of SP. If stack is not full then use PUSH.

PUSH Operation :

SP ← SP + 1 increment stack pointer

M[SP] ← DR write item on top of the stack

If (SP = 0) then (Full ← 1), check if stack is full

EMPTY ← 0 mark the stack not empty.

Pop Operation : If Empty = 0

DR ← M[SP] Read item from the top of stack

SP ← SP – 1 Decrement stack pointer

It (SP = 0) then (empty ← 1) check if stack is empty.

Full ← 0 Mark the stack not full.

Memory Stack : A stack can exist as a stand alone unit or can be implemented in a random access memory attached to a CPU. The implementation of a stack in the CPU is done by assigning a portion of memory to a stack operation and using a processor register as a stack pointer. Portion of computer memory partitioned into three segments.

Program

Data and

Stack

The program counter PC points at the address of next instrument in the program.

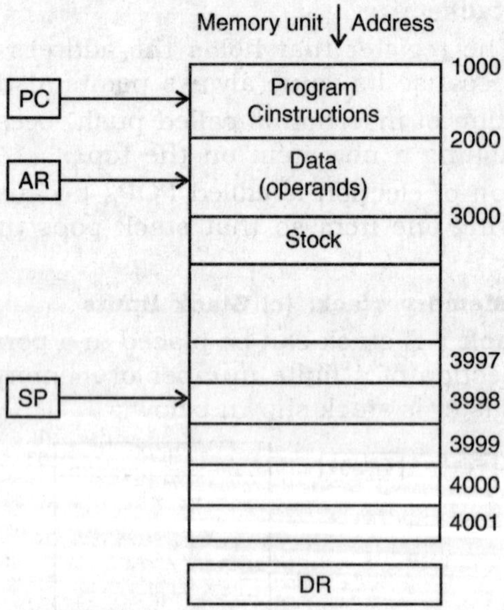

Fig. 2.13.

The address register array points at an array of data the stack points SP points at top of stack.

Stack Limits : Most computers do not provide hardware to check for stack overflow (Full stack) underflow (empty stack). The stack limits can be by using two processor register are to hold the upper limit and other to hold the lower limit. After a push operation,

SP is compared with upper limit register and after a operation, SP is compared with the lower limit register.

Q.3. What is reverse publish notation ? Change the following in reverse polish notation ?

(a) A + B × C

(b) (A + B) × C

(c) A$ B × C – D + E/F/(G + H)

(d) A – B (C× DBE)

Solution: A stack organization is very effective for evaluating arithmetic expressions. The post fix notation referred to as reverse polish notation (RPN). Places the operator after the operands

e.g., A + B ← in fix notation

+ AB ← Prefix of polish notation

AB + ← Post fix or reverse polish notation

The reverse polish notation is in a form suitable for stack manipulation. The expression A × B + C × D is written in RPN AB × CD +

(a) ABC × +

(b) AB + C×

(c) AB $ C × D EF/GH + / +

(d) ABC DE $ × –

Q.4. What is program interrupt ? Describe all functionality of the interrupt.

Solution: Program interrupt refers to the transfer of a program control from a currently running program to another service program as a result of external or internal generated request. Control returns to the original program after the service program is executed.

Interrupt Procedure is main Functionality :

(a) The interrupt is usually initiated by an internal or external signal rather than from the execution of an instruction

(b) The address of interrupt service program is determined by the hardware rather than from the address field of an instruction.

(c) The interrupt procedure usually stores all the information necessary to define the state of the CPU rather than storing only the program counter. After a program has been interrupted and the service routine been executed the CPU must return to exactly the same state that it was when the interrupt occurred. The state of the CPU at the end of the execute. Cycle is determined from:

(i) The content of program counter

(ii) The content of processor register

(iii) The content of certain status conditions

Q.5. How many interrupts are there in program control describe briefly ?

Solution: There are three major types of interrupts that cause a break in the normal execution of a program :

(a) External interrupts

(b) Internal interrupts

(c) Software interrupts

(a) External interrupts : External interrupts come from input out (I/O) devices, from a timing devices from a circuit monitoring. The power supply or from any other external source.

(b) Internal interrupts : Internal interrupts arise from illegal erroveous use of an instruction or data. Internal interrupts are also called traps. Examples of internal interrupt caused by internal error conditions arc.

Register overflow, attempt to divide by zero, an invalid operation code, stack over flow and protection violation. These error condition usually occur as a result of a permeative termination of the instruction execution.

(c) Software interrupt : A software interrupt is initiated by executing an instruction. Software interrupt is a special call instruction that behaves like an interrupt rather than a subroutine call. It can be used by the programmer to initiate an interrupt procedure at any desired point in the program. The most common use of software interrupt is associated with a supervisor call instruction. This instruction provides means for switching from CPU user mode to supervisor mode.

Q.6. What are the differences between external and internal interrupts ?

Solution: The difference between internal and external interrupt is that

(*i*) The internal interrupt is initiated by some exceptional condition caused by the program itself rather than by an external event.

(*ii*) Internal interrupts are synchronous with the program while external interrupts are asynchronous.

(*iii*) If the program is rerun, the internal interrupts will occur in the same place each time.

(*iv*) External interrupts depend on external conditions that are independent of the program being executed at the time.

(*v*) External or internal interrupts are initiated from signals that occur in the hardware of the CPU.

Q.7. What is the difference between user mode and supervisor mode ?

Solution: A program written by a user must run in the user mode when an input or output transfer is required, the supervisor mode is requested by means of a supervisor call instruction. This instruction causes a software interrupt that stores the old CPU state and brings in a new PSW that belongs to the supervisor mode.

Q.8. Define the following terms :

(a) Program status word (PSW)

(b) RISC

(c) CISC

Solution: (a) PSW : (Program Status Word) The collection of all status bit conditions in the CPU is sometimes called a program status word or PSW. The PSW is stored in a separate hardware register and contains the status information that characterizes the state of the CPU.

(b) RISC : (Reduced Instruction Set Computer) A computer with instructions with simple constructs, so they can be executed much faster within the CPU without having to use memory as often. This type of computer is classified as a reduced instruction set computer or RISC.

RISC Characteristics :

1. Relatively few instructions
2. Relatively few addressing modes
3. Memory access limited to load and store instruction,
4. All operations done within the registers of the CPU.

5. Fixed length, easily decoded instruction formed.
6. Single cycle instruction execution.
7. Hardware rather than micro Programmed control.

CISC : A computer with large number of instruction is classified as a complex instruction set computer.

Characteristics of CISC :

1. A large number of instructions typically from 100 to 250 instructions.
2. Some instructions that perform specialized tasks and are used infrequently.
3. A large variety of addressing mode typically from 5 to 20 different modes.
4. Variable length instruction format.
5. Instruction that manipulate operands in memory.

Q.9. A bus-organized CPU similar to figure given below has 16 registers with 32 bits in each, an ALU and a destination decoder.

(a) How many multiplexers are there in the bus A and what is the size of each multiplexer.

(b) How many selection inputs are needed for B MUX and MUXB.

(c) How many inputs and outputs are there in the decoder ?

(d) How many inputs and outputs are there in the ALU for data, including input and output carries ?

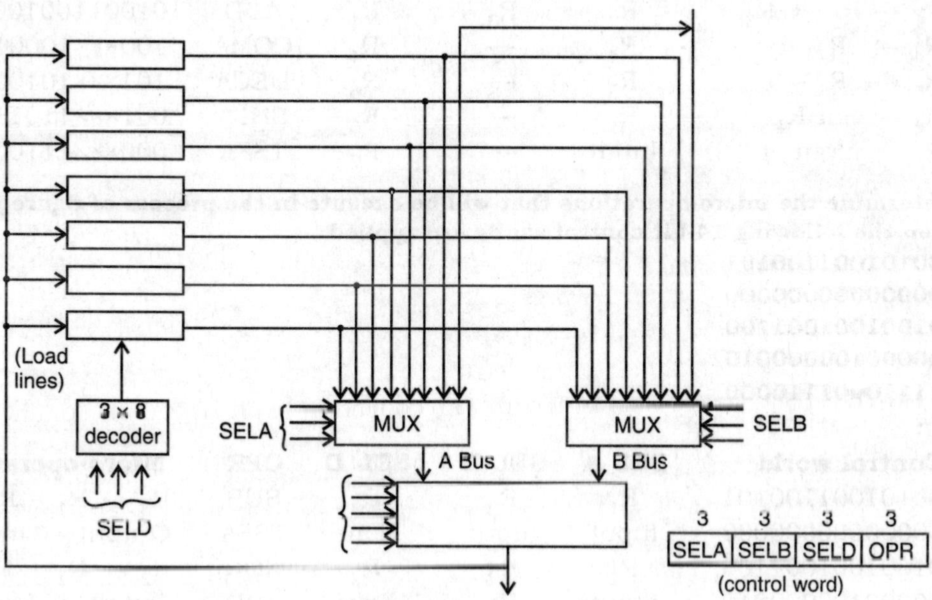

Fig. 2.14.

(e) Formulate a control word for the system assuming that the ALU has 35 operations.

Solution:

(a) 32 multiplexers, each of size 16 × 1

(b) 4 inputs each, to select one of 16 registers.

(c) 4–to-16 line decoder.

(d) 32 + 32 + 1 = 65 data input lines 32 + 1 = 33 data output lines

(e)	4	4	4	6	= 18 bits
	SEL A	SEL B	SEL D	OPR	

Q.10. The bus system of figure given in question 8. Has the following propagation delay times :

30 ns for the signals to propagate through the multiplexer

80 ns to perform the ADD operation in the ALU

20 ns delay in the destination decoder, and

10 ns to clock the data into the destination register

What is the minimum cycle time that can be used for the clock ?

Solution: 30 + 80 + 10 = 120 n sec (The recorder signals propagate at the same as the mixes)

Q.11. Specify the control word that must be applied to the processor of figure given in Q. 8 implement the following micro-operations.

(a) $R_1 \leftarrow R_2 + R_3$

(b) $R_4 \leftarrow \overline{R_4}$

(c) $R_5 \leftarrow R_5 - 1$

(d) $R_6 \leftarrow$ Shl R_1

(e) $R_7 \leftarrow$ input

Solution:

	SEL A	SEL B	SEL D	OPR	Control word
(a) $R_1 \leftarrow R_2 + R_3$	R_2	R_3	R_1	ADD	01001100100010
(b) $R_4 \leftarrow \overline{R_4}$	R_4	—	R_4	COMA	100×××10001110
(c) $R_5 \leftarrow R_5 - 1$	R_5	—	R_6	DECA	101×××10100110
(d) $R_6 \leftarrow$ Shl R_1	R_1	—	R_6	SHLA	001×××11011000
(e) $R_7 \leftarrow$ input	Input	—	R_7	TSFA	000×××11100000

Q.12. Determine the micro-operations that will be execute in the process of figure given in Q.8. When the following 14-bit control words arc applied.

(a) 00101001100101

(b) 00000000000000

(c) 01001001001100

(d) 00000100000010

(d) 11110001110000

Solution:

Control world	SEL A	SEL B	SEL D	OPR	Micro-operation
(a) 00101001100101	R_1	R_2	R_3	SUB	$R_3 \leftarrow R_1 - R_2$
(b) 00000000000000	Input	Input	None	TSFA	Output ← Input
(c) 01001001001100	R_2	R_2	R_2	XOR	$R_2 \leftarrow R_2 + R_2$
(d) 00000100000010	Input	R_1	None	ADD	Output ← Input + R_1
(d) 11110001110000	R_7	R_4	R_3	SHRA	$R_3 \leftarrow$ Shr R_7

Q.13. Let SP = 000000 in the stack of figure given below. How many items are there in stack it

(a) FULL = 1 and EMPTY = 0 ?

(b) FULL = 0 and EMPTY = 1 ?

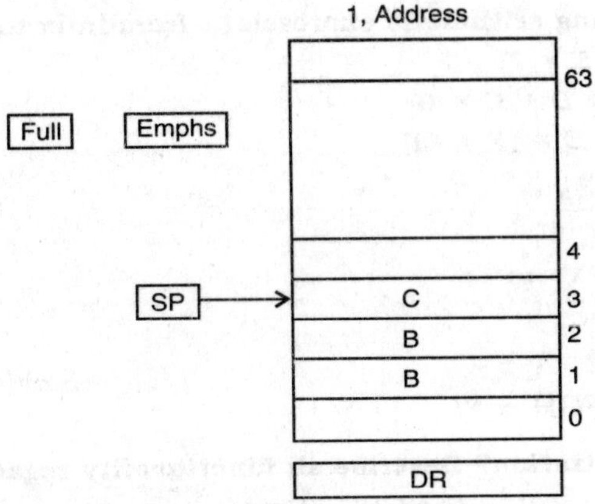

Fig. 2.15.

Solution: (*a*) Stack full with 64 items

(*b*) Stack Empty.

Q.14. A stack organized such that SP always points at the next empty location on the stack. This means that SP can be initiated to 4000 in figure given below and first item in the stack is storeding location 4000. List the micro-operations for the push and pop operations.

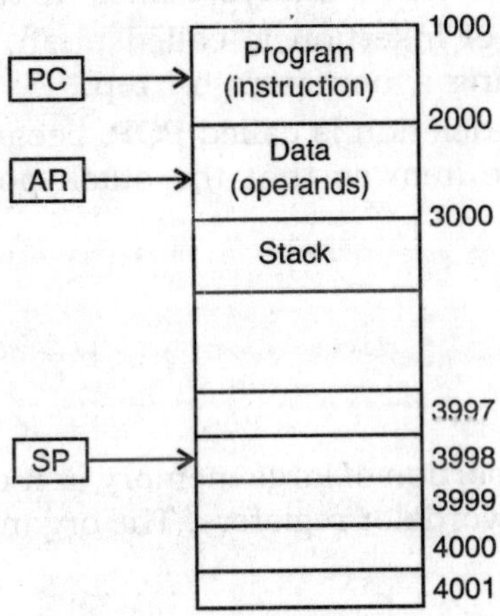

Fig. 2.16. (Computer memory with Program, data dn stack segments)

Solution:

PUSH : M[SP] ← DR

SP ← SP – 1

POP : SPH

DR ← M[SP]

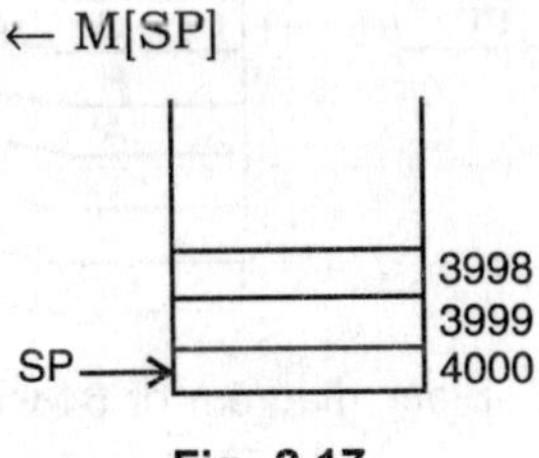

Fig. 2.17.

Q.15. Convert the following arithmetic expressions from infix to reverse polish notation.

(a) $A \times B + C \times D + E \times F$

(b) $A \times B + A \times CB \times D + C \times G$

(c) $A + B \times [C \times D + E \times (F + G)]$

(d) $\dfrac{A \times [B + C \times (D + E)]}{F \times (G + H)}$

Solution: *(a)* AB * CD * EX* ++

(b) AB * ABD* CE* + * +

(c) FG + E* CD* + B* A +

(d) ABCDE + α + * FGH ≠ */

Q.16. What is stack organization? Describe all functionality regarding to stack?

Solution: Stack is very useful in the CPU of most computers. Stack is also called last in first out (LIFO). A stack is a storage device that stores information in such a manner that the item stored last is the first item retrieved. The stack in digital computers is essentially a memory unit with an address register that can count only after an initial value is loaded into it.

Various function of stack

(*a*) **Stack pointer.** The register that holds the address for the stack is called a *stack pointer* (SP) because its value always points at the top item in the stack.

(*b*) PUSH. The operation of insertion is called push, because it can be thought of as the result of pushing a new item on top.

(*c*) POP. The operation of deletion is called POP, because it can be thought of as the result of removing one item so that the stack pops up.

Q.17. Define the following :

(*a*) Register stack

(*b*) Memory stack

(*c*) Stack limits.

Solution: (*a*) Register Stack.

A stack can be placed in a portion of large memory or it can be organized as a collection of a finite number of memory words or registers. The organization of a 64-word register is stack shown below :

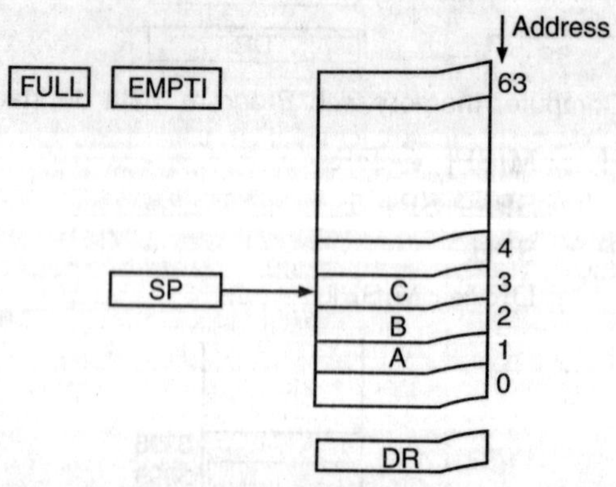

Fig. 2.18. Block diagram of 64-word stack

The stack point register SP contains a binary numbers, whose value is equal to the address of the word that is currently on top of the stack. So, in above block diagram, the content of SP is now 3. To remove the top item. The stack is popped by reading the memory word at address 3 and decrementing the content of SP. If tack is not full then use PUSH.

Push Operation

SP ← SP + 1 Increment stack pointer
M[SP] ← DR Write item on top of the stack
If (SP = 0) then (Full ← 1), check if stack is full
EMPTY ← 0 Mark the stack not empty.

POP OPERATION : If Impty = 0

DR ← M[SP] **Read item from the top of stack**
SP ← SP – 1 **Decrement stack pointer**
If (SP = 0) then (empty ← 1) **Check if stack is empty**
FULL ← 0 **Mark the stack not full**

(*B*) Memory Stack

A stack can exist as a stand alone unit or can be implemented in a random access memory attached to a CPU. The implementation of a stack in the CPU is done by assigning a portion of memory to a stack operation and using a processor register as a stack pointer. Portion of computer memory partitioned into three segments.

Program,
Data and
Stack.

The program counter PC points at the address of next instruction in the program.

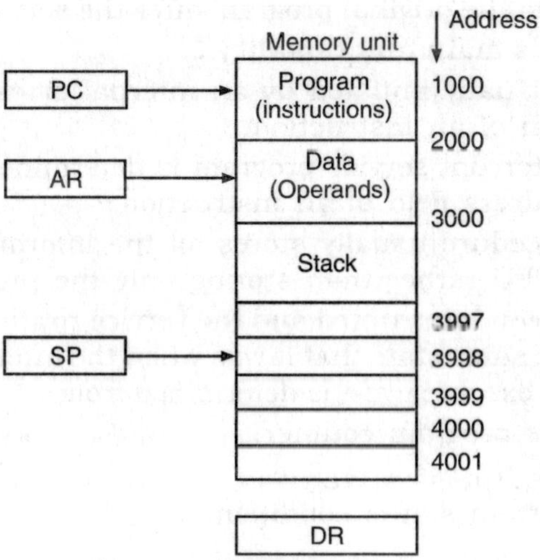

Fig. 2.19.

The address register array points at an array of data. The stack point SP points at top o stack.

(*c*) Stack Limits

Most computers do not provide hardware to check for stack overflow (full stack) or underflow (empty stack). The stack limits can be checked by using two processor register are to hold the upper limit and other to hold the lower limit. After a push operation, SP is compared with upper limit register and after a pop operation, SP is compared with the lower limit register.

Q.18. What is reverse polish notation? Change the following in reverse polish notation?

(a) A + B * C)
(b) (A + B) * C
(c) A $ B * C – D + E/F/(G + H)
(d) NA – B(C * D $ E).

Solution: A stack organization is very effective for evaluating arithmetic expressions. The post fix notation, referred to as reverse polish notation (RPN), places the operator after the operands.

e.g.,

A + *B* — In fix notation
+ *AB* — Prefix or polish notation
AB + — Postfix or reverse polish notation.

The reverse polish notation is in a form suitable for stack manipulation. The expression *A* * *B* + *C* * *D* is written in RPN *AB* * *CD* * + :

(*a*) *ABC* * +
(*b*) *AB* + *C* *
(*c*) *AB* $ *C* * *D* – *EF*/ *GH* + / +
(*d*) *ABCDE* $ * –

Q.19. What is program interrupt ? Describe all functionality of the interrupt.

Solution: Program interrupt refers to the transfer of a program control from a currently running program to another service program as a result of an external or internal generated request. Control returns to the original program after the service program is executed.

Interrupt procedure is main functionality :

(*a*) The interrupt is usually initiated by an internal or external signal rather than from the execution of an instruction.
(*b*) The address of interrupt service program is determined by the hardware rather than from the address field of an instruction.
(*c*) The interrupt procedure usually stores all the information necessary to define the state of the CPU rather than storing only the program counter.

After a program has been interrupted and the service routine been executed, the CPU must return to exactly the same state that it was when the interrupt occured. The state of the CPU at the end of the execute cycle is determined from :

(*i*) The content of the program counter.
(*ii*) The content of all processor register.
(*iii*) The content of certain status conditions.

Q.20. How many interrupts are there in program control, describe briefly?

Solution: There are three major types of interrupts that cause a break in the normal execution of a program :

(*a*) External interrupts.
(*b*) Internal interrupts.
(*c*) Software interrupts.

(a) External Interrupts

External interrupts come from input-output (input/output) devices. from a timing devices, from a circuit monitoring the power supply or from any other external source.

(b) Internal Interrupts

Internal interrupts arise from illegal or erroneous use of an instruction or data. Internal interrupts are also called **traps**. Examples of interrupt caused by internal error conditions are :

Register over flow, attempt to divide by zero, an invalid operation code, stack over flow and protection violation. These error condition usually occur as a result of a premature termination of the instruction execution.

(c) Software Interrupts

A software interrupt is initiated by executing an instruction. Software interrupt is a special call instruction that behaves like an interrupt rather than a subroutine call. It can be used by the programmer to initiate an interrupt procedure at any desired point in the program. The most common use of software interrupt is associated with a supervisor call instruction. This instruction provides means for switching from CPU user mode to supervisor mode.

Q.21. What are the differences between external and internal interrupts?

Solution: The difference between internal and external interrupts is that :

(i) The internal interrupt is initiated by some exceptional condition caused by the program itself rather than by an external event.

(ii) Internal interrupts are synchronous with the program while external interrupts are asynchronous.

(iii) If the program is rerun, the internal interrupts will occur in the same place each time.

(iv) External interrupts depend on external conditions that are independent of the program being executed at the time.

(v) External or internal interrupts are initiated from signals that occur in the hardware of the CPU.

Q.22. What is the difference between user mode and supervisor mode ?

Solution: A program written by a user must run in the user mode. When an input or output transfer is required, the supervisor mode is requested by means of a supervisor call instruction. This instruction causes a software interrupt that stores the old CPU state and brings in a new PSW that belongs to the supervisor mode.

Q.23. Define following terms :

(a) Program status word

(b) RISC

(c) CISC.

Solution: (a) PSW (Program Status Word)

The collection of all status bit conditions in the CPU is sometimes called a program status word or PSW. The PSW is stored in a seprate hardware register and contains the status information that characterizes the state of the CPU.

(b) RISC (Reduced Instruction Set Computer)

A computer with fewer instructs with simple constructs, so they can be executed much faster within the CPU without having to use memory as often. This type of computer is classified as a reduced instruction set computer or RISC.

RISC characteristics :

1. Relatively few instructions.
2. Relatively few addressing modes.
3. Memory access limited to load and store instructions.
4. All operations done within the registers of the CPU.
5. Fixed length, easily decoded instruction format.
6. Single cycle instruction execution.
7. Hardware rather than micro programed control.

(c) CISC

A computer with large number of instructions is classified as a complex instruction set computer.

Characteristics of CISC

1. A large number of instructions typically from 100 to 250 instructions.
2. Some instructions that perform specialized tasks and are used infrequently.
3. A large variety of addressing mode typically from 5 to 20 different modes.
4. Variable length instruction format.
5. Instructions that manipulate operands in memory.

Q.24. A bus-organized CPU similar to figure given below has 16 registers with 32-bits in each, an ALU, and a destination decoder.

(a) How meiny multiplexers are there in the A bus, and what is the size of each multiplexer?

(b) How many selection inputs are needed for MUX A and MUX B?

(c) How many inputs and outputs are there in the decoder?

(d) How many inputs and outputs are there in the ALU for data, including input and output carries?

(e) Formulate a control word for the system assuming that the ALU has 35 operations?

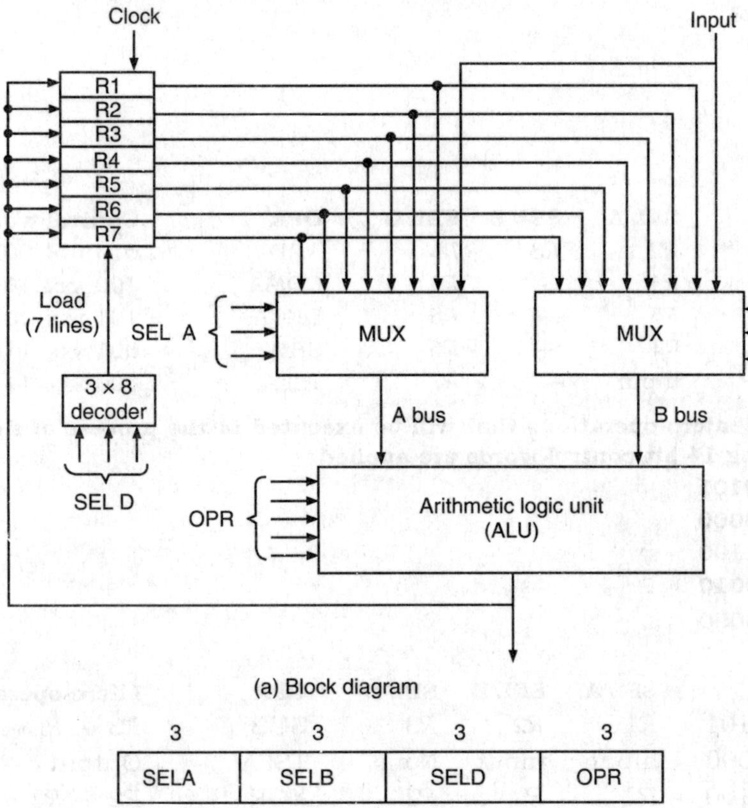

Fig. 2.20. Register set with common ALU

Solution:

(*a*) 32 multiplexers, each of size.

(*b*) 5 inputs each, to select one of 16 registers.

(*c*) 4-to-16-line decoder.

(*d*) 32 + 32 + 1 = 65 data input lines.

32 + 1 = 33 data output lines.

(*e*) 5 5 4 6 = 20 bits

SEL A	SEL B	SEL D	OPR

Q.25. The bus system of figure given in Q. 9 has the following propagation delay times :

30 ns for the signals to propagate through the multiplexers,

80 ns to perform the ADD operation in the ALU,

20 ns delay in the destination decoder, and

10 ns to clock the data into the destination register.

What is the minimum cycle time that can be used for the clock?

Solution: 30 + 80 + 10 = 20 nsec.

(The decoder signals propagate at the same as the muxes).

Q.26. Specify the control word that must be applied to the processor of figure given in Q.9 implement the following micro-operations.

(a) R1 ← R2 + R3
(b) R4 ← R4
(c) R5 ← R5 – 1
(d) R6 ← Shl R1
(e) R7 ← input

Solution:

	SEL A	SEL B	SEL D	OPR	Control word
(a) R1 ← R2 + R3	R2	R3	R1	ADD	010 011 001 00010
(b) R4 ← R4	R4	—	R4	COMA	100 xxx 100 00000
(c) R5 ← R5 – 1	R5	—	R5	DECA	101 xxx 101 00110
(d) R6 ← Shi R1	R1	—	R6	SHUT	001 xxx 110 11000
(e) R7 ← Input	Input	—	R7	TSFA	000 xxx 111 00000

Q.27. Determine the micro-operations that will be executed in the process of figure given Q.9 when the following 14-bit control words are applied.

(a) 00101001100101
(b) 00000000000000
(c) 01001001001100
(d) 00000100000010
(e) 11110001110000

Solution :

	Control word	SEL A	SEL B	SEL D	OPR	Micro-operation
(a)	001 010 011 00101	R1	R2	R3	SUB	R3 ← R1 – R2
(b)	000 000 000 00000	Input	Input	None	TSFA	Output ← Input
(c)	010 010 010 01100	R2	R2	R2	XOR	R2 ← R2 + R2
(d)	000 001000 00010	Input	R1	None	ADD	Output ← Input + R1
(e)	111 100 011 10000	R7	R4	R3	SHRA	R3 ← Shr27

Q.28. Let SP = 000000 in the stack of figure given below. Haw many items are there in the stack if :

(a) FULL = 1 and EMTY = 0?
(b) FULL = 0 and EMTY = 1?

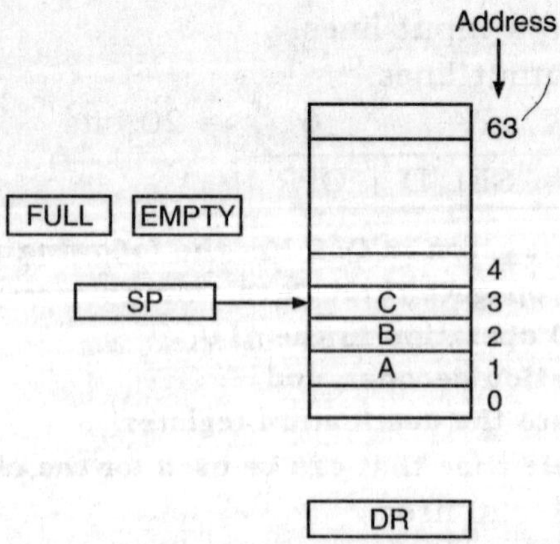

Fig. 2.21. Block diagram of a 64-word stack

Solution:

(*a*) Stack full with 64 items.

(*b*) Stack empty.

Q.29. A stack is organized such that SP always points at the next empty location on the stack. This means that SP can be initialized to 4000 in figure given below and the first item in the stack is stored in location 4000. List the micro-operations for the push and pop operations.

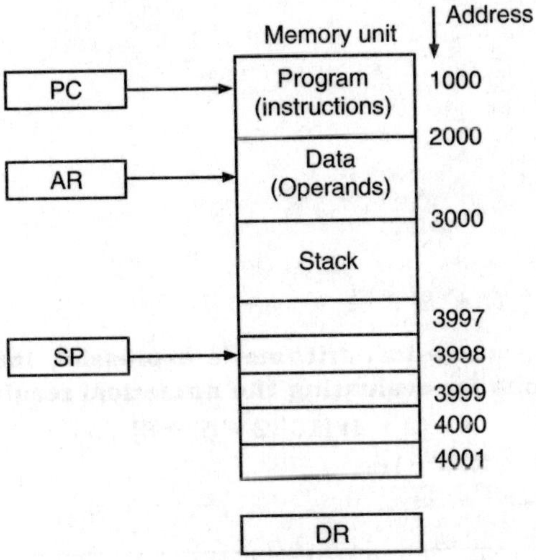

Fig. 2.22. Computer memory with program, data and stack segments

Solution:

Push : $M[SP] \leftarrow DR$

$SP \leftarrow SP - 1$

POP : $SP \leftarrow SP + 1$

$DR \leftarrow M[SP]$

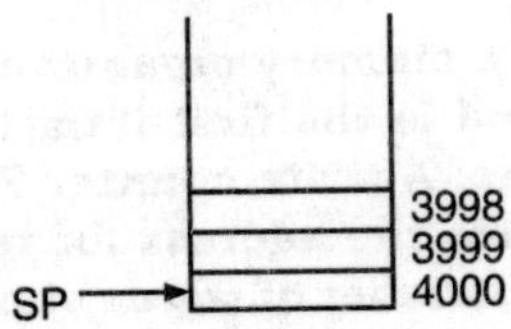

Fig 2.23.

Q.30. Convert the following arithmetic expressions from infix to reverse polish notation.

(*a*) $A * B + C * D + E * F$

(*b*) $A * B + A * (B * D + C * E)$

(*c*) $A + B * [C * D + E * (F + G)]$

(*d*) $\dfrac{A * [B + C * (D + E)]}{F * (G + H)}$

Solution:

(*a*) $AB * CD * E * + +$

(*b*) $AB * ABD * CE * + * +$

(*c*) $FG + E * CD * + B * A +$

(*d*) $ABCDE + * + * FGH + * /$

Q.31. Convert the following arithmetic expressions from reverse polish notation to infix notation.

(*a*) *ABCDE* + * – /

(*b*) *ABCDE* * / – +

(*c*) *ABC* * / *D* – *EF* / +

(*d*) *ABCDEFG* + * + * + *

Solution:

(*a*) $\dfrac{A}{B-(D+E)*C}$

(*b*) $A + B - \dfrac{C}{D*E}$

(*c*) $\dfrac{A}{B*C} - D + \dfrac{E}{F}$

(*d*) $(((F + G) * E + D) * C + B) * A$

Q.32. Convert the following numerical arithmetic expression into reverse polish notation and show the stack operations for evaluating the numerical result.

(3 + 4) [10 (2 + 6) + 8]

Solution: (3 + 4) [10 (2 + 6) + 8] = 616

RPN : 34 + 26 + 10 * 8 + *

				6		10		8		
	4		2	2	8	8	80	80	88	
3	3	7	7	7	7	7	7	7	7	616
3	4	+	2	6	+	10	*	8	+	*

Fig. 2.24.

Q.33. A first-in, first-out (FIFO) has a memory organization that stores information in such a manner that the item that is stored is the first item that is retrieved. Show how a FIFO memory operates with three counters. A write counter ????? holds the address for writing into memory. A read counter RC holds the address for reading from memory. An availables storage counter ASC indicataaes the number of words stored in FIFO. ASC is incremented for every word stored and decremented for every item that is retrieved.

Solution:

WRITE (if not full) :

M [WC] ← DR

WC ← WC + 1

ASC ← ASC + 1

READ : (if not empty)

DR ← M [RC]

RC ← RC + 1

ASC ← ASC – 1

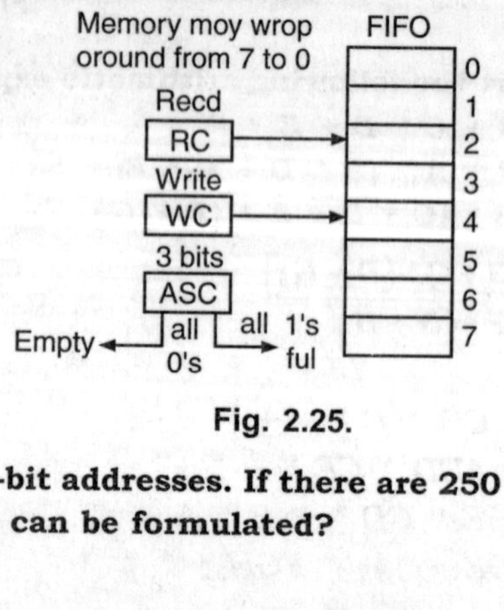

Fig. 2.25.

Q.34. A computer has 32-bit instructions and 12-bit addresses. If there are 250 two address instructions, how many one-address instructions can be formulated?

Solution:

8	12	12
Opcode	Address	Address

= 32 bits

(Two address instruction)

$$2^8 = 256 \text{ combinations.}$$

$$256 - 250 = \text{S combinations can be used for one address.}$$

Opcode	Address
6×2^{12}	

(one address instructions)

Maximum number of one address instruction = $6 \times 2^{12} = 24{,}576$

Q.35. The memory unit of a computer has 256K words of 32-bits each. The compter has an instruction format with four fields : an operation code field, a mode field to specify one of seven addressing modes, a register address field to specify one of 60 processor register, and a memory address. Specify the instruction format and the number of bits in each field if the in instruction is in one memory word.

Solution: $256\text{ K} = 2^8 \times 2^{10} = 2^{18}$ Address = 18 bits

Mode = 3 bits

Register = 6 bits

27 bits

Opcode 5

Opcode	Mode	Register	Address
5	3	6	18

32 bits

Q.36. A two-word instruction is stored in memory at an address designated by the symbol *W*. The address field of the instruction (stored at *W* + 1) is designated by the symbol *Y*. The operand used during the execution of instruction is stored at an address symbolized by *Z*. An index register contains the value of *X*. State how *Z* is calculated from the other addresses if the addressing mode of the instruction is :

(*a*) direct

(*b*) indirect

(*c*) relative

(*d*) indexed

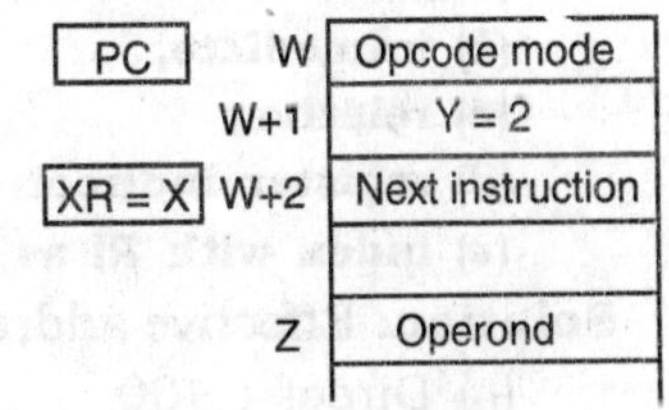

Fig. 2.26.

Solution: Z = Effective address

(*a*) Direct : $Z = Y$

(*b*) Indirect : $Z = M(Y)$

(*c*) Relative : $Z = Y + W + 2$

(*d*) Indexed : $Z = Y + X$

Q.37. A relative mode branch type of instruction stored in memory at an address equivalent to decimal 750. The branch is made to an address equivalent to decimal 500.

(*a*) What should be the value of the relative address field of the instruction (in decimal)?

(*b*) Determine the relative address value in binary using 12 bits. (Why must the number be in 2's complement?)

(*c*) Determine the binary value in PC after the fetch phase and calculate the binary value of 500. Then show that the binary value in PC plus the relative address calculated in part (*b*) is equal to the binary value of 500.

Solution:

(*a*) Relative address = 500 – 751 = – 251

(*b*) 251 = 000011111011; – 251 = 111100000101

(*c*) *PC* = 751 = 001011101111; 500 = 000111110100

PC = 751 = 001011101111

RA = – 251 = + 111100000101

EA = 500 = 000111110100

Q.38. How many times does the control unit refer to memory when it fetches and executes an indirect addressing modes instruction if the instruction is

(*a*) a computational type requiring and operand from memory;

(*b*) a branch type.

Solution: Assuming one word per instruction or operard.

(*a*) Computational type

Fetch instruction.

Fetch effective address.

Fetch operand

3 memory references

(*b*) Branch Type

Fetch instruction

Fetch effective addresses and transfer to *PC*

2 memory references

Q.39. What must the address field of an indexed addressing mode instruction be to make it the same as a register indirect mode instruction?

Solution: The address part of the indexed mode instruction must be set to zero,

Q.40. An instruction is stored at location 300 with its address field at location 301. The address field has the value 400. A processor register *R*1 contains the number 200. Evaluate the effective address if the addressing mode of the instruction is

(*a*) direct;

(*b*) immediate;

(*c*) relative;

(*d*) register indirect;

(*e*) index with *R*1 as the index register.

Solution: Effective address

(*a*) Direct : 400

(*b*) Immediate : 301

(*c*) Relative : 302 + 400 = 702

(*d*) Register indirect : 200

(*e*) Indexed : 200 + 400 = 600

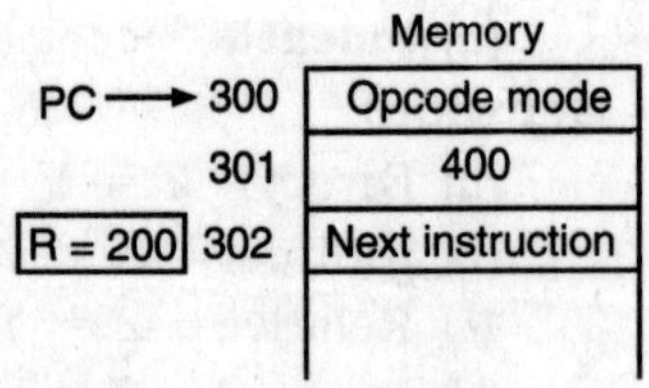

Fig. 2.27.

Q.41. Assuming an 8-bit computer, show the multiple precision addition of the two 32-bit unsigned numbers listed below using the add with carry instruction. Each byte is expressed as a two-digit hexadecimal number.

(6*E*, C3 56 7*A*) + (13 55 6*B* 8*F*)

Solution:	1 = *C*	0 = *C*	1 = *C*	0 = Reset initial carry
	6*E*	C3	56	7*A*
	??	??	??	??

Q.42. Perform the logic AND, OR and XOR with the two binary strings 10011100 and 10101010.

Solution :

10011100	10011100	10011100
10101010 AND	10101010 OR	10101010 XOR
10001000	11111110	00110110

Q.43. Given the 16-bit value 1001101011001101, what operation must be performed in order to :

(a) clear to 0 the first eight bits?

(b) set to 1 the last eight bits ?

(c) complement the middle eight bits?

Solution:

(a) AND with : 0000001111111

(b) OR with : 0000000011111111

(c) XOR with : 0000111111110000

Q.44. An 8- bit register contains the value 01111011 and the carry bit is equal to 1. Perform the eight shift operations given by the instructions listed in Table given below. Each time, start from the initial value given above,

Table 2.5. Typical Shift Instructions

Name	Mnemonic
Logical shift right	SHR
Logical shift left	SHL
Arithmetic shift right	SHRA
Arithmetic shift left	SHLA
Rotate right	ROR
Rotate left	ROL
Rotate right through carry	RORC
Rotate left through carr	ROLC

Solution:

Initial	:	01111011	$C = 1$
SHR	:	00111101	
SHL	:	11110110	
SHRA	:	00111101	
SHLA	:	11110110	(Overflow)
ROL	.	11110110	
RORC	:	10111101	
ROLC	:	11110111	

Q.45. Represent the following signed numbers in binary using eight bits + 83, – 83, + 68, – 68.

(a) Perform the addition (– 83) + (+ 68) in binary and interpret the result obtained.

(b) Perfrom the subtraction (– 68) – (+ 83) in binary and indicate if there is an.

(c) Shift binary – 68 once to the right and give the value of the shifted number in decimal.

(d) Shift binary – 83 once to the left and indicate if there is an overflow.

Solution:

+ 83 = 01010011– 83 = 10101101

+ 68 = 01000100 – 68 = 10111100

		10 carries
	– 83	10101101
(*a*)	+ 68	+ 01000100
	– 15	11110001
	– 68	10111100
(*b*)	– 83	+ 10101101
	– 151	01101001

(*c*) – 68 = 10111100

– 34 = 11011110

(*d*) – 83 = 10101101

– 166 ≠ 01011010 (overflow)

Q.46. Show that the circuit labeled 'check for zero output'in figure given below is an 8-bot NOR gate.

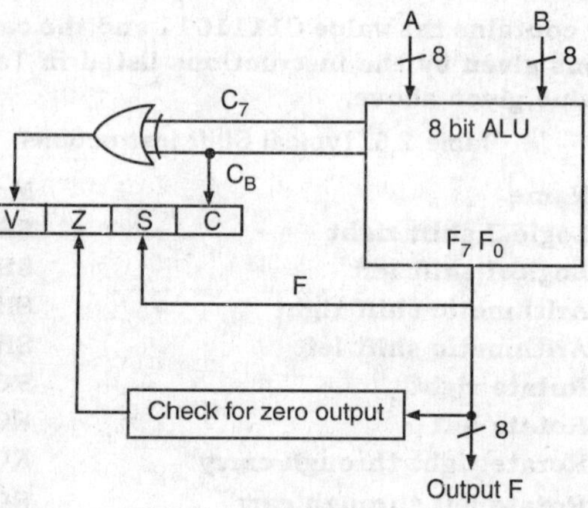

Fig. 2.28.

Solution : $Z = F_0' F_1' F_2' F_3' F_4' F_5' F_6' F_7' = (F_0 + F_1 + F_3 + F_4 + F_5 + F_6 + F_7)'$.

Q.47. An 8-bit computer has a register *R*. Determine the values of status bits *C, S, Z,* and *V* figure given; in Q. 30, after each of the following instructions. The initial value of register *R* in each case is hexadecimal 72. The numbers below are also in hexadecimal.

(*a*) Add immediate operand *C*6 to *R*.

(*b*) Add immediate operand *IE* to *R*.

(c) Subtract immediate operand 9*A* from *R*.

(*d*) AND immediate operand 8*D* to *R*.

(*e*) Exclusive-*OR R* with *R*.

Solution:

(a) 72 01110010

*C*6 11000110

138 00111000

$C = 1$ $S = 0$ $Z = 0$ $V = 0$

(*b*) 72 01110010

IE 00011110

138 10010000

$C = 0$ $S = 1$ $Z = 0$ $V = 1$

(*c*) 9*A* = 10011010

01100110 2's complement

72	01110010
*D*8	11011000

C = 0 *S* = 1 *Z* = 0 *V* = 0 (Borrow = 1)

(*d*)

72	01110010
8*D*	10001101
00	00000000

C = 0 *S* = 0 *Z* = 1 *V* = 0

(*e*) *C* = 0 *S* = 0 *Z* = 1 *V* = 0

Q.48. Two unsigned numbers *A* and *B* are compared by subtracting *A* – *B*. The carry status bit is considered as a borrow bit after a compare instruction in most commercial computers, so that C = 1 if *A* < *B*. Show that the relative magnitude of A and *B* ean be determined from inspection of status bits C and *Z* as specified in the tables.

Relation	Condition of Status Bits
A > *B*	C = 0 and *Z* = 0
A ≥ *B*	C = 0
A < *B*	C = 1
A ≤ *B*	C = 1 or *Z* = 1
A = *B*	*Z* = 1
A ≠ *B*	*Z* = 0

TABLE 2.6. For Conditional Branch Instruct

Mnemonic	Branch condition	Tested condition
BZ	Branch if zeroZ = 1	
BNZ	Branch if not zero	*Z* = 0
BC	Branch if carry	*C* = 1
BNC	Branch if no carry	C = 0
BP	Branch if plus	S = 0
BM	Branch if minus	S = 1
BV	Branch if overflow	*V* = 1
BNV	Branch if not overflow	*V* = 0
	Unsigned compare conditions (*A* – *B*)	
BHI	Branch if higher	*A* > *B*
BHE	Branch if higher or equal	*A* ≤ *B*
BLO	Branch if lower	*A* < *B*
BLOE	Branch if lower or equal	*A* ≤ *B*
BE	Branch if equal	*A* = *B*
BNE	Branch if not equal	*A* ≠ *B*
	Signed Compare Conditions (*A* – B)	
BGT	Branch if greater that	*A* > *B*
BGE	Branch if greater or equal	*A* ≥ *B*
BLT	Branch if less than	*A* < *B*
BLE	Branch if less or equal	*A* ≤ *B*
BE	Branch if equal	*A* = *B*
BNE	Branch if not equal	*A* ≠ *B*

Solution: $C = 1$ if $A < B$, therefore $C = 0$ if $A \geq B$

$Z = 1$ if $A = B$, therefore $Z = 1$ıı $A \neq B$

For $A > B$ we must have $A \leq B$ provided $A \neq B$

or

For $A \leq B$ we must have $A < B$ or $A = B$

or $C = 1$ or $Z = 1$ $(C + Z) = 1$

Q.49. Two signed numbers *A* and *B* represented in signed – 2's complement form are compared by subtracting *A* – *B*. Status bits *S*, *Z* and *V* are set or cleared depending on the result of the operation, (Note that there is a sign reversal if an overflow occurs). Show that the relative magnitude of *A* and *B* can be determinded from inspection of the status bits as specified below. (See also Table for conditional branch instruction given in Q. 32).

Relation	Condition of Status Bits
$A > B$	$(S \oplus V) = 0$ and $Z = 0$
$A > B$	$(S \oplus V) = 0$
$A < B$	$(S \oplus V) = 1$
$A < B$	$(S \oplus V) = 1$ or $Z = 1$
$A = B$	$Z = 1$
$A \neq B$	$Z = 0$

Solution: $A > B$ implies that $A - B \geq 0$ (positive or zero)

Sign $S = 0$ if no overflow (positive)

or $S = 1$ if overflow (sign reversal)

Boolean expression :

$S' V' + SV = 1$

or $(S + V) = 0$

$A < B$ is the complement of $A \geq B$ $(A - B)$ negative

then $S = 1$ if $V = 0$

or $S = 0$ if $V = 1(S \oplus V) = 1$

$A > B$ implies $A \geq B$ but not $A = B$

$(S \oplus V) = 0$ and $Z = 0$

$A \leq B$ implies $A < B$ or $A = B$

$(S \oplus V) = 1$ or $Z = 1$

Q.50. It is necessary to design a digital circuit with four inputs *C*, *S*, *Z* and *V* and 10 outputs, one for each of the branch conditions listed in Probs. 32 and 33. (The equal and unequal conditions are common to both tables). Draw the logic diagram of the circuit using two OR gates, one XOR gate, and five inverters.

Solution:

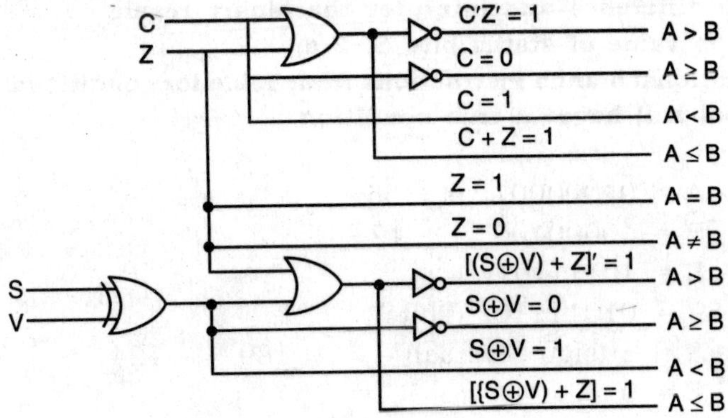

Fig. 2.29.

Q.51. Considser the two 8-bit numbers *A* = 01000001 and *B* = 10000100

(*a*) Give the decimal equivalent of each number assuming that

(*i*) they unsignsed, and

(*ii*) they are signed.

(*b*) Add the two binary numbers and interpret the sum assuming that the numbers are

(*i*) unsigned, and

(*ii*) signed.

(*c*) Determine the values of the *C*, *Z*, *S* and *V* status bits after the addition.

(*d*) List the conditional branch instructions from Table for conditional branch instruction in Q.32.

Solution:

		Unsigned	Signed
(*a*)			
	A = 01000001	65	+ 65
	B = 10000100	132	– 124
(*b*)	$A + B$ = 11000101	197	– 59

(*c*) $C = 0 \quad Z = 0 \quad S = 1 \quad V = 0$

(*d*) BNC BNZ BM BNV

Q.52. The program in a computer compares two unsigned numbers *A* and *B* by performing a subtaction *A* – *B* and updating the status bits. Let *A* = 01000001 and *B* = 10000100.

(*a*) Evaluate the difference and interpret the binary result.

(*b*) Determine the values of statuo bits *C* (borrow) and *Z*.

(*c*) List the conditional branch instructions from Table for conditional branch instrucution in Q. 8.26 that a will have a true condition.

Solution:

(*a*) A = 01000001 = 65

B = 10000100 = 132

A – B = 10111101 = – 67 (2's complement of 01000011)

(*b*) C (borrow) = 1, Z = 0 65 < 132

$A < B$

(*c*) BL, BLE, BNE

Q.53. The program in a computer compares two signed numbers *A* and *B* by performing the subtraction *A* – *B* and updating the status bits. Let *A* = 01000001 and *B* = 10000100.

(a) Evaluate the difference and interpret the binary result.

(b) Determine the value of status bits S, Z ar ' V.

(c) List the conditional branch instructions from Table for conditional branch instruction in Q. 32. that will haves a true condition.

Solution:

(a)

$A = 01000001 = +65$

$B = 10000100 = -124$

$A - B = 10111101$

$+189 = 010111101$ (9bits)

(b)

$S = 1$ (sign reversal) $+189 > -124$

$Z = 0$

$V = 1$ (overflow) $65 > -124$

$A > B$

(c) BGT, BGE, BNE

Q.54. The content of the top of a memory stack is 5320. The content of the stack pointer SP is 3560. A two-word call subroutine instruction is located in memory at address 1120 followed by the address field of 6720 at location 1121. What are the content of PC, SP and the top of the stack :

(a) Before the call instruction is fetched from memory?

(b) After the call instruction is executed ?

(c) After the return from subroutine?

Solution:

	PC	SP	Top of Stack
Initial	1120	3560	5320
Afer CALL	6720	3559	1122
After RETURN	1122	3560	5320

Q.55. What are the basic differences between a branch instruction, a call subroutine instruction, and program interrupt?

Solution: Branch instruction – Branch without being able to return

Subroutine call – Branch to subroutine and then return to calling program.

Program intesrrupt – Hardware initiated branch with possibility to return

Q.56. A computer responds to an interrupt request signal by pushing onto the stack the contents of PC and the current 'PSW' (program status word). It reads a new ????? from memory from a location given by an interrupt address symbolized by ?????. The first address of the service program is taken from memory at location IAD + 1.

(a) List the sequence of micro-operations for the interrupt cycle.

(b) List the sequence of micro-operations for the return from interrupt instructions.

Solution :

(a) $SP \leftarrow SP - 1$

$M(CP) \leftarrow PSW$

$SP \leftarrow SP - 1$

$M(SP) \leftarrow PC$

$TR \leftarrow IAD$ (TR is a temporary register)

$PSW \leftarrow M(TR)$

$TR \leftarrow TR + 1$

$PC \leftarrow M(TR)$

Go to fetch phase

(b) $PC \leftarrow M(SP)$

$SP \leftarrow SP + 1$

$PSW \leftarrow M(SP)$

$SP \leftarrow SP + 1$

Q.57. Three computers use register windows with the following characteristics. Determine the window size and the total number of registers in each computer.

	Computer 1	Computer 2	Computer 3
Global registers	10	8	16
Local registers	10	8	16
Common registers	6	8	16
Number of windows	8	4	16

Solution: Window Size = $L + 2C + G$

Computer 1 : 10 + 12 + 10 = 32

Computer 2 : 8 + 16 + 8 = 32

Computer 3 : 16 + 32 + 16 = 64

Register file = $(L + C)\ W + G$

Computer 1 : (10 + 6) 8 + 10 = 16 × 8 + 10 = 138

Computer 2 : (8 + 8) 4 + 8 = 16 × 4 + 8 = 72

Computer 3 : (16 + 16) 16 + 16 = 32 × 16 + 16 = 528

Q.58. Give an example of a RISC *I* instructions that will perfrom the following operations.

(a) Decrement a register

(b) Complement a register

(c) Negate a register

(d) Clear a register to 0

(e) Divide a signed number by 4

(f) No operation

Solution :

(a)	SUB *R*22, # 1, *R*22	*R*22 ← *R*22 – 1 (Subtract 1)
(b)	XOR *R*22, # – 1, *R*22	*R*22 ← *R*22 ⊕ all l's (*x* ? 1 = *x*')
(c)	SUB *R*0, *R*22, *R*22	*R*22 ← 0 – *R*22
(d)	ADD *R*0, *R*0, *R*22	*R*22 ← 0 + 0
(e)	SRA *R*22, # 2, *R*22	Arithmetic shift right twice
(f)	OR *R*1, *R*1, *R*1	*R*1 ← *R*1 v *R*1
or	ADD *R*1, *R*0, *R*1	*R*1 ← *R*1 + 0
or	SLL *R*1, # 0, *R*1	shift left 0 times

Q.59. Write the RISC *I* instructions in assembly language that wil *I* cause a jump to address 3200 if the *Z* (zero) status bit is equal to 1.

(a) Using immediate mode

(b) Using a relative address mode (assume that PC = 3400)

Solution:

(a) JMP *Z*, # 3200, (*RO*) PC ← 0 + 3200

(b) JMPR *Z*, – 200 PC ← 3400 + (– 200)

SUMMARY

1. (*a*) A CPU can be defined as a general purpose instruction set processor responsible for program execution.
   (*b*) A CPU consists of address bus, data bus and control bus.
   (*c*) A computer with one CPU is called microprocessor and with more than one CPU in called multiprocessor.
   (*d*) The address bus in used to transfer addresses from the CPU to main memory or to I/O devices.
   (*e*) Data bus is the main path by which information is transferred to and from the CPU
   (*f*) Control bus is used by CPU to control various devices connected and to synchronise their operations with those of the CPU.
2. (*a*) A control unit take the instructions one by one to execute. It takes data from input devices and store it in memory. And also sends data from memory to the output device.
   (*b*) All arithmetic and logical operations are carried out by Arithmetic Logical Unit
   (*c*) A control unit and the arithmetic logical unit together is known as CPU
3. (*a*) The accumulator is the main register of the ALU.
   (*b*) In execution of the most of the instructions the accumulator is used to store a input data or output result.
   (*c*) Instructions register holds the opcode of the current instruction.
   (*d*) Memory address register holds the address of the current instructions.
4. A accumulator based CPU consists of (a) data processing unit (b) program control unit and (c) memory and input/output interface unit.
   (*a*) (*i*) In the data processing unit, data is processed to get some results.
   (*ii*) The accumulator in the main operand register of the ALU.
   (*b*) (*i*) Program control unit controls various parts of CPU.
   (*ii*) Program counter holds the address of the next instructions to be read from memory after the current instruction is executed.
   (*iii*) Instruction register holds the opcode of the current instruction.
   (*iv*) Control circuits hold the responsibility of every operation of the CPU.
   (*c*) (*i*) Data registers of the memory and input/output interface unit acts as a buffer between the CPU and main memory.
   (*ii*) Address register contains the address of the present instructions obtained from the program control unit.
5. (*a*) Stack pointer and flag register are special registers of CPU.
   (*b*) Stack pointers holds the address of the most recently entered item into the stack.
   (*c*) Flag registers indicates the status which depends on the results of the operation.
6. (*a*) Micro-operations are the operations executed on data stored in registers.
   (*b*) A set of micro-operations specified by an instruction is known as macro-operation.
7. (*a*) A sequence of operations involved in processing an instruction constitutes an instruction cycle.
   (*b*) Fetch cycle in defined as the time required for getting the instruction code from main memory to CPU
   (*c*) Executed cycle is the time required to decode and execute an instruction.
   (*d*) Fetch cycle requires a fixed time slot and execute cycle requires variable time slot.
8. (*a*) A word length indicates the number of bits the CPU can process at a time.
   (*b*) A memory size indicates the total storage capacity of the CPU.
   (*c*) Word length is the indication of bit length of each register.
9. A computer is said to be operated based on stored program concept if it stores the instructions as well as data of a program in main memory when they are waiting to execute.

10. A stored program in main memory is executed instruction after instruction in a successive method by using program counter.

EXERCISE

(I) Fill in the Blanks (Solved)

1. SIMD stands for .... instruction stream, multiple data stream.
2. .... control is faster than microprogrammed control.

(II) True or False (Solved)

1. Indexing is one of the addressing mode.
2. MOV X is used to upload instruction in the processor.

(III) Fill in the Blanks (Unsolved)

1. In stack the values are stored in .... form.
2. .... is a visualisation of processing segments collection.

(IV) True or False (Unsolved)

1. Stack organized processor does not use address field for instructor ADD or MUL.
2. PSW stands for program stack word.

(V) Detailed Questions

1. A relative mode branch type of instruction is stored in memory at an address equivalent to decimal 750. The branch is made to an address equivalent to decimal 500.
   (*a*) What should be the value of the relative address field of the instruction in decimal?
   (*b*) Determine the relative address value in binary using 12 bits.
2. Write a program to evaluate the arithmetic statement
   (*a*) Using a general purpose register with three address instructions.
   (*b*) Using an accumulator type computer with one address instruction.
3. Perform the logic AND, OR , and XOR with the two binary strings 11011101 and 11001111.

ANSWERS

(I) Fill in the Blanks (Solved)

1. single
2. Hardwired

(II) True or False (Solved)

1. True
2. False

□□□

Chapter

3

Data Representation

Chapter Outline

- Introduction
- Binary Number System
- Bit and Byte
- K & M
- Number System
- Fixed-Point Representation
- IEEE Standard 754 Floating Point Numbers
- Decimal to Any Base Conversion
- Any Base to Decimal Conversion
- Decimal to Binary
- Binary to Decimal
- Decimal to Octal
- Octal to Decimal
- Octal to Binary
- Binary to Octal
- Octal to Hexa Decimal
- Hexa Decimal to Octal
- Addition and Multiplication Table
- Booth Multiplication Table

3.0 INTRODUCTION

When working with any kind of digital electronics in which numbers are being represented, it is important to understand the different ways numbers are represented in these systems. Almost without fail, numbers are represented by two voltage levels which can represent a one or a zero (an interesting exception to this rule is the new memory device recently announced by Intel which uses one of four possible voltage levels, thereby increasing the amount of information that can be stored in a given space). The number system based on ones and zeroes is called the binary system (because there are only two possible digits). Before discussing the binary system, a review of the decimal (ten possible digits) system is in order, because many of the concepts of the binary system will be easier to understand when introduced alongside their decimal counterpart.

3.1 BINARY NUMBER SYSTEM

A binary number is made up of only 0s and 1s.

This is $1 \times 8 + 1 \times 4 + 0 \times 2 + 1 + 1 \times (1/2) + 0 \times (1/4) + 1 \times (1/8)$(=13.625 in Decimal)

Similar to the decimal system, numbers can be placed to the left or right of the point, to indicate values greater than one or less than one. For binary numbers:

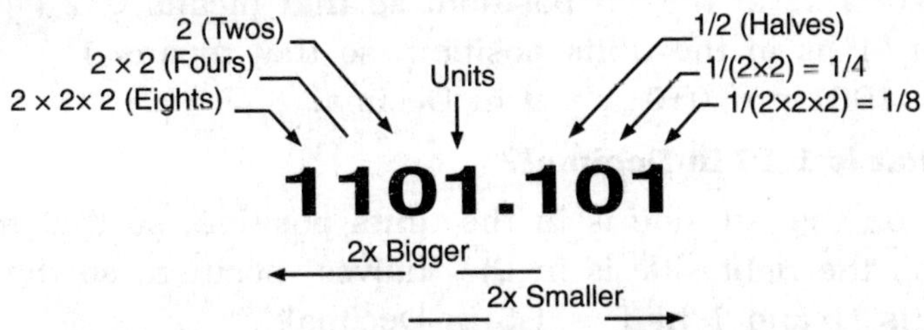

The number just to the left of the point is a whole number, we call this place units. As we move left, every number place gets 2 times bigger.

The first digit on the right of the point means halves (1/2). As we move further right, every number place gets 2 times smaller (one half as big).

Different Values

Because you can only have 0s or 1s, this is how you count using binary:

Decimal:	0	1	2	3	4	5	6	7	8	9	10	11	12	13	14	15
Binary:	0	1	10	11	100	101	110	111	1000	1001	1010	1011	1100	1101	1110	1111

"Binary is as easy as 1, 10, 11."

Here are some more equivalent values:

Decimal:	20	25	30	40	50	100	200	500
Binary:	10100	11001	11110	101000	110010	1100100	11001000	111110100

Definition of Binary

The word **binary** comes from "Bi-" meaning two. We see "bi-" in words such as "bicycle" (two wheels) or "binocular" (two eyes).

When you say a binary number, pronounce each digit (example, the binary number "101" is spoken as "one zero one", or sometimes "one-oh-one"). This way people don't get confused with the decimal number.

Bits

A single binary digit (like "0" or "1") is called a "bit". For example **11010** is five bits long.

The word **bit** is made up from the words "**binary digit**"

How to show that a number is binary

To show that a number is a binary number, follow it with a little 2 like this: $\mathbf{101_2}$

This way people won't think it is the decimal number "101" (one hundred and one).

Examples

Example 1: What is 1111_2 in Decimal?

- The "1" on the left is in the "2×2×2" position, so that means 1×2×2×2 (=8)
- The next "1" is in the "2×2" position, so that means 1×2×2 (=4)
- The next "1" is in the "2" position, so that means 1×2 (=2)
- The last "1" is in the units position, so that means 1
- Answer: 1111 = 8+4+2+1 = 15 in Decimal

Example 2: What is 1001_2 in Decimal?

- The "1" on the left is in the "2×2×2" position, so that means 1×2×2×2 (=8)
- The "0" is in the "2×2" position, so that means 0×2×2 (=0)

- The next "0" is in the "2" position, so that means 0×2 (=0)
- The last "1" is in the units position, so that means 1
- Answer: 1001 = 8+0+0+1 = 9 in Decimal

Example 3: What is 1.12 in Decimal?

- The "1" on the left side is in the units position, so that means 1.
- The 1 on the right side is in the "halves" position, so that means 1×(1/2)
- So, 1.1 is "1 and 1 half" = 1.5 in Decimal

Example 4: What is 10.112 in Decimal?

- The "1" is in the "2" position, so that means 1×2 (=2)
- The "0" is in the units position, so that means 0
- The "1" on the right of the point is in the "halves" position, so that means 1×(1/2)
- The last "1" on the right side is in the "quarters" position, so that means 1×(1/4)
- So, 10.11 is 2+0+1/2+1/4 = 2.75 in Decimal

"There are 10 kinds of people in the world, those who understand binary numbers, and those who don't."

3.2 BIT & BYTE

Computer uses the binary system. Any physical system that can exist in two distinct states (e.g., 0-1, on-off, hi-lo, yes-no, up-down, north-south, etc.) has the potential of being used to represent numbers or characters.

A binary digit is called a **bit**. There are two possible states in a bit, usually expressed as 0 and 1.

A series of eight bits strung together makes a **byte**, much as 12 makes a dozen. With 8 bits, or 8 binary digits, there exist 2^8=256 possible combinations. The following Table shows some of these combinations. (The number enclosed in parentheses represents the decimal equivalent.)

00000000	(0)	00010000	(16)	0010000	(32)	...	01110000	(112)
00000001	(1)	00010001	(17)	00100001	(33)	...	01110001	(113)
00000010	(2)	00010010	(18)	00100010	(34)	...	01110010	(114)
00000011	(3)	00010011	(19)	00100011	(35)	...	01110011	(115)
00000100	(4)	00010100	(20)	00100100	(36)	...	01110100	(116)
00000101	(5)	00010101	(21)	00100101	(37)	...	01110101	(117)
00000110	(6)	00010110	(22)	00100110	(38)	...	01110110	(118)
00000111	(7)	00010111	(23)	00100111	(39)	...	01110111	(119)
00001000	(8)	00011000	(24)	00101000	(40)	...	01111000	(120)
00001001	(9)	00011001	(25)	00101001	(41)	...	01111001	(121)
00001010	(10)	00011010	(26)	00101010	(42)	...	01111010	(122)
00001011	(11)	00011011	(27)	00101011	(43)	...	01111011	(123)
00001100	(12)	00011100	(28)	00101100	(44)	...	01111100	(124)
00001101	(13)	00011101	(29)	00101101	(45)	...	01111101	(125)
00001110	(14)	00011110	(30)	00101110	(46)	...	01111110	(126)
00001111	(15)	00011111	(31)	00101111	(47)	...	01111111	(127)

:

(continued)

:

10000000	(128)	10010000	(144)	10100000	(160)	...	11110000	(240)
10000001	(129)	10010001	(145)	10100001	(161)	...	11110001	(241)
10000010	(130)	10010010	(146)	10100010	(162)	...	11110010	(242)
10000011	(131)	10010011	(147)	10100011	(163)	...	11110011	(243)
10000100	(132)	10010100	(148)	10100100	(164)	...	11110100	(244)
10000101	(133)	10010101	(149)	10100101	(165)	...	11110101	(245)
10000110	(134)	10010110	(150)	10100110	(166)	...	11110110	(246)
10000111	(135)	10010111	(151)	10100111	(167)	...	11110111	(247)
10001000	(136)	10011000	(152)	10101000	(168)	...	11111000	(248)
10001001	(137)	10011001	(153)	10101001	(169)	...	11111001	(249)
10001010	(138)	10011010	(154)	10101010	(170)	...	11111010	(250)
10001011	(139)	10011011	(155)	10101011	(171)	...	11111011	(251)
10001100	(140)	10011100	(156)	10101100	(172)	...	11111100	(252)
10001101	(141)	10011101	(157)	10101101	(173)	...	11111101	(253)
10001110	(142)	10011110	(158)	10101110	(174)	...	11111110	(254)
10001111	(143)	10011111	(159)	10101111	(175)	...	11111111	(255)

3.3 K & M

2^10=1024 is commonly referred to as a "K". It is approximately equal to one thousand. Thus, 1 Kbyte is 1024 bytes. Likewise, 1024K is referred to as a "Meg". It is approximately equal to a million. 1 Mega byte is 1024*1024=1,048,576 bytes. If you remember that 1 byte equals one alphabetical letter, you can develop a good feel for size.

3.4 NUMBER SYSTEM

You may regard each digit as a box that can hold a number. In the binary system, there can be only two choices for this number -- either a "0" or a "1". In the octal system, there can be eight possibilities:

"0", "1", "2", "3", "4", "5", "6", "7".

In the decimal system, there are ten different numbers that can enter the digit box: "0", "1", "2", "3", "4", "5", "6", "7", "8", "9".

In the hexadecimal system, we allow 16 numbers:

"0", "1", "2", "3", "4", "5", "6", "7", "8", "9", "A", "B", "C", "D", "E", and "F".

As demonstrated by the following table, there is a direct correspondence between the binary system and the octal system, with three binary digits corresponding to one octal digit. Likewise, four binary digits translate directly into one hexadecimal digit. In computer usage, hexadecimal notation is especially common because it easily replaces the binary notation, which is too long and human mistakes in transcribing the binary numbers are too easily made.

Base Conversion Table

BIN	OCT	HEX	DEC
0000	00	0	0
0001	01	1	1

0010	02	2	2
0011	03	3	3
0100	04	4	4
0101	05	5	5
0110	06	6	6
0111	07	7	7
1000	10	8	8
1001	11	9	9
1010	12	A	10
1011	13	B	11
1100	14	C	12
1101	15	D	13
1110	16	E	14
1111	17	F	15

3.4.1 Convert From Any Base To Decimal

Let's think more carefully what a decimal number means. For example, 1234 means that there are four boxes (digits); and there are 4 one's in the right-most box (least significant digit), 3 ten's in the next box, 2 hundred's in the next box, and finally 1 thousand's in the left-most box (most significant digit). The total is 1234:

Original Number:	1	2	3	4
	|	|	|	|
How Many Tokens:	1	2	3	4
Digit/Token Value:	1000	100	10	1
Value:	1000 +	200 +	30 +	4 = 1234

or simply, 1*1000 + 2*100 + 3*10 + 4*1 = 1234

Thus, each digit has a value: 10^0=1 for the least significant digit, increasing to 10^1=10, 10^2=100, 10^3=1000, and so forth.

Likewise, the least significant digit in a hexadecimal number has a value of 16^0=1 for the least significant digit, increasing to 16^1=16 for the next digit, 16^2=256 for the next, 16^3=4096 for the next, and so forth. Thus, 1234 means that there are four boxes (digits); and there are 4 one's in the right-most box (least significant digit), 3 sixteen's in the next box, 2 256's in the next, and 1 4096's in the left-most box (most significant digit). The total is:

1*4096 + 2*256 + 3*16 + 4*1 = 4660

Example: Convert the hexadecimal number 4B3 to decimal notation. What about the decimal equivalent of the hexadecimal number 4B3.3?

Solution:

Original Number:	4	B	3 .	3
	|	|	|	|
How Many Tokens:	4	11	3	3
Digit/Token Value:	256	16	1	0.0625
Value:	1024 +	176 +	3 +	0.1875 = 1203.1875

Example. Convert 234.14 expressed in an octal notation to decimal.

Solution:

Original Number:	2	3	4	.	1	4
	|	|	|		|	|
How Many Tokens:	2	3	4		1	4
Digit/Token Value:	64	8	1		0.125	0.015625
Value:	128 +	24 +	4 +		0.125 +	0.0625 = **156.1875**

Another way is to think of a cash register with different slots, each holding bills of a different denomination.

3.4.2 Convert From Decimal to Any Base

Again, let's think about what you do to obtain each digit. As an example, let's start with a decimal number 1234 and convert it to decimal notation. To extract the last digit, you move the decimal point left by one digit, which means that you divide the given number by its base 10.

$$1234/10 = 123 + 4/10$$

The remainder of 4 is the last digit. To extract the next last digit, you again move the decimal point left by one digit and see what drops out.

$$123/10 = 12 + 3/10$$

The remainder of 3 is the next last digit. You repeat this process until there is nothing left. Then you stop. In summary, you do the following:

```
              Quotient Remainder
-------------------
1234/10 =      123     4 ------+
 123/10 =       12     3 ----+ |
  12/10 =        1     2 --+ | |   (Stop when the quotient is 0.)
   1/10 =        0     1 -+| | |
                          | | | |
                          1 2 3 4  (Base 10)
```

Now, let's try a nontrivial example. Let's express a decimal number 1341 in binary notation. Note that the desired base is 2, so we repeatedly divide the given decimal number by 2.

```
          Quotlent Remainder
-------------------
1234/2 =   670    1 ----------------------+
 670/2 =   335    0 --------------------+ |
 335/2 =   167    1 ------------------+ | |
 167/2 =    00    1 ----------------+ | | |
  83/2 =    41    1 --------------+ | | | |
  41/2 =    20    1 ------------+ | | | | |
  20/2 =    10    1 ----------+ | | | | | |
  10/2 =     5    0 --------+ | | | | | | |
   5/2 =     2    0 ------+ | | | | | | | |
   2/2 =     1    1 ----+ | | | | | | | | |
   1/2 =     0    1 --+ | | | | | | | | | |  Stop when the
                      | | | | | | | | | | |
                      1 0 1 0 0 1 1 1 1 0 1  (BIN; Base 2)
```

Let's express the same decimal number 1341 in octal notation.

```
                 Quotient Remainder
        ------------------
1341/8 =     167    5 ----------+
 167/8 =      20    7 --------+ |
  20/8 =       2    2 ------+ | |  (Stop when the quotient is 0)
   2/8 =       0    2 ----+ | | |
                          | | | |
                          2 4 7 5  (OCT; Base 8)
```

Let's express the same decimal number 1341 in hexadecimal notation.

```
                 Quotient Remainder
        ------------------
1341/16 =     83    13 ----------+
  83/16 =      3     3 --------+ |
   5/16 =      5     5 ------+ | |  (Stop when the quotient is 0)
                             | | |
                             5 3 D  (HEX; Base 16)
```

Example: Convert the decimal number 3315 to hexadecimal notation. What about the hexadecimal equivalent of the decimal number 3315.3?

Solution:

```
                 Quotient Remainder
        ------------------
3315/16 =    207     3 ----------+
 207/16 =     12    15 --------+ |
  12/16 =      0    12 ------+ | |  (Stop when the quotient is 0)
                             | | |
                             C F 3  (HEX; Base 16)
```

```
                                          (HEX; Base 16)
           Product Integer Part          0.4 C  C  C ...
     ----------------------------            |  |  |  |
     0.3*16  =  4.8      4         ---------+  |  |  |
     0.8*16  =  12.8    12         ------------+  |  |
     0.8*16  =  12.8    12         ---------------+  |
     0.8*16  =  12.8    12         ------------------+ |
         :                         --------------------+
         :
     Thus, 3315.3 (DEC) --> CF3.4CCC... (HEX)
```

Note that from the base conversion Table, you can easily get the binary notation from the hexadecimal number by grouping four binary digits per hexadecimal digit, or from or the octal number by grouping three binary digits per octal digit, and vice-versa.

HEX	5	3	D	
BIN	0101	0011	1101	
OCT	2	4	7	5
BIN	010	100	111	101

Finally, the fractional part is a decimal number can also be converted to any base by repeatedly multiplying the given number by the target base. Example: Convert a decimal number 0.1234 to binary notation

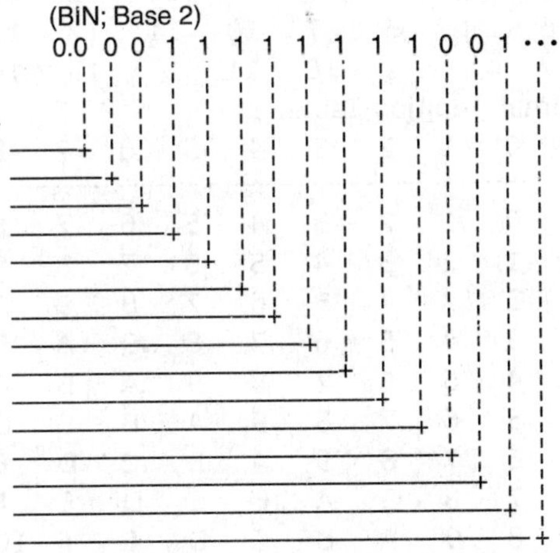

		Product	Integer Part
0.1234*2	=	0.2468	0
0.2468*2	=	0.4936	0
0.4936*2	=	0.9872	0
0.9872*2	=	1.9744	1
0.9744*2	=	1.9488	1
0.9488*2	=	1.8976	1
0.8976*2	=	1.7952	1
0.7952*2	=	1.5904	1
0.5904*2	=	1.1908	1
0.1808*2	=	0.3616	0
0.3616*2	=	0.7232	0
0.7232*2	=	1.4464	1

3.4.3 Addition and Multiplication Tables

You generate the addition tables in bases other then 10 by following the same rule you do in base 10. The resulting tables have the appearance of shifting the columns to the left by one in each subsequent rows. Note how simple the addition and multiplication tables are for the binary system; addition operation is simply the bit-wise XOR operation with carry, and multiplication is simply the logical AND operation.

Decimal Addition Table:

	0	1	2	3	4	5	6	7	8	9
0	0	1	2	3	4	5	6	7	8	9
1	1	2	3	4	5	6	7	8	9	10
2	2	3	4	5	6	7	8	9	10	11
3	3	4	5	6	7	8	9	10	11	12
4	4	5	6	7	8	9	10	11	12	13
5	5	6	7	8	9	10	11	12	13	14
6	6	7	8	9	10	11	12	13	14	15
7	7	8	9	10	11	12	13	14	15	16
8	8	9	10	11	12	13	14	15	16	17
9	9	10	11	12	13	14	15	16	17	18

Binary Addition Table:

	0	1
0	0	1
1	1	10

Octal Addition Table:

	0	1	2	3	4	5	6	7
0	0	1	2	3	4	5	6	7
1	1	2	3	4	5	6	7	10
2	2	3	4	5	6	7	10	11
3	3	4	5	6	7	10	11	12
4	4	5	6	7	10	[illegible]	[illegible]	[illegible]

5	5	6	7	10	11	12	13	14
6	6	7	10	11	12	13	14	15
7	7	10	11	12	13	14	15	16

Hexadecimal Addition Table:

	0	1	2	3	4	5	6	7	8	9	A	B	C	D	E	F
0	0	1	2	3	4	5	6	7	8	9	A	B	C	D	E	F
1	1	2	3	4	5	6	7	8	9	A	B	C	D	E	F	10
2	2	3	4	5	6	7	8	9	A	B	C	D	E	F	10	11
3	3	4	5	6	7	8	9	A	B	C	D	E	F	10	11	12
4	4	5	6	7	8	9	A	B	C	D	E	F	10	11	12	13
5	5	6	7	8	9	A	B	C	D	E	F	10	11	12	13	14
6	6	7	8	9	A	B	C	D	E	F	10	11	12	13	14	15
7	7	8	9	A	B	C	D	E	F	10	11	12	13	14	15	16
8	8	9	A	B	C	D	E	F	10	11	12	13	14	15	16	17
9	9	A	B	C	D	E	F	10	11	12	13	14	15	16	17	18
A	A	B	C	D	E	F	10	11	12	13	14	15	16	17	18	19
B	B	C	D	E	F	10	11	12	13	14	15	16	17	18	19	1A
C	C	D	E	F	10	11	12	13	14	15	16	17	18	19	1A	1B
D	D	E	F	10	11	12	13	14	15	16	17	18	19	1A	1B	1C
E	E	F	10	11	12	13	14	15	16	17	18	19	1A	1B	1C	1D
F	F	10	11	12	13	14	15	16	17	18	19	1A	1B	1C	1D	1E

3.4.4 Arithmetic Operations

You do arithematic with hexadecimal numbers or numbers in any base in exactly the same way you do with decimal numbers, except that the addition and multiplcation tables you employ to base your calculations are a bit different. Substraction is equivalent to adding a negative number, and division is equivalent to multiplying by the inverse.

Example: Find the sum of two hexadecimal integers 123 and DEF.

Solution:

From the above hexadecimal addition table, we see that:

$$3 + F = 12,\ 2 + E = 10, \text{ and } 1 + D = E$$

```
        123
      + DEF
      --------
carry   11
        E02
      ---------
sum     F12
```

Example: Find the product of two hexadecimal integers 123 and DEF.

Solution:

Step 1: We break down the second multiplier into single digits.

$$123*DEF = 123*(D00+E0+F)$$
$$= (123*D)*100 + (123*E)*10 + (123*F)$$

Step 2: We find the product in parentheses.

From the above hexadecimal multiplication table, we see that:

1*D = D, 2*D = 1A, 3*D = 27; thus,

```
123*D = (100 + 20 + 3)*D
      = 1*D*100 + 2*D*10 + 3*D
      = D*100 + 1A*10 + 27
      = D00 + 1A0 + 27
      = EC7
```

Likewise,

```
123*E = (100+20+3)*E
      = 1*E*100 + 2*E*10 + 3*E
      = E*100 + 1C*10  + 2A
      = E00 + 1C0 + 2A
      = FEA
123*F = (100 + 20 + 3)*F
      = 1*F*100 + 2*F*10 + 3*F
      = F*100 + 1E*10 + 2D
      = F00 + 1E0 + 2D = 110D
```

Or, in elementary school style:

```
   123       123        123
 x   D     x   E      x    F
 -------   --------   ----------
    27        2A        2D
   1A        1C        1E
  D         E         F
 -------   -------    --------
 EC7       FEA        110D
```

Step 3: We sum up the individual products.

```
123*DEF = (123*D)*100 + (123*E)*10 + (123*F)
        = EC7*100 + FEA*10 + 110D
        = EC700 + FEA0 + 110D
        = FD6AD
```

Or, in elementary school style:

```
    123
 x  DEF
  -------
   110D
   FEA
  EC7
  -------
 FD6AD
```

3.5 FIXED-POINT REPRESENTATION

3.5.1 Signed 2's complement

A fixed point number (an integer) can be represented by signed 2's complement, defined as:

- If $X \geq 0$ it is represented in binary form, but the MSB must be 0.
- If $X < 0$ it is represented by 2's complement of its magnitude $|X|$, but the MSB must be 1.

The MSB can therefore be used as the sign bit. Both cases can be represented by the following:

$$2^n + X = \begin{cases} 2^n + X = X, & \text{if } X \geq 0 \\ 2^n - |X|, & \text{if } X < 0 \end{cases}$$

Note:

- Assume there are a fixed number of n-bits available in hardware for representing a given value.
- When $X > 0$ 2^n| in the expressions above corresponds to a 1 in the $(n + 1)$th bit, which is always ignored in n-bit representation as the $(n + 1)$th bit does not exist. The remaining X is in binary form.
- When $X < 0$ $2^n - |X|$ is the definition of 2's complement of X .

Range of integer values representable

Example: $n = 3$, $2^{n-1} = 4$|, $2^{n-1} - 1 = 2^2 - 1 = 3$

	0	1	2	3	4	5	6	7
X	000	001	010	011	100	101	110	111
X	000	111	110	101	100	011	010	001

In this example, the range of values representable by 3-bit signed 2's complement is from 0 to $2^2 - 1 = 3$ for positive values and from $-2^2 = -4$ to-1 for negative values. In general, we note:

- When $X \geq 0$, MSB = 0| represents the plus sign, and the remaining n-1 bits represent the magnitudes in the range $0 \leq X \leq (2^{n-1} - 1)$|.
- When $X < 0$, MBS = 1| represents the minus sign, and the remaining n-1 bits represent the magnitude in the range $1 \leq (|X| = -X) \leq 2^{n-1}$, *i.e.* $-2^{n-1} \leq X \leq 1$| . The MSB is thus used as the sign-bit to indicate whether the number is positive or negative.
- The overall range for signed-complement representation is: $-2^{n-1} \leq X \leq (2^{n-1} - 1)$;

Any value beyond this range cannot be properly represented as the MSB no longer correctly indicates whether or $X \geq 0$ or $X < 0$.

- Signed-2's complement representation simplifies the operations when dealing with negative operands and subtractions.

-2^{n-1} .... -1 0 1 .... $2^{n-1} - 1$

When these 2^n| values are represented by signed 2's complement, MSB = 1 for the 2^{n-1} values on the left (from -2^{n-1} to -1), MSB = 0 for the 2^{n-1} values on the right (from 0 to $2^{n-1} - 1$).

Examples:

Range of value reperesentable

$n = 4$ (–23 = –8) ~ (23 –1 = 7)

$n = 6$ (–25 = –32) ~ (25 –1 = 31)

Singed 2's compl (n = 3)	decimal value
000	0
001	1
010	2
011	3
100	–4
101	–3
110	–2
111	–1

Value	5-bit Register	9-bit Register
+9	01001	000001001
–9	10111	11111011
+12	01100	000001100
–12	10100	111110100
+16	10000(*)	000010000
–16	10000	111110000

* This is an overflow, as when n = 5, the range of values representable is from –16 to 15.

3.6 IEEE STANDARD 754 FLOATING POINT NUMBERS

IEEE Standard 754 floating-point is the most common representation today for real numbers on computers, including Intel-based PC's, Macintoshes, and most Unix platforms. This article gives a brief overview of IEEE floating-point and its representation. Discussion of arithmetic implementation may be found in the book mentioned at the bottom of this article.

3.6.1 What Are Floating Point Numbers?

There are several ways to represent real numbers on computers. Fixed point places a radix point somewhere in the middle of the digits, and is equivalent to using integers that represent portions of some unit. For example, one might represent 1/100ths of a unit; if you have four decimal digits, you could represent 10.82, or 00.01. Another approach is to use rationals, and represent every number as the ratio of two integers

Floating-point representation - the most common solution - basically represents reals in scientific notation. Scientific notation represents numbers as a base number and an exponent. For example, 123.456 could be represented as 1.23456 × 102. In hexadecimal, the number 123.abc might be represented as 1.23abc × 162.

Floating-point solves a number of representation problems. Fixed-point has a fixed window of representation, which limits it from representing very large or very small numbers. Also, fixed-point is prone to a loss of precision when two large numbers are divided.

Floating-point, on the other hand, employs a sort of "sliding window" of precision appropriate to the scale of the number. This allows it to represent numbers from 1,000,000,000,000 to 0.0000000000000001 with ease.

3.6.2 Storage Layout

IEEE floating point numbers have three basic components: the sign, the exponent, and the mantissa. The mantissa is composed of the fraction and an implicit leading digit (explained below). The exponent base (2) is implicit and need not be stored.

The following figure shows the layout for single (32-bit) and double (64-bit) precision floating-point values. The number of bits for each field are shown (bit ranges are in square brackets):

	Sign	Exponent	Fraction	Bias
Single Precision	1 [31]	8 [30-23]	23 [22-00]	127
Double Precision	1 [63]	11 [62-52]	52 [51-00]	1023

The Sign Bit

The sign bit is as simple as it gets. 0 denotes a positive number; 1 denotes a negative number. Flipping the value of this bit flips the sign of the number.

The Exponent

The exponent field needs to represent both positive and negative exponents. To do this, a bias is added to the actual exponent in order to get the stored exponent. For IEEE single-precision floats, this value is 127. Thus, an exponent of zero means that 127 is stored in the exponent field. A stored value of 200 indicates an exponent of (200-127), or 73. For reasons discussed later, exponents of -127 (all 0s) and +128 (all 1s) are reserved for special numbers.

For double precision, the exponent field is 11 bits, and has a bias of 1023.

The Mantissa

The mantissa, also known as the significant, represents the precision bits of the number. It is composed of an implicit leading bit and the fraction bits.

To find out the value of the implicit leading bit, consider that any number can be expressed in scientific notation in many different ways. For example, the number five can be represented as any of these:

5.00 × 100
0.05 × 102
5000 × 10-3

In order to maximize the quantity of representable numbers, floating-point numbers are typically stored in normalized form. This basically puts the radix point after the first non-zero digit. In normalized form, five is represented as 5.0 × 100.

A nice little optimization is available to us in base two, since the only possible non-zero digit is 1. Thus, we can just assume a leading digit of 1, and don't need to represent it explicitly. As a result, the mantissa has effectively 24-bits of resolution, by way of 23 fraction bits.

Putting it All Together

So, to sum up:

1. The sign bit is 0 for positive, 1 for negative.
2. The exponent's base is two.
3. The exponent field contains 127 plus the true exponent for single-precision, or 1023 plus the true exponent for double precision.

4. The first bit of the mantissa is typically assumed to be 1.f, where f is the field of fraction bits.

3.6.3 Ranges of Floating-Point Numbers

Let's consider single-precision floats for a second. Note that we're taking essentially a 32-bit number and re-jiggering the fields to cover a much broader range. Something has to give, and it's precision. For example, regular 32-bit integers, with all precision centered around zero, can precisely store integers with 32-bits of resolution. Single-precision floating-point, on the other hand, is unable to match this resolution with its 24-bits. It does, however, approximate this value by effectively truncating from the lower end. For example:

```
    11110000 11001100 10101010 00001111  // 32-bit integer
=   +1.1110000 11001100 10101010 x 231    // Single-Precision Float
=   11110000 11001100 10101010 00000000  // Corresponding Value
```

This approximates the 32-bit value, but doesn't yield an exact representation. On the other hand, besides the ability to represent fractional components (which integers lack completely), the floating-point value can represent numbers around 2127, compared to 32-bit integers maximum value around 232.

The range of positive floating point numbers can be split into normalized numbers (which preserve the full precision of the mantissa), and denormalized numbers (discussed later) which use only a portion of the fractions's precision.

	Denormalized	Normalized	Approximate Decimal
Single Precision	$\pm 2^{-149}$ to $(1-2^{-23}) \times 2^{-126}$	$\pm 2^{-126}$ to $(2-2^{-23}) \times 2^{127}$	$\pm \sim 10^{-44.85}$ to $\sim 10^{38.53}$
Double Precision	$\pm 2^{-1074}$ to $(1-2^{-52}) \times 2^{-1022}$	$\pm 2^{-1022}$ to $(2-2^{-52}) \times 2^{1023}$	$\pm \sim 10^{-323.3}$ to $\sim 10^{308.3}$

Since the sign of floating point numbers is given by a special leading bit, the range for negative numbers is given by the negation of the above values.

There are five distinct numerical ranges that single-precision floating-point numbers are not able to represent:

1. Negative numbers less than – (2 – 2 – 23) × 2127 (*negative overflow*)
2. Negative numbers greater than – 2 – 149 (*negative underflow*)
3. Zero
4. Positive numbers less than 2 – 149 (*positive underflow*)
5. Positive numbers greater than (2– 2 – 23) × 2127 (*positive overflow*)

Overflow means that values have grown too large for the representation, much in the same way that you can overflow integers. Underflow is a less serious problem because is just denotes a loss of precision, which is guaranteed to be closely approximated by zero.

Here's a Table of the effective range (excluding infinite values) of IEEE floating-point numbers:

	Binary	Decimal
Single	± (2-2-23) × 2127	~ ± 1038.53
Double	± (2-2-52) × 21023	~ ± 10308.25

Note that the extreme values occur (regardless of sign) when the exponent is at the maximum value for finite numbers (2127 for single-precision, 21023 for double), and the mantissa is filled with 1s (including the normalizing 1 bit).

3.6.4 Special Values

IEEE reserves exponent field values of all 0s and all 1s to denote special values in the floating-point scheme.

Zero

As mentioned above, zero is not directly representable in the straight format, due to the assumption of a leading 1 (we'd need to specify a true zero mantissa to yield a value of zero). Zero is a special value denoted with an exponent field of zero and a fraction field of zero. Note that – 0 and + 0 are distinct values, though they both compare as equal.

Denormalized

If the exponent is all 0s, but the fraction is non-zero (else it would be interpreted as zero), then the value is a denormalized number, which does not have an assumed leading 1 before the binary point. Thus, this represents a number (–1)s × 0.f × 2–126, where s is the sign bit and f is the fraction. For double precision, denormalized numbers are of the form (–1)s × 0.f × 2–1022. From this you can interpret zero as a special type of denormalized number.

Infinity

The values + infinity and – infinity are denoted with an exponent of all 1s and a fraction of all 0s. The sign bit distinguishes between negative infinity and positive infinity. Being able to denote infinity as a specific value is useful because it allows operations to continue past overflow situations. Operations with infinite values are well defined in IEEE floating point.

Not A Number

The value NaN (Not a Number) is used to represent a value that does not represent a real number. NaN's are represented by a bit pattern with an exponent of all 1s and a non-zero fraction. There are two categories of NaN: QNaN (Quiet NaN) and SNaN (Signalling NaN).

A QNaN is a NaN with the most significant fraction bit set. QNaN's propagate freely through most arithmetic operations. These values pop out of an operation when the result is not mathematically defined.

An SNaN is a NaN with the most significant fraction bit clear. It is used to signal an exception when used in operations. SNaN's can be handy to assign to uninitialized variables to trap premature usage.

Semantically, QNaN's denote indeterminate operations, while SNaN's denote invalid operations.

3.6.5 Special Operations

Operations on special numbers are well-defined by IEEE. In the simplest case, any operation with a NaN yields a NaN result. Other operations are as follows:

Operation	Result
n ÷ ±Infinity	0
±Infinity × ±Infinity	±Infinity
±nonzero ÷ 0	±Infinity
Infinity + Infinity	Infinity
±0 ÷ ±0	NaN
Infinity - Infinity	NaN
±Infinity ÷ ±Infinity	NaN
±Infinity × 0	NaN

SOLVED PROBLEMS

Q.1. Represent the binary number 1010111101110111 in octa form or hexadecimal form ?

Solution: Binary number = 1010 1111 0111 0111

Hexadecimal = *A* *F* 7 7 = $(AF77)_{16}$

Octal form = 1010111101110111

⇒ 1 2 7 5 6 7

= $(127567)_8$.

Q.2. Describe the alpha numeric representation.

Solution: An alpha numeric character set is a set of elements that includes the 10 decimal digits, the 26 letters of the alphabets and a number of special characters, such as, $, + and = etc. such a set contains 32 and 64 elements (if only upper case letters are included) or between 64 to 128 (if both upper case and lower case letters are included). In the first case, the binary code will require six bits and in second case seven bits the standard alphanumeric binary code is the ASCII (American Standard Code for Information Interchange) which uses seven bits to code 128 characters.

Some examples in alphanumeric

a = 97 (ASCII)

A = 65 (ASCII)

O = 48 (ASCII).

Q.3. Find the value of number 12389 in

(*a*) 10's complement form **(*b*) 9's complement form**

(*c*) 1's complement form **(*d*) 2's complement form.**

Solution: Given number = 12389

9's complement = 99999 – 12389 = 87610

10's complement = 100000 – 12389 = 87611

1's complement binary of 1238 is

= 11000001100101

1's complement = 00111110011010

2's complement = 00111110011011

Q.4. Describe the fixed point representation.

Solution: Positive integers including zero can be represented as unsigned numbers. However to represent negative integers left most position (bit) is 1.

The fixed point method assumes that the binary point is always fixed in one position. Two position most widely used are :

(*i*) A binary point in the extreme left of the register to make the stored number a fraction and

(*ii*) A binary point in the extreme right to make the stored number and integer.

Q.5. What do you mean by overflow ? Describe the overflow detection.

Solution: When the two numbers of *n* digits each are added and the sum occupies *n* digits, we say over flow occurred.

The detection of an overflow after addition of two binary numbers depends on whether the numbers are considered to be signed or unsigned when two unsigned numbers are added, an overflow is detected from the end carry out of the most significant position. In the case of sign numbers, the left most bit always represent the sign, and negative numbers are in 2's complement form.

When two signed numbers are added the sign bit is treated as part of the number and the end carry does not indicate an overflow.

An overflow cannot occur after an addition if one number is positive and other is negative, since adding a positive number to a negative number produces a result that is smaller than larger of the two original numbers.

An overflow occur if the two numbers added are both positive or both negative.

+ 70	01000110
+ 80	01010000
+ 150	10010110

Since sum of two number is +150, it exceeds the capacity of the 8-bit register.

−70	10111010
−80	10110000
−150	01001010

Overflow Detection : An overflow condition can be detected by observing the carry into the sign bit position and carry out of the sign bit position. If these two carry are not equal, an overflow condition is produced. If two carries are applied to an exclusive *–OR* gate, an overflow will be detected when the output of gate is equal to 1.

Q.6. Represent the decimal number

***(a)* 27, *(b)* 396, *(c)* 4096**

Into binary form in

***(i)* Binary code** ***(ii)* BCD code**

***(iii)* Excess-3 code** ***(iv)* Gray code**

***(v)* Octal code** ***(vi)* Hexadecimal code.**

Solution: *(a)* *(i)* 27 = 11011

(ii) 27 = 00100111

(iii) 27 = 01011010

(iv) 00110100

(v) 27 = $(33)_8$ = 011011

(vi) 27 = $(1B)_{16}$ = 00011011

(b) *(i)* 396 = 110001100

(ii) 396 = 001110010110

(iii) 396 = 011011001001

(iv) 396 = 001011010101

(v) 396 = $(614)_8$ = 110001100

(vi) 396 = $(186)_{16}$ = 000110001100

(c) *(i)* 4096 = 1000000000000

(ii) 4096 = 0100000010010110

(iii) 4096 = 0111001111001001

(iv) 4096 = 0110000011010101

(v) 4096 = $(10000)_8$ = 001000000000000

(vi) 4096 = $(1000)_{16}$ = 0001000000000000

Q.7. What is floating point notation ? What is benefit of these ?

Solution: The usual method used by computers to represent real number is floating point notation. The key concept is that a real number is represented by a numbers called a mantissa, times a base raised to an integer power called an exponent. The base is usually fixed and mantissa and exponent vary to represent different real numbers.

e.g. If the base is fixed at 10, the number 378.53 could be represented as 37853×10^{-2}. The mantissa is 37853 and the exponent is –2. Other possible representations are $.37853 \times 10^3$. A real number is represented by a 32-bit string consisting of 24-bit mantissa followed by an 8-bit exponent.

The base is fixed at 10. Both the mantissa and exponent are two's complement binary integers.

e.g. 24 bit binary representation of 38753 is 000000001001011101100001 and 8 it two's complement binary representation of –2 is 11111110. So the representation of 387.53 is 0000 0000 1001 0111 0110 0001.

The advantage of floating point notation is that it can be used to represent numbers with extremely small absolute values. For example, in the notation presented priviously, the largest number can be represented is $(2^{23} - 1) \times 10^{127}$, which is very large number indeed.

The small positive number that can be represented is 10^{-128} which is quite small.

Q.8. Convert the following binary number to decimal :

(*a*) 101110, (*b*) 1110101, (*c*) 110110100.

Solution: (*a*) $(101110)_2 = 32 + 8 + 4 + 2 = 46$

(*b*) $(1110101)_2 = 64 + 32 + 16 + 4 + 1 = 117$

(*c*) $(110110100)_2 = 256 + 128 + 32 + 16 + 4 = 436.$

Q.9. Convert the following numbers with the indicated bases to decimal :

(*a*) $(12121)_3$ (*b*) $(4310)_5$

(*c*) $(50)_7$ (*d*) $(198)_{12}$.

Solution: (*a*) $(12121)_3 = 3^4 + 2 \times 3^3 + 3^2 + 2 \times 3 + 1 = 81 + 54 + 9 + 6 + 1 = 151.$

(*b*) $(4310)_5 = 4 \times 5^3 + 3 \times 5^2 + 5 = 500 + 75 + 5 = 580.$

(*c*) $(50)_7 = 5 \times 7 = 35.$

(*d*) $(198)_{12} = 12^2 + 9 + 12 + 8 = 144 + 108 + 8 = 260.$

Q.10. Convert the following decimal numbers to binary :

(*a*) 1231, (*b*) 673, (*c*) 1998.

Solution: (*a*) $(1231) = 1024 + 128 + 64 + 15 = 2^{10} + 2^7 + 2^6 + 2^3 + 2^2 + 1$

$= (10011001111)_2$

(*b*) $673 = 512 + 128 + 32 + 1 = 2^9 + 2^7 + 2^5 + 1 = (1010100001)_2$

(*c*) $1998 = 1024 + 512 + 256 + 128 + 64 + 8 + 4 + 2$

$= 2^{10} + 2^9 + 2^8 + 2^7 + 2^6 + 2^3 + 2^2 + 1 = (11111001110)_2.$

Q.11. Convert the following decimal numbers to the bases indicated.

(*a*) 7562 to octal

(*b*) 1998 to hexadecimal

(*c*) 175 to binary.

Solution: (*a*) $(7562)_{10} = (16612)_8$

(*b*) $(1938)_{10} = (792)_{16}$

(*c*) $(175)_{10} = (10101111)_2.$

Q.12. Convert the hexadecimal number F3A7C2 to binary and octal.

Solution: $(F3A7C2)_{16} = (111100111010011111000010)_2 = (74723702)_8$.

Q.13. What is the radix of the numbers if the solution to the quadratic equation $x^2 - 10x + 31 = 0$ is $x = 5$ and $x = 8$?

Solution: $(x^2 - 10x + 31) = [(x - 5)(x - 8)]_{10} = x^2 - (5 + 8)_{10}x + (40)_{10}x$

$(10)_r = (13)_r$

Radix = 13

Also, $(31)_r = 3 \times 13 + 1 = (40)_{10}$

Radix = 13.

Q.14. Show the value of all bits of a 12-bit register that hold the number equivalent to the decimal 215 in

(a) binary

(b) binary coded octal

(c) binary coded hexadecimal

(d) binary coded decimal.

Solution: $(215)_{10} = 128 + 64 + 16 + 7 = (11010111)_2$

(a) 000 011 010 111

(b) 000 011 010 111 binary coded octal
  0 3 2 7

(c) 0000 1101 0111 binary coded hexadecimal
  0 0 7

(d) 0010 0001 0101 binary coded decimal
  2 1 5

Q.15. Show the bit configuration of a 24-bit register when its content represents the decimal equivalent of 295 :

(a) binary

(b) in BCD

(c) in ASCII using eight bits with even parity.

Solution: $(295)_{10} = 256 + 32 + 7 = (100100111)_2$

(a) 0000 C000 0000 0001 0010 0111

(b) 0000 0000 0000 0010 1001 0101

(c) 10110010 00111001 00110101.

Q.16. Decode the following ASCII code :

1001010 10C1111 1001000 1001110 0100000 1000100 1001111 100010

Solution: JOHN DOE.

Q.17. Obtain the 9's complement of the following eight digit decimal numbers :

(a) 12349876 **(b) 00980100**

(c) 90009951 **(d) 00000000.**

Solution: (a) 87650123 (b) 99019899

(c) 09990048 (d) 99999999.

Q.18. Obtain the 10's complement of the following six digit decimal numbers :

(a) 876100 **(b) 909343**

(c) 900000 **(d) 000000.**

Solution: (*a*) 876100 (*b*) 909343

(*c*) 900000 (*d*) 000000.

Q.19. Obtain the 1's and 2's complements of the following six digit decimal numbers :

(*a*) 10101110 **(*b*) 10000001**

(*c*) 10000000 **(*d*) 00000001**

(*e*) 00000000.

Solution: 1's Complement :

(*a*) 01010001 (*b*) 01111110

(*c*) 01111111 (*d*) 11111110

(*e*) 11111111.

2's Complement :

(*a*) 01010010 (*b*) 01111111

(*c*) 10000000 (*d*) 11111111

(*e*) 00000000.

Q.20. Perform the subtraction with the following unsigned decimal numbers by taking the 10's complement of the subtrahend.

(*a*) 5250 - 1321 **(*b*) 1753 - 8640**

(*c*) 20 - 100 **(*d*) 1200 - 250.**

Solution: (*a*)

```
   5250
   8629
1) 3929
```

(*b*)

```
   1753
   1360
0) 3113
  –6887  10's complement
```

(*c*)

```
    020
  + 900
0)  920
   –080  10's complement
```

(*d*)

```
   1200
 | 9750
0) 0950
```

Q.21. Perform the subtraction with the following unsigned binary numbers by taking the 2's complement of the subtrahend.

(*a*) 11010 - 10000 **(*b*) 11010 - 1181**

(*c*) 11010 - 110000 **(*d*) 1010100 - 1010100.**

Solution: (*a*)

```
    11010
    10000
 1) 01010
(26 – 16 = 10)
```

(*b*)

```
    11010
    10011
 1) 01101
(26 – 13 = 13)
```

(*c*)
```
       000100
       010000
    ---------
     1) 010100
  (4 – 48 = –44)
```

(*d*)
```
      1010110
      0101100
    ---------
    1) 0000000
  (84 – 84 = 0)
```

Q.22. Perform the arithmetic operations (+42) + (–13) and (–42) – (–13) in binary using signed 2's complement representation for negative numbers.

Solution:

+42 = 0101010		+13 = 0001101	
–42 = 1010110		–13 = 1110011	
(+42) = 0101010		(–42) = 1010110	
(–13) = 1110011		(+13) = 0001101	
(+29) = 0011101		(–29) = 1100011	

Q.23. Represent the binary number 1010 1111 0111 0111 in Octal form or Hexadecimal form ?

Solution: Binary number = 1010 1111 0111 0111

Hexadecimal = A F 77

Octal form = 1010 111 101 110 111

$\Rightarrow$ 1 2 7 5 6 7 = $(127567)_8$.

Q.24. Describe the alpha numeric representation ?

Solution: An alpha numeric character set is a set of elements that includes the 10 decimal digits, the 26 letters of the alphabet and a number of special characters, such as $, +, and =. Such a set contains 32 and 64 elements (if only upper case letters are included) or between 64 to 128 (if both upper case and lower case letters are included). In the first case the binary code will require six bits and in second case seven bits the standard alpha numeric binary code is the ASCII (American Standard Code for Information Interchange) which uses seven bits to code 128 characters.

Some Example in Alpha Numeric

a = 97 (in ASCII)

A = 65 (in ASCII)

O = 48 (in ASCII)

Q.25. Find the value of number 12389 in

(*a*) 10's complement form

(*b*) 9's complement form

(*c*) 1's complement form

(*d*) 2's complement form

Solution: Given number = 12389

9's complement = 99999 – 12389 = 87610

10's complement = 100000 – 12389 = 87611

1's complement binary number of 12389 is = 11000001100101

1's complement = 00111110011010

2's complement = 00111110011010

1

00111110011011

Q.26. Describe the fixed-point representation ?

Solution: Positive integer including zero can be represented as unsigned. However we represent negative integers leftmost position (bit) is 1.

The fixed point method assumes that the binary point is always fixed in one position. The two position most widely used are :

(*i*) a binary point in the extreme left of the register to make the stored number of fraction, and

(*ii*) a binary point in the extreme right of the register to make the stored number and integer.

Q.27. What do you mean by overflow ? Describe the over flow detection.

Solution: When two numbers of a digits each are added and the sum occupies *n* digits, we say over flow occurred.

The detection of an overflow after the addition of two binary numbers depends on whether the numbers are considered to be signed or unsigned when two unsigned number is added, an overflow is detected from the end carry out of the most significant position. In the case of sign numbers, the leftmost bit always represent the sign, and negative numbers are in 2's complement form.

When two signed numbers are added, the sign bit is treated as part of the number and the end carry does not indicate an overflow.

An overflow cannot occur after an addition if one number is positive and other is negative, since adding a positive number to a negative number produces a result that is smaller than larger of the two original numbers.

An overflow occur, if the two numbers added are both positive or both negative.

+ 70	0 1000110
+ 80	0 1010000
+ 150	1 0010110

Since sum of two number is + 150, it exceeds the capacity of the 8-bit registers.

Carries : 10

– 70	1 0111010
– 80	1 0110000
– 150	0 1001010

Over Flow Detection : An over flow condition can be detected by observing the carry into the sign bit position and carry out of the sign bit position. If these two carry are not equal, an overflow condition is produced. If two carries are applied to an exclusive — OR gate, an overflow will be detected when the output of the gate is equal to 1.

Q.28. Represent the decimal number

(*a*) 27

(*b*) 396

(*c*) 4096

in binary form in

(*i*) Binary code

***(ii)* d**

***(iii)* d**

***(iv)* Gray code**

***(v)* Octal code**

***(vi)* Hexadecimal code**

Solution:

(a) (*i*) 27 = 11011

(*ii*) 27 = 00100111

(*iii*) 27 = 01011010

(*iv*) 0011 0100

(*v*) 27 = $(33)_8$ = 011 011

(*vi*) 27 = $(1B)_{16}$ = 0001 1011

(b) (*i*) 396 = 11000 1100

(*ii*) 396 = 0011 1001 0110

(*iii*) 396 = 0110 1100 1001

(*iv*) 396 = 0010 1101 0101

(*v*) 396 = $(614)_8$ = 110 001 100

(*vi*) 396 = $(18C)_{16}$ = 0001 1000 1100

(c) (*i*) 4096 = 10000 0000 0000

(*ii*) 4096 = 0100 0000 1001 0110

(*iii*) 4096 = 0111 0011 1100 1001

(*iv*) 4096 = 0111 0011 1100 1001

(*v*) 4096 = $(10000)_8$ = 001 000 000 000000

(*vi*) 4096 = $(1000)_{16}$ = 0001 0000 0000 0000

Q.29. What is floating point notation ? What is benefit of these ?

Solution: The usual method used by computers to represent real number is floating point notation. The key concept is that a real number is represented by a number, called a mantissa, times a base raised to an integer power called an *exponent.* The base is usually fixed and the mantissa and exponent vary to represent different real numbers.

e.g., if the base is fixed at 10, the number 378.53 could be represented as 38753 × 10^{-2}, The mantissa is 38753 and the exponent is – 2. Other possible representations are .38753 × 10^3 and 387.53 × 10^0.

A real number is represented by a 32 bit string consisting of a 24-bit mantissa followed by an 8-bit exponent. The base is fixed at 10. Both the mantissa and exponent are two's complement binary integers.

e.g., 24 bit binary representation of 38753 is 000000001001011111000001 and 8 bit two's complement binary representation of – 2 is 11111110. So, the representation of 387.53 is 0000 0000 1001 0111 0110 0001.

The advantage of floating point notation is that it can be used to represent numbers with extremely large or extremely small absolute values. For example, in the notation presented previously, the largest number can be represented is $(2^{23} - 1) \times 10^{127}$. Which is very large number indeed.

Q.30. Convert the following binary numbers to decimal :

101110; 1110101; and 110110100

Solution: $(101110)_2$ = 32 + 8 + 4 + 2 = 46

$(1110101)_2 = 64 + 32 + 16 + 4 + 1 = 117$
$(110110100)_2 = 256 + 128 + 32 + 16 + 4 = 436.$

Q.31. Convert the following numbers with the indicated bases to decimal :

$(12121)_3$; $(4310)_5$; $(50)_7$; and $(138)_{12}$

Solution: $(12121)_3 = 3^4 + 2 \times 3^3 + 3^2 + 2 \times 3 + 1 = 81 + 54 + 9 + 6 + 1 = 151$
$(4310)_5 = 4 \times 5^3 + 3 \times 5^2 + 5 = 500 + 75 + 5 = 580$
$(50)_7 = 5 \times 7$
$(198)_{12} = 12^2 + 9 \times 12 + 8 = 144 + 108 + 8 = 260.$

Q.32. Convert the following decimal numbers to binary :

1231; 673 and 1998

Solution: $(1231)_{10} = 1024 + 128 + 64 + 15 = 2^{10} + 2^7 + 2^6 + 2^3 + 2^2 + 1$
$= (10011001111)_2$
$(673)_{10} = 512 + 128 + 32 + 1 = 2^9 + 2^7 + 2^5 + 1 = (1010100001)_2$
$(1998)_{10} = 1024 + 512 + 256 + 128 + 64 + 8 + 4 + 2$
$= 2^{10} + 2^9 + 2^8 + 2^6 + 2^3 + 2^2 + 1 = (11111001110)_2$

Q.33. Convert the following decimal numbers to the bases indicated.

(a) 7562 to octal
(b) 1938 to hexadecimal
(c) *175 to binary*

Solution:

(a) $(7562)_{10} = (16612)_8$
(b) $(1938)_{10} = (792)_{19}$
(c) $(175)_{10} = (10101111)_2$

Q.34. Convert the hexadecimal number F3A7C2 to binary and octal.

Solution: $(F3A7C2)_{16} = (11110011\ 10100111\ 1100\ 0010)_2 = (74723702)_8$

Q.35. What is the radix of the numbers if the solution to the quadratic equation

$x^2 - 10x + 31 = 0$

is $x = 5$ and $x = 8$?

Solution: $(x^2 - 10x + 31)_r = [(x - 5)(x - 8)]_{10} = x^2 - (5 + 8)_{10}\, x + (40)_{10}\, x$

$\therefore$ $(10)_r = (13)_{10}$
Radix = 13

Also $(31)_r = 3 \times 13 + 1 - (40)_{10}$
Radix = 13.

Q.36. Show the value of all bits of a 12-bit register that hold the number equivalent to decimal 215 in

(a) binary
(b) binary-coded octal;
(c) binary-coded hexadecimal
(d) *binary coded decimal (BCD).*

Solution: (215) 1D = $128 + 64 + 16 + 7 = (110\ 10\ 111)_2$

(a) 000011010111 Binary
(b) 000 011 010 111 Binary coded octal
0 3 2 7

(*c*) 0000 1101 0111 Binary coded hexadecimal
0 2 7

(*d*) 0010 0001 0101 Binary coded decimal
2 1 5

Q.37. Show the bit configuration of a 24-bit register when its content represent the decimal equivalent of 295 :

(*a*) in binary

(*b*) in BCD

(*c*) in ASCII using eight bits with even parity.

Solution: $(295)_{10} = 256 + 32 + 7 + (100100111)_2$

(*a*) 0000 0000 0000 0001 0010 0111

(*b*) 0000 0000 0000 0010 1001 0101

(*c*) 10110010 00111001 00110101

Q.38. Decode the following ASCII Code :

1001010 1001111 1001000 1001110 0100000 1000100 1001111 1000101

Solution: JOHN DOE

Q.39. Obtain the 9's complement of the following eight-digit decimal numbers :

12349876; 00980100; 90009951 and 00000000

Solution: 87650123; 99019899; 09990048; 99999999

Q.40. Obtain the 10's complement of the following six-digit decimal numbers :

123900, 090657; 100000; and 000000

Solution: 876100; 909343; 900000; 000000.

Q.41. Obtain the 1's and 2's complements of the following eight-digit binary numbers :

10101110, 10000001, 10000000, 00000001; and 00000000.

Solution: 01010001; 01111110; 01111111; 11111110; 11111111

01010010; 01111111; 10000000; 11111111; 00000000

Q.42. Perform the subtraction with the following unsigned decimal numbers by taking the 10's complement of the subtrahend.

(*a*) 5250 – 1321

(*b*) 1753 – 8640

(*c*) 20 – 100

(*d*) 1200 – 100

Solution:

(*a*)
```
   5250
 + 8629
 1) 3929
```

(*b*)
```
   1753
 + 1350
 0) 3113
 - 6887   10's complement
```

(*c*)
```
       020
     + 900
    0) 920
     - 080    10's complement
```
(*d*)
```
      1200
    + 9750
   1) 0950
```

Q.43. Perform the subtraction with the following unsigned binary numbers by taking the 2's complement of the subtrahend.

(*a*) 11010 – 10000

(*b*) 11010 – 1101

(*c*) 100 – 110000

(*d*) 1010100 – 1010100

Solution:

(*a*)
```
       11010
       10000
    1) 01010
   (26 – 16 = 10)
```
(*b*)
```
       11010
       10011
    0) 01101
   (26 – 13 = 13)
```
(*c*)
```
      000100
      010000
   1) 010100
    – 101100
   (4 – 48 = – 44)
```
(*d*)
```
      1010110
      0101100
   1) 0000000
   (84 – 84 = 0)
```

Q.44. Perform the arithmetic operations (+ 42) + (– 13) and (– 42) – (– 13) in binary using signed – 2's complement representation for negative numbers.

Solution:
```
     + 42 = 0101010          + 13 = 0001101
     – 42 = 1010110          – 13 = 1110011

   (+ 42)   0101010        (– 42)   1010110
   (–13)    1110011        (+13)    0001101
   (+ 29)   0011101        (– 29)   1100011
```

Q.45. Perform the arithmetic operations (+70) + (+80) and (–70) + (+80) with binary numbers in signed 2's complement representation. Use eight bits to accommodate each number together with its sign. Show that overflow occurs in both cases, that the last two carries are unequal and that there is a sign reversal.

Solution: 01 ← last two carries → 10

+ 70	01000110	– 70	10111010
+ 80	01010000	– 80	10110000
+ 150	10010110	– 150	01101010
greater ↑ than + 1271	↑ negative	less than ↑ – 128	↑ Positive

Q.46. Perform the following arithmetic operations with the decimal numbers using signed – 10's complement representation for negative numbers.

***(a)* (– 638) + (+ 758)**

***(b)* (– 638) – (+ 185)**

Solution:

(*a*) (– 638)	9362
(+ 785)	+ 0785
(+ 147)	0147
(*b*) (– 638)	9362
(+ 185)	+ 9815
(+ 823)	9177

Q.47. A 36-bit floating-point binary number has eight bits plus sign for the exponent and 26 bits plus sign for the mantissa. The mantissa is a normalized fraction. Numbers in the mantissa and exponent are in signed magnitude representation. What are the largest and smallest positive quantities that can be represented, excluding zero ?

Solution:

	Mantissa	Exponent	
	+ 26 bits	S 8 bits	36 bits
Largest :	+ 0.1111...1	+ 11111111	
	$1 - 2^{-26}$	+ 255	$(1 - 2^{-26}) \times 2^{+255}$
Smallest :	+ 0.1000...0	– 11111111	
(normalized)	2^{-1}	– 255	2^{-256}

Q.48. Represent the number (+ 46.5) 10 as a floating-point number with 24 bits.

Solution: The normalized tractions mantissa has 16 bits and the exponents has 8 bits.

$$46.5 = 32 + 8 + 4 + 2 + 0.5 = (101110.12)_2$$

Sign		
0	101110100000000	00000110
		(+ 6)
	24 – bit mantissa	8 – bit exponent

Q.49. The Gray code is sometimes called a reflected code because the bit values are elected on both sides of any 2^n value. For example, as shown in Table given below the values of the three low-order bits are reflected over a line drawn between 7 and 8. Using this property of the Gray code, obtain :

***(a)* The Gray code numbers for 16 through 31 as a continuation of Table given below.**

***(b)* The excess-3 Gray code for decimals 10 to 19.**

Table. 3.1 Bit Gray Code

Binary code	Decimal equivalent	Binary Code	Decimal equivalent
0000	0	1100	8
0001	1	1101	9
0011	2	1111	10
0010	3	1110	11
0110	4	1010	12
0111	5	1011	13
0101	6	1001	14
0100	7	*1000*	*15*

(*a*) (*b*)

Solution:

Decimal	Gray code
16	11000
17	11001
18	11011
19	11010
20	11110
21	11111
22	11101
23	11100
24	10100
25	10101
26	10111
27	10110
28	10010
29	10011
30	10001
31	10000

Decimal	Excess-3	Gray
9	0010	1010
10	0110	1010
11	0110	1110
12	0110	1101
13	0110	1101
14	0110	1100
15	0110	0100
16	0110	0101
17	0110	0111
18	0110	0110
19	0110	0010
20	0111	0010

Q.50. Represent decimal number 8620 in

(*a*) **BC D;**
(*b*) **excess-3 code**
(*c*) **2421 code;**
(*d*) **as a binary number.**

Solution: Decimal number = 8620

(*a*) BCD 1000 0110 0010 0000
(*b*) xs-3 1011 1001 0101 0011
(*c*) 2421 1110 1100 0010 0000
(*d*) Binary 10000110101100 (8192 + 256 + 28 + 32 + 8 + 4)

Q.51. List the 10 BCD digits with an even parity in the leftmost position (total of five bits per digit). Repeat an odd-parity bit.

Solution:

Decimal	*BCD with even parity*	*BCD with odd parity*
0	0000	10000
1	10001	00001
2	10010	00011
3	00011	10100
4	10100	00101
5	00101	10110
6	00110	10110
7	10111	00111
8	11000	01000
9	0100	11001

Q.52. Represent decimal 3984 in the 2421 code. Complement all bits of the coded number and show that the result is the 9's complement of 3984 in the 2421 code.

Solution:

3984 = 0011 1111 1110 0100

1100 0000 0001 1011 = 6015

Q.53. Derive the circuits for a 3-bit parity generator and 4-bit parity checker using an even parity bit. (The circuits is given below, use odd parity).

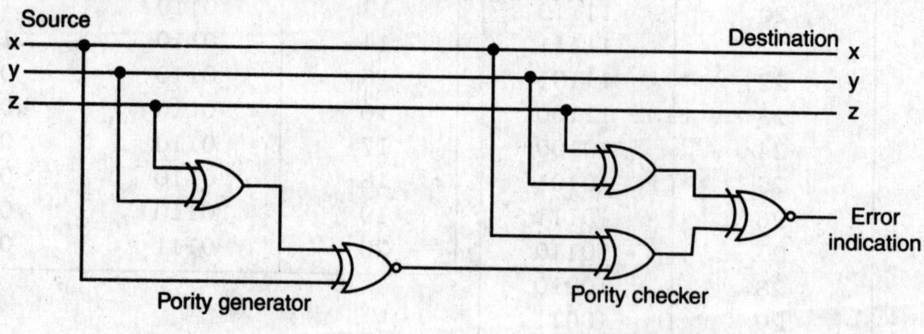

Fig. 3.1. Error detection with odd parity bit

Solution: Same as in figure given above without the complemented circles in the outputs of the gates.

$$P = x \oplus y \oplus z$$

$$\text{Error} = x \oplus y \oplus z \oplus P.$$

SUMMARY

1. Numbers are represented by a string of digit symbols
2. The binary number system uses the base 2, the octal 8 and hexadecimal 16.
3. Remainder Method is the most commonly used method for carrying out conversions from decimal system to other systems. It involves the following steps:
   (*a*) Divide the given number of decimal system by the radix R of the proposed system. From this, get a quotient Q_1 and a remainder R_1.
   (*b*) Dividing the quotient Q_1 by the radix *R* again to get quotient Q_2 and a remainder R_2.
   (*c*) Divide the quotient Q_2 by the radix R again to get quotient Q_3 and a remainder R_3.
4. American Standard Code for Information Interchange uses seven bits to code 128 characters

5. Floating-point representation has inherent demerits in the sense that the same number can be represented in many ways. For example, and all represents the same number. This necessitates a representation that is unique or normal for m and can be adopted by each system.
6. The mantissa is said to be normalized if the digits to the right of radix point is not zero, that is, no lending zero appear in the magnitude part of the number. Hence, for above number 0.3´1015 is the normal form for m or we can say that 0.3´1015 is a normalized floating point number.
7. Complements are the operations employed for the simplification of the subtraction and logical manipulation as well as for negative numbers representation.
8. In a base-r system, two types of complements are possible : (*i*) the r's complement and (*ii*) (*r* – 1)'s complement. For example: for binary numbers there are 2's and 1's complements and decimal numbers have 10's and 9's complements.
9. A set of elements that comprises of the 10 decimal digits, the 26 letters of alphabets, a number of special characters, such as $, @, and = etc. form a alphanumeric character set. The standard alphanumeric binary code is the ASCII (American Standard Code for Information Interchange), which uses seven bits to code 128 characters.
10. The decimal digits in ASCII can be converted to BCD (Binary Coded Decimal) by removing the three high-orders bits, 011. Since registers can only hold binary information, the code must be in binary form.

EXERCISE

(I) Fill in the Blanks (Solved)

1. $(11011.1001)_2$ = (..)$_8$
2. ASCII stands for ....

(II) True or False (Solved)

1. Memory read micro-operation takes 10 clock cycles.
2. A microprogram is a set of micro-operations.

(III) Fill in the Blanks (Unsolved)

1. The channel through which CPU and memory accomplish using ....
2. Mirco-operation for adding content of memory location 5 to the content of register R1 is ....
3. The 10's complement of (5432)10 is ....
4. $(9876)_{10}$ – $(5432)_{10}$ using 9's complement is ....
5. Acronym IEEE stands for ....

(IV) True or False (Unsolved)

1. Accumulator is a decoder.
2. Control unit is a part of ALU.

(V) Detailed Questions

1. Find the decimal equivalent of the following binary numbers :
   (*i*) 101011 (*ii*) 111101101 (*iii*) 100001 (*iv*) 111111
2. Find the decimal equivalent of the following binary numbers :
   (*i*) 0.01110 (*ii*) 111.00110 (*iii*) 1101.0011 (*iv*) 1111.0001
3. What is the advantage of using hexadecimal numbers?
4. Convert the following binary numbers into decimal and hexadecimal equivalents :
   (*i*) 45.525 (*ii*) 87.1250

5. In a number system, the symbols used are 0, 1, 2, 3, 4. What is the base of the system?
6. Find the 1's and 2's complement of the following binary numbers :
   (*i*) 1010110 (*ii*) 0110110 (*iii*) 1101.0011 (*iv*) 1101.1100
7. Find the 10's complement of (9456)10.
8. Perform the subtraction of following decimal numbers using (a) 10's complement (b) 9's complement. Check the answer by direct subtraction.
   (*i*) 5432-1234 (*ii*) 876.45-283.4 (*iii*) 5438-9432 (*iv*) 124.43-876.85
9. Write first 20 decimal numbers in base 5.
10. Perform the arithmetic operations with binary numbers given below and with negatives numbers unsigned 2's complement representation. Use seven bits to accommodate each number together with its sign
    (*a*) (+ 15) + (+ 14), (*b*) (– 15) + (– 14), (*c*) (– 15) + (+ 14)
11. Describe the hardware algorithm for addition with an example
12. Multiply the following numbers using Booth multiplication algorithm
    (*a*) + 14 * – 13, (*b*) – 14 * + 3; use 5 bit register
13. Design 2–bit multiplier by 3 bit array multiplier
14. Divide the number 0110000010 by 10001
15. Describe the flow chart of division algorithm
16. Why floating point number operations are move complicated than fixed point numbers operations.
17. Describe addition algorithm for addition of floating point numbers with example.
18. Perform operations ash r A and ash l A where the content of register A is 0111100001100000
19. Show the Block diagram of BCD adder. Hence show the block diagram of adding decimal numbers using BCD address.
20. Describe strobe control in Asynchronous Data transfer.
21. What is hand shaking ? What advantage it has over strobe control ?
22. What are different modes of data transfer to and from peripherals ?
23. What is need of input/output processor ? Give block diagram of CPU with input/output processor.

Answers

(I) Fill in the Blanks (Solved)

1. 33.41
2. American Standard Code for Information Interchange.

(II) True or False (Solved)

1. False
2. True

Chapter

4

Instruction Set

Chapter Outline

4.0 INTRODUCTION

Computers provide an extensive set of instructions to give the user the flexibility to carry out various computational tasks. The instruction set of different computers differ from each other mostly in the operands are determined from the address and mode field. The basic set of operations available in a typical computer is covered in this and in the next section.

Most of computer instructions can be classified into three categories:

1. Data transfer instructions.
2. Data manipulation instructions.
3. Program control instructions.

In the next three sections we will briefly describe those types and list some sample instructions based on the instruction format of the ATMEL Microcontroller instruction set.

4.1 INSTRUCTION SET

An instruction on our simple computer consists of one 12-bit word. The leading four bits form the operation code (opcode) which specifies the action to be taken, and the remaining 8 bits, when used, indicate the memory address of one of the instruction's operands. For those instructions that have two operands, the other operand is always contained within the accumulator.

Fig. 4.1. ATMEL AVR Microcontroller

4.1.1 Data transfer instructions

Data transfer instructions move data in a computer from one place to another without changing the data content. The most common transfers are between memory and processor registers, between processor registers and input or output, and between the processor registers themselves. The following is a list of some data transfer instructions used in many computers.

Table 4.1. Typical Data Transfer instructions

Name	Mnemonic
Load	LD
Store	ST
Move	MOV
Exchange	XCH
Input	IN
OutputOUT	
Push	PUSH
Pop	POP

Selected ATMEL Microcontroller Data Transfer instructions are listed below:

- **LD** – Load Indirect from Data Space to Register

  Example:

  ```
  clr r 27        ; Clear X high byte
  ldi r 26,$60    ; Set X low byte to $60
  ```

- **SPM** – Store Program Memory
- **ST (STD)** – Store Indirect From Register to Data Space

  Example:

  ```
  clr r 27        ; Clear X high byte
  ldi r 26,$60    ; Set X low byte to $60
  st X+,r0        ; Store r 0 in data ;space loc. ;$60(X post ;inc)
  st X,r1         ; Store r1 in data ;space loc. $61
  ldi r 26,$63    ; Set X low byte to $63
  ```

```
st X,r 2      ; Store r 2 in data ;space loc. $63
st -X,r 3     ; Store r 3 in data ;space loc.
;$62(X pre dec)
```

- **MOV** – Copy Register
- **MOVW**– Copy RegisterWord
- **IN** - Load an I/O Location to Register
- **OUT** – Store Register to I/O Location
- **PUSH** – Push Register on Stack
- **POP** – Pop Register from Stack

4.1.2 Data Manipulation Instructions

Data manipulation instructions perform operations on data and provide the computational capabilities for the computer. They are usually divided into three basic types:

1. Arithmetic instructions.
2. Logical and bit manipulation instructions.
3. Shift instructions.

4.1.3 Arithmetic instructions

The four basic arithmetic operations are addition, subtraction, multiplication and division. A list of typical arithmetic instructions is given in Table 4.2. The increment instruction adds one to the value stored in a register or memory word. One common characteristic of the increment operations is that a binary number of all 1's when incremented produces a result of all 0's. The decrement instruction subtracts 1 from a value stored in a register or memory word. A number of all 0's when decremented produces a result of all ones.

Table 4.2. Typical Arithmetic instructions

Name	Mnemonic
Increment	INC
DecrementDEC	
Add	ADD
Subtract	SUB
Multiply	MUL
Divide	DIV
Add with carry	ADDC
Subtract with borrow	SUBB
Negate (2's Complement)	NEG

Selected ATMEL Microcontroller Arithmetic instructions are listed below:

- **INC** – Increment

  Example:

  ```
  clr r 22 ; clear r 22
  inc r 22 ; increment r 22
  ```

- **DEC** – Decrement

Example:

```
ldi r17,$10      ; Load constant in r17
add r1,r 2       ; Add r 2 to r1
dec r17          ; Decrement r17
```

- **ADC** – Add with Carry
- **ADD** – Add without Carry

Example:

```
add r1,r 2       ; Add r 2 to r1 (r1=r1+r 2)
add r 28,r 28 ; Add r 28 to itself (r 28=r 28+r 28)
```

- **ADIW** – Add Immediate to Word
- **SUB** – Subtract without Carry

Example:

```
sub r13,r12    ; Subtract r12 from r13
```

- **SUBI** – Subtract Immediate
- **MUL** – Multiply Unsigned

Example:

```
mul r 5,r 4    ; Multiply unsigned r 5 and r 4
movw r 4,r 0 ; Copy result back in r 5:r 4
```

- **MULS** – Multiply Signed
- **NEG** – Two's Complement

Logical and bit manipulation instructions

Logical instructions perform binary operations on strings of bits stored in registers. Some typical logical and bit manipulation instructions are listed in Table 4.3.

Table 4.3. Typical Logical and bit manipulation instructions

Name	Mnemonic
Clear	CLR
Complement	COM
AND	AND
OR	OR
Exclusive-OR	XOR
Clear CarryCLRC	
Set Carry	SETC
Complement Carry	COMC
Enable interrupt	EI
Disable interrupt	DI

Selected ATMEL Microcontroller Logical and bit manipulation instructions are listed below:

- **CBI** – Clear Bit in I/O Register

Example:

```
cbi $12,7       ; Clear bit 7 in Port D
```

- **CBR** – Clear Bits in Register
- **COM** – One's Complement

Example:

```
com r4          ; Take one's complement of r4
```

- **AND** – Logical AND

Example:

```
and r 2,r 3     ; Bitwise and r2 and r3, result in ;r2
```

- **ANDI** – Logical AND with Immediate
- **OR** – Logical OR

Example:

```
or r15,r16      ; Do bitwise or between registers
```

- **ORI** – Logical OR with Immediate
- **EOR** – Exclusive OR
- **CLC** – Clear Carry Flag

Example:

```
add r 0,r 0     ; Add r 0 to itself
clc             ; Clear Carry Flag
```

- **SEC** – Set Carry Flag
- **SEI** – Set Global Interrupt Flag
- **CLI** – Clear Global Interrupt Flag

4.1.4 Shift Instructions

Shifts are operations in which the bits of a word are moved to the left or right. Shift instructions may specify either logical shifts, arithmetic shifts, or rotate-type operations. Table 4.4 lists four types of shift instructions and four types of rotate instructions.

Table 4.4. Typical Shift and Rotate instructions

Name	Mnemonic
Logical shift right	SHR
Logical shift left	SHL
Arithmetic shift right	SHRA
Arithmetic shift left	SHLA
Rotate right	ROR
Rotate left	ROL
Rotate right through carry	RORC
Rotate left through carry	ROLC

Selected ATMEL Microcontroller Shift and Rotate instructions are listed below:

- **LSL** – Logical Shift Left

Example:

```
add r 0,r 4     ; Add r4 to r 0
lsl r 0                 ; Multiply r 0 by 2
```

- **LSR** – Logical Shift Right
- **ASR** – Arithmetic Shift Right

Example:

```
ldi r16,$10     ; Load decimal 16 into r16
asr r16         ; r16=r16 / 2
```

```
ldi r17,$FC     ; Load -4 in r17
asr r17         ; r17=r17/2
```

- **ROL –** Rotate Left trough Carry
- **ROR –** Rotate Right through Carry

Example:

```
lsr r19         ; Divide r19:r18 by two
ror r18         ; r19:r18 is an unsigned two-byte; integer
```

4.1.5 Program Control Instructions

Program flow can be altered by instructions that modify the value of the program counter: important feature of a digital computer provides a control over the program flow and capability for branching to different program segments. Typical program control instructions are shown in Table 4.5.

Table 4.5. Typical program control instructions branch and jump instructions may be conditional or unconditional.

Name	Mnemonic
Branch	BR
Jump	JMP
Skip	SKP
Call	CALL
Return	RET
Compare (by subtraction)	CMP
Test (by ANDing)	

An unconditional branch instruction causes a branch to the specific address without any conditions, *e.g.*:

```
jmp    DisplayGreeting
```

The conditional branch specifies a condition, e.g. branch if zero: only when the condition is met, the program counter is loaded with the branch address, *e.g.*:

Compare and test instructions can be used in setting conditions for subsequent conditional branch instructions.

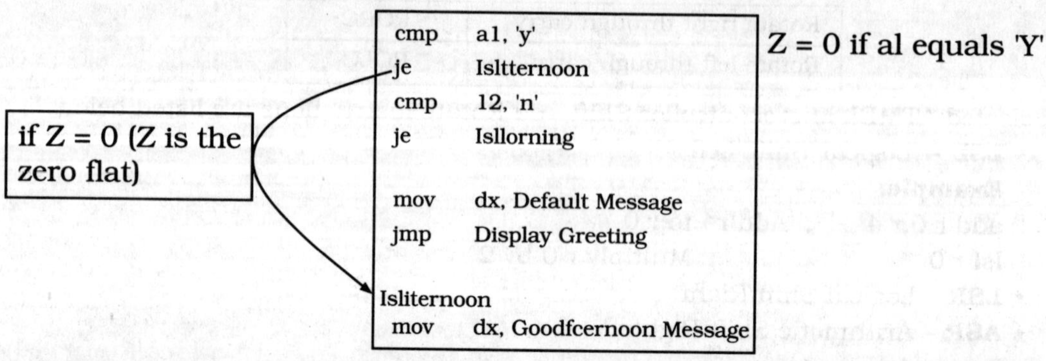

Compare performs an arithmetic subtraction: result is not saved – only status bit conditions are set as a result of operation. Similarly, test performs logical AND of two operands and updates certain status bits.

Selected ATMEL Microcontroller Shift and Rotate instructions are listed below:

- **BRCC** – Branch if Carry Cleared

  Example:

  ```
  add r 22,r 23 ; Add r 23 to r 22
  brcc nocarry  ; Branch if carry cleared
  ...
  nocarry: nop ; Branch destination (do nothing)
  ```

- **BRCS** – Branch if Carry Set
- **BRID** – Branch if Global Interrupt is Disabled
- **BRIE** – Branch if Global Interrupt is Enabled
- **BRVC** – Branch if Overflow Cleared

  Example:

  ```
  add r 3,r 4    ; Add r 4 to r 3
  brvc noover    ; Branch if no overflow
  ...
  noover: nop    ; Branch destination (do nothing)
  ```

- **BRVS** – Branch if Overflow Set
- **JMP** – Jump

  Example:

  ```
  mov r1,r 0     ; Copy  0 to r1
  jmp farplc     ; Unconditional jump
  ...
  farplc:  nop   ; Jump destination (do nothing)
  ```

- **SBI** – Set Bit in I/O Register
- **SBIC** – Skip if Bit in I/O Register is Cleared
- **CALL** – Long Call to a Subroutine
- **RCALL** – Relative Call to Subroutine
- **RET** – Return from Subroutine
- **CP** – Compare

  Example:

  ```
  cp r4,r19      ; Compare r4 with r19
  brne noteq     ; Branch if r4 <> r19
  ...
  noteq:              nop
  ```

- **Test**– Test fpor Zero or inus;

4.1.6 Status Bit Conditions

The Status register stores the values of the status bit (status register is composed of the status). Bit of the status register are modified as a result of an operation perfored in the ALU. Fig. 4.2 shows the block diagram of an 8-bit ALU with a 4-stauts register.

Status bit can be checked after ALU operation to determine certain relationships that exist between the valuves of A and B. V indicates overflow *i.e.,* for 8-bit ALU the result is greate than 127 or less than –127. If Z is set, the result is zero: we can use, *e.g.* XOR operation to compare to number (the result is zero if A = B) and Z indicates the result of comparison. A single bit in A can be checked with a mask that contains 1 in that particular bit position (others being 0's) and by using AND operations.

C (carry) is set to 1 if carry C8 is 1.
C is cleared to D if C8 is 0.

S (sign) is set to 1 the highest-order bit F7 is 1. S is set to 0 if F7 is 0.

Z (zero) is set to 1 if the output of ALU is 0 (all 0's). Z is set to 0 otherwise.

V (overflow) is set to 1 is XOR of the last two carries is equal to 1 and cleared to 0 otherwise.

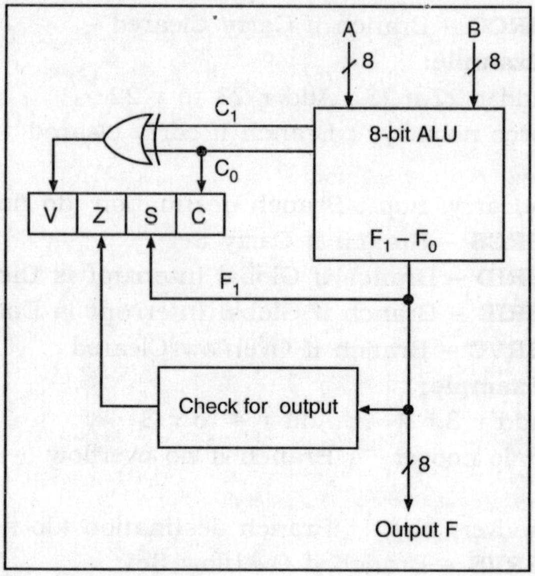

Fig. 4.2. Status register bits

4.1.7 Conditional Branch Instruction

Conditional branch instructions use the status bits for checking conditions for branching. Table 4.6. gives a list of the most common branch instruction.

Table 4.6. Conditional branch instructions some conditional branch instructions of ATMEL AVR have been listed in section 1.6.3

Mnemoice	Branch condition	Tested condition
BZ	Branch if zero	Z = 1
BNZ	Branch if not zero	Z = 0
BC	Branch if carry	C = 1
BNC	Branch if no carry	C = 0
BP	Branch if plus	S = 0
BM	Branch if minus	S = 1
BV	Branch if overlflow	V = 1
BNV	Branch if no overflow	V = 0
Unsigned compare conditions (A – B)		
BHI	Branch if higher	A > B
BHE	Branch if higher or equal	A ≥ B
BLO	Branch if lower	A < B
BLOE	Branch if lower to equal	A ≤ B
BE	Branch if equal	A = B
BNE	Branch if not equal	A ≠ B
Signed compare conditions (A – B)		
BGT	Branch if greater than	A > B
BGE	Branch if greater or equal	A ≥ B
BLT	Branch if less than	A < B
BLE	Branch if less or equal	A ≤ B
BE	Branch if equal	A = B
BNE	Branch if not equal	A ≠ B

80×80 Z = 0 if al equals Y

```
cmp   al, 'y'
je    Isktner noon
cmp   al, 'x'
je    al, 'y'
mov   dx, Default Message
jmp
```

Numerical Example

Let A = 11110000 and B = 00010100. Perform A – B, update the status bits, and list the conditional branch instructions.

1. The ALU takes the 2's complement of B and adds it to A.

   A : 11110000

   B + 1 : 11101100

2. The status bits are:

   C = 1, S = 1, V = 0, and Z = 0

3. For unsigned numbers:

   A = 240, and B = 20

   Then we have that A>B and A [* No filter found for the requested operation. | In-line.šTMTM *]B, *i.e.* the instructions that will cause a branch are BHI, BHE, and BNE.

4. For signed numbers:

   A = –16, and B = 20

Then we have that A<B and A [* No filter found for the requested operation. | In-line.šTMTM *]B, *i.e.* the instructions that will cause a branch are BLT, BLE, and BNE.

4.1.8 Subroutine Call and Return

For subroutine calls, different computers can use a different temporary location for storing the return address, some computers use the first memory location of the subroutine (like the "Basic Computer"), some store the return address in a fixed memory location, and some computers use a processor register. Stack memory is yet another possibility (the most efficient way): when a succession of subroutines is called (nested calls), the sequential return addresses can be pushed into the stack. The 'return from subroutine' instruction pops the return address (and assigns to program counter) from the top of the stack: we always have the return address for the last called subroutine.

Subroutine call (Stack based) micro-operations;

$SP \leftarrow SP - 1$	Decrement stack pointer
$M[SP] \leftarrow PC$	Push content of PC onto the stack
$PC \leftarrow$ effective address	Transfer control to the subroutine

Subroutine return microoperations:

$PC \leftarrow M[SP]$	Pop stack and transfer to PC
$SP \leftarrow SP + 1$	Increment stack pointer

4.2 INSTRUCTION FORMAT & TYPES

The format of an instruction is usually depicted in a rectangular box symbolizing the bits of the instruction as they appear in memory words or in a control register. The bits of the instruction are divided into groups called fields. The most common fields found in instruction formats are:

1. Operation code specifying the operation: add, subtract, complement, etc.
2. Address field designating a memory address or a register
3. Mode field for specifying the way for determining the effective address of an operand.

The number of address fields in the instruction format of a computer depends on the internal organization of its registers. E.g.: MIPS (a RISC microprocessor architecture developed by MIPS Computer Systems Inc.).

Computers may have instructions of several different lengths containing varying number of addresses. The number of address fields in the instruction format of a computer depends on the internal organization of its register. Most computers fall into one of three types of CPU organizations:

1. Single accumulator organization (uses **one** address field).
2. General register organization(uses **two or three** address fields).
3. Stack organization (uses **zero** address field).

Example:

To illustrate the influence of the number of addresses on computer programs, we will evaluate the arithmetic statement

X = (A + B)*(C + D)

using zero, one, two, or three address instructions.

1. Using three address instructions

```
ADD   R1, A, B    R1 ←M[A] + M[B]
ADD   R2, C, D    R2 ←M[C] + M[D]
MUL   X, R1, R2   X ←R1*R
```

2. Using two address instructions

```
MOV ←R1, A    R1←M[A]
ADD ←R1, B    R1←R1 + M[B]
MOV ←R2, C    R2←M[C]
ADD ←R2, D    R2←R2 + M[D]
MUL ←R1, R2   R1←R1* R2
MOV←X, R1     X ←R1
```

3. Using one address instructions

```
LOAD    A    AC←M[A]
ADD     B    AC←AC + M[B]
STORE   T    M[T]←AC
LOAD    C    AC← M[C]
ADD     D    AC←AC + M[D]
MUL     T    AC←AC*M[T]
STORE   X    M[X]←AC
```

4. Using zero address instructions (RPN: AB+CD+*)

```
PUSH   A   TOS ← A
PUSH   B   TOS ← B
ADD        TOS ← (A + B)
PUSH   C   TOS ← C
PUSH   D   TOS ← D
ADD        TOS ← (C + D)
MUL        TOS ← (C + D) * (A + B
POP    X   M[X] ← TOS
```

Classifying Instructions According to Number of Operands

Possible to have anywhere from 0 to 4 or more operands. Fewer operands make for shorter but less flexible instructions since some of the operands are implicit.

- Zero Address - any stack architecture - *e.g.* IJVM Mic -1
- One Address - accumulator is implicit operand - *e.g.* PDP-8, IAS
- Two Address - one operand is both source and destination - *e.g.* Intel 80x86
- Three Address - two source operands and one destination operand

- VAX has 3 address instructions (a good example of a CISC architecture)
- ARC has 3 register instructions (a good example of RISC architecture)

Four Address - 4th operand is either address of next instruction (obsolete technique) or for some reason the instruction requires more than three operands (*e.g.* integer division produces quotient and remainder)

4.3 PROGRAMMING CONSIDERATIONS

4.3.1 Machine Language

We write the computer program in a high level language. But it is not possible for the system to understand it as the internal hardware of the computer can understand only the binary language.

- To write a program for a computer we need to specify, directly or indirectly, a sequence of machine instructions. Machine instructions inside the computer form a binary pattern, which is difficult to understand and manipulate.
- The instruction set of the basic computer, whose hardware organization was explored earlier is used to program a computer. The 25 instructions of the basic computer are given in Table 4.7 to provide an easy reference for the programming examples that follow.

Table 4.7. Computer Instructions

Symbol	Hexadecimal code	Description
AND	0 or 8	AND M to AC
ADD	1 or 9	Add M to AC, carry to E
LDA	2 or A	Load AC from M
STA	3 or B	Store AC in M
BUN	4 or C	Branch unconditionally to m
BSA	5 or D	Save return address in m and branch to m+1
ISZ	6 or E	Increment M and skip if zero
CLA	7800	Clear AC
CLE	7400	Clear E
CMA	7200	Complement AC
CME	7100	Complement E
CIR	7080	Circulate right E and AC
CIL	7040	Circulate left E and AC
INC	7020	Increment AC
SPA	7010	Skip if AC is positive
SNA	7008	Skip if AC is negative
SZA	7004	Skip if AC is zero
SZE	7002	Skip if E is zero
HLT	7001	Halt computer
INP	F800	Input information and clear flag
OUT	F400	Output information and clear flag
SKI	F200	Skip if input flag is on
SKO	F100	Skip if output flag is on
ION	F080	Turn interrupt on
IOF	F040	Turn interrupt off

- A program is a list of instructions to tell the computer to do need processing on the data. We can write a program for a computer in various types of programming languages, but this program can be executed by the computer only when it is in binary form. If it is written in some other language it has to be translated to the binary form so that it can be executed by the computer.
- A symbolic instruction and its binary equivalent has a one-to-one relationship between them. An assembly language can be thought of a machine level language writing 16 bits for each instruction. Because there are several digits, it becomes tedious. By writing the equivalent hexadecimal code, we can reduce to four digits. We can go one step further and replace each hexadecimal address by a symbolic address and each hexadecimal operand by a decimal operand. We find it convenient because, we generally do not know exactly the numeric memory location of operands at the time of writing a program. As we know that there is a set of rules for a programming language, we must confirm with all format and rules of the language if we want our programs to be translated correctly. A line of code is the fundamental unit of an assembly language program.

Table 4.8. Binary Program to Add Two Numbers

Location	Instruction Code			
0	0010	0000	0000	0100
1	0001	0000	0000	0101
10	0011	0000	0000	0110
11	0111	0000	0000	0001
100	0000	0000	0101	0011
101	1111	1111	1110	1001
110	0000	0000	0000	0000

4.3.2 Assembly Language

We all know that a program written to instruct a computer is not understood by the system directly. But it has first to be converted into equivalent machine language. That's why a programming language is defined by a set of rules. If users want their programs to be translated correctly they must conform to all formats and rules of the language. Every digital computer has its own separate assembly language. The rules for writing assembly language programs are available from the computer manufacturer in the form of documents and manuals.

A line of code is the fundamental unit of an assembly language program. The specific language is defined by a group of rules. This group describes the symbols that can be used. It also tells how a line of code can be made from them. We will now give the rules to write assembly language programs for the basic computer.

4.3.2.1 Rules for the Assembly Language

- A line of code of an assembly language program is divided in three columns called fields. The fields describe the following information.
  1. The label: may be empty or it may specify a symbolic address.
  2. The instruction: specifies a machine instruction or a pseudo-instruction.
  3. The comment: may be empty or it may include a comment.

- A symbolic address has one, two, or three alphanumeric characters. But it cannot have more than three alphanumeric characters. The first character is an alphabet; the next two may be alphabets or numeric digits. We can choose the symbols arbitrarily. A symbolic address in the label field is terminated by a comma to make it a label. The instruction field specifies one of the following items:
    1. A memory-reference instruction (MRI)
    2. A register-reference (*i.e.* input-output instruction) (non-MRI)
    3. A pseudo-instruction with or without an operand
- A memory-reference instruction occupies two or three symbols. These symbols are separated by spaces. The first must be a three-letter symbol defining an MRI operation code. The second one is a symbolic address. The third symbol, which is optional, is the letter I. It is a direct address instruction, if I is missing otherwise it is an indirect address instruction.
- A non-MRI is an instruction that does not have an address part. A non-MRI is found in the instruction field of a program by any one of the three-letter symbols for the register-reference and input-output instructions.
- The following is an illustration of the symbols that may be placed in the instruction field of a program.

    CLA non-MRI
    ADD OPR direct address MRI
    ADD PTR I indirect address MRI
- The first three-letter symbol in each line must be one of the instruction symbols of the computer. A memory-reference instruction, such as MUL, must be followed by a symbolic address. The letter I may or may not be present.
- The memory location of an operand is determined by a symbolic address in the instruction field. This location is mentioned somewhere in the program by appearing again as a label in the first column. If we want to translate program from assembly language to a machine language, each symbolic address that is mentioned in the instruction field must occur again in the label field.
- A pseudo-instruction is an instruction to the assembler giving information about some phase of the translation (it is not a machine instruction). Four pseudo-instructions that are recognized by the assembler are listed in Table 4.9. The assembler is informed by the origin (ORG) pseudo-instruction that the instruction or operand in the following line is to be placed in a memory location specified by the number next to ORG.

Table 4.7. Definition of Pseudo-instructions

Symbol	Information for the Assembler
ORG N	Hexadecimal number N is the memory location for the instruction or operand listed in the following line
END	Denotes the end of symbolic program
DEC N	Signed decimal number N to be converted to binary
HEX N	Hexadecimal number N to be converted to binary

- To inform the assembler that the program is terminated the END symbol is placed at the end of the program. The radix is given by the other two pseudo-instructions. They also describe the operand and tell the assembler how to convert the listed number to a binary one.

We reserve the third field in a program for comments. A line of code may or may not have a comment. But if there is a comment, it must be preceded by a slash for the assembler to recognize the beginning of a comment field. Comments are useful for explaining the program and are helpful in understanding the step-by-step procedure taken by the program. Comments are used for explanation and are not neglected during the binary translation process.

An Example

The program of Table 4.10 is an example of an assembly language program. The first line has the pseudo instruction ORG to define the origin of the program at memory location (100)16.

The next six lines define machine instructions, and the last four have pseudo-instructions. Three symbolic addresses have been used and each is listed in column 1 as a label and in column 2 as an address of a memory-reference instruction. Three of the pseudo-instructions specify operands, and the last one signifies the END of the program.

When the program is converted into the binary code and executed by the computer it performs a subtraction between two numbers. We can perform subtraction operation by adding the minuend to the 2's complement of the subtrahend. We know that subtrahend is a negative number, we convert it into a binary number as signed 2's complement representation because we dictate that all negative numbers be in their 2's complement form. Thus, – 23 converts to + 23 and the difference is 83 + (2's complement of -23) = 83 + 23=106.

Table 4.10. Language Program to Subtract Two Numbers

	ORG 100	/Origin of program is location 100
	LDA SUB	/Load subtrahed to AC
	CMA	/Complement AC
	INC	/ Increment AC
	ADD MIN	/Add minuend to AC
	STA DIF	/ Store difference
	HILT	/ Halt computer
MIN,	DEC 83	/Minuend
SUB,	DEC–23	/Subtrahend
DIF,	HEX 0	/Difference stored here
	END	/End of symbolic program

4.3.3 Translation to Binary

The conversion of the symbolic program into binary is done by a program, which is called an assembler. The tasks performed by the assembler can be easily understood if we perform the translation on paper first.

- The translation of the symbolic program of Table 4.11 into an equivalent binary code may be done by scanning the program and replacing the symbols by their machine code binary equivalent.

- Starting from the first line, we encounter an ORG pseudo-instruction. This tells us to start the binary program from hexadecimal-location 100.
- The second line has two symbols. It must be a memory-reference instruction to be placed in location 100. Since the letter I is missing, the first bit of the instruction must be 0. The symbolic name of the operation is LDA. We find that the first hexadecimal digit of the instruction should be 2. The value of the address part is obtained from the address symbol SUB. We scan the label column and find this symbol in line 9. To determine its hexadecimal value we note that line 2 contains an instruction for location 100 and every other line specifies a machine instruction or are operand for sequential memory locations.
- Counting lines, we find that label SUB in line 9 corresponds to memory location 107. So the hexadecimal address of the instruction LDA must be 107. When the two parts of the instruction are assembled, we obtain the hexadecimal code 2107. The other lines representing machine instructions are translated in a similar fashion and their hexadecimal code is listed in Table 4.11.
- The pseudo-instruction DEC is specified by two lines in the symbolic program. A third line gives a zero with the help of a HEX pseudo-instruction. We can use DEC as well. Now we convert (83)10 to binary and place in location 106 in hexadecimal equivalent. Decimal -23 is a negative number and so convert it into binary in signed-2's complement form.

Table 4.11. Listing of Translated Program of Table 4.10

	ORG 100	/Origin of program is location 100
	LDA SUB	/Load subtrahed to AC
	CMA	/Complement AC
	INC	/ Increment AC
	ADD MIN	/Add minuend to AC
	STA DIF	/ Store difference
	HILT	/ Halt computer
MIN,	DEC 83	/Minuend
SUB,	DEC–23	/Subtrahend
DIF,	HEX 0	/Difference stored here
	END	/End of symbolic program

- The hexadecimal equivalent of the binary number is stored in location 107. The END symbol specifies the end of the symbolic program.

We can simplify the translation process if we scan the entire symbolic program two times. In the first scan no translation. We assign a memory location to each machine instruction and operand. The address value of labels is obtained by the location assignment. It also facilitates the translation process in the second scan. Hence in Table 4.12 assign location 100 to the first instruction after ORG. Now sequential locations are assigned for each line of code that has a machine instruction (or operand up to the end of the program). After the first scan, we attach with each label its location number and form a table that defines the hexadecimal value of each symbolic address. The address symbol table for the above program is:

Table 4.12. Listing of Translated Program of Table 4.10

Hexadecimal code Location	Content		Symbolic program
			ORG 100
100	2107		LDA SUB
101	7200		CMA
102	7020		INC
103	1106		ADD MIN
104	3108		STA DIF
105	7001		HLT
106	0053	MIN,	DEC 83
107	FFE9	SUB,	DEC – 23
108	0000	DIF,	HEX 0
	END		

The hexadecimal equivalent of the binary number is stored in location 107. The END symbol specifies the end of the symbolic program.

We can simplify the translation process if we scan the entire symbolic program two times. In the first scan no translation. We assign a memory location to each machine instruction and operand. The address value of labels is obtained by the location assignment. It also facilitates the translation process in the second scan. Hence in Table 4.12 assign location 100 to the first instruction after ORG. Now assign sequential locations are assigned for each line of code that has a machine instruction (or operand up to the end of the program). After the first scan, we attach with each label its location number and form a table that defines the hexadecimal value of each symbolic address. The address symbol table for the above program is:

Address symbol	Hexadecimal address
MIN	106
SUB	107
DIF	108

4.3.4 The Assembler

An assembler is nothing but a software program that takes input as the program in assembly language and its output is the binary equivalent of the program. The input is called the source program and the resulting binary program is called the object program. Basically an assembler works on the program by considering it as a character string and produces the binary equivalent.

4.3.4.1 How the program is stored in the memory ?

Before the work can be done on a program, it should be stored in a memory. We as user write the symbolic program on a computer. Then with the help of a program called loader, this symbolic program is taken into the memory. Since the program consists of symbols, its representation in memory must use an alphanumeric character code. Usually each character is represented by an 8-bit code, in the basic computer. The high-order bit is always 0 and the other seven bits are as described by ASCII code. Table 4.13 gives the hexadecimal equivalent of the character set. Each character is given two hexadecimal digits. So each character can be easily converted to their equivalent 8-bit code. The last entry in the table does not print a character, it looks after the physical movement of the cursor in the terminal. When the return key is depressed, the code for CR is produced. Therefore "carriage" goes to its initial position and we can start typing a new line.

We store a line of code in consecutive memory locations. Two characters in each location. Since a memory word has a capacity of 16-bits we can store two characters stored in each word. A comma delimits a label symbol. Now we see how the operation code and addresses are terminated. They are terminated with a space and the end of the line is recognized by the CR code. For example, the following line or code: PL3, LDA SUB I is stored in seven consecutive memory locations, as shown in Table 4.14. The label PL3 occupies two words and is terminated by the code for comma (2C). The instruction field in the line of code may have one or more symbols. Each symbol is terminated by the code for space (20) except for the last symbol, which is terminated by the code of carriage return (0D). If the line of code has a comment, the assembler recognizes it by the code for a slash (2F). The assembler neglects all characters in the comment field and keeps checking for a CR code. When this code is encountered, it replaces the space code after the last symbol in the line of code.

Table 4.13. Hexadecimal Character Code

Character	Code	Character	Code	Character	Code
A	41	Q	51	6	36
B	42	R	52	7	37
C	43	S	53	8	38
D	44	T	54	9	39
E	45	U	55	space	20
F	46	V	56	(	28
G	47	W	57	)	29
H	48	X	58	*	2A
I	49	Y	59	+	2B
J	4A	Z	5A	,	2C
K	4B	0	30	–	2D
L	4C	1	31	.	2E
M	4D	2	32	/	2F
N	4E	3	33	=	3D
O	4F	4	34	CR	0D (Carriage return)
P	50	5	35		

Table 4.14. Computer Representation of the line of code: PL3, LDA SUB I

Memory Word	Symbol	Hexadecimal code		Binary representaion			
1	P L	50	4C	0101	0000	0100	1100
2	3,	33	2C	0011	0011	0010	1100
3	L D	4C	44	0100	1100	0100	0100
4	A	41	20	0100	0001	0010	0000
5	S U	53	55	0101	0011	0101	0101
6	B	42	20	0100	0010	0010	0000
7	1 CR	49	0D	0100	1001	0000	1101

The user's symbolic language program in ASCII is input for the assembler program. The assembler scans this input twice to produce the equivalent binary program. The binary program constitutes the output generated by the assembler. We will now describe briefly the major tasks that must be performed by the assembler during the translation process.

4.3.4.2 First Pass

The functioning of the first pass of an assembler can be understood in the following steps:

- Entire symbolic program is scanned by a two-pass assembler twice. After the first pass, it generates a table that correlates all user-defined symbols with their equivalent value in binary. The binary translation is done during the second pass.
- To keep track of the location of instructions, the assembler uses a memory word called a location counter (abbreviated LC). The content of LC stores the value of the memory location assigned to the instruction or operand presently being processed.
- The ORG pseudo-instruction initializes the location counter to the value of the first location. Since instructions are stored in sequential locations, the content of LC is incremented by 1 after processing each line of code. To avoid ambiguity in case ORG is missing, the assembler sets the location counter to 0 initially.
- The flowchart of Fig. 4.3 describes the tasks performed by the assembler during the first pass. LC is initially set to 0. A line of symbolic code is analyzed to determine if it has a label (by the presence of a comma).
- If the line of code has no label, the assembler checks the symbol in the instruction field. If it contains an ORG pseudo-instruction, the assembler sets LC to the number that follows ORG and goes back to process the next line.
- If the line has an END pseudo-instruction, the assembler terminates the first pass and goes to the second pass. (Note that a line with ORG or END should not have a label.)

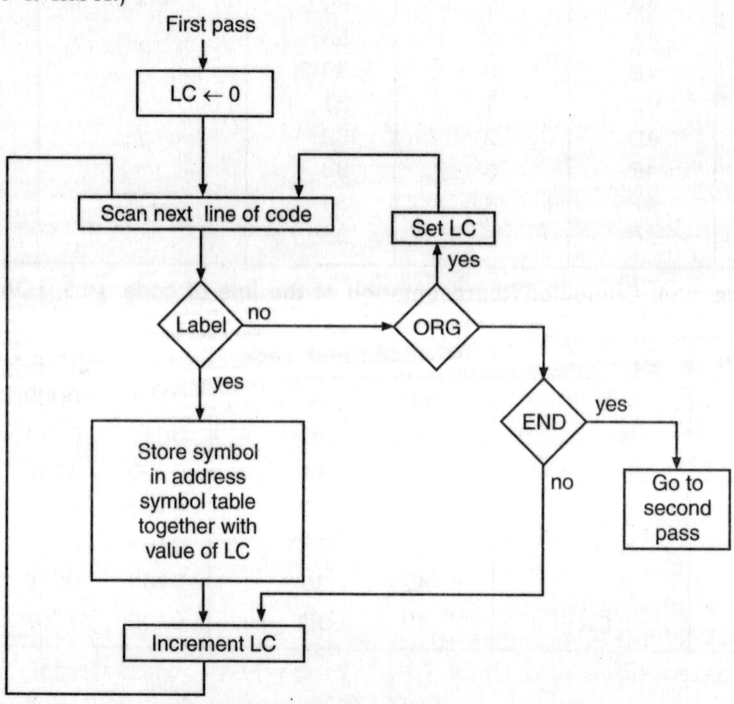

Fig. 4.3. Flowchart for first pass of assembler

- If the line of code contains a label, it is stored in the address symbol table together with its binary equivalent number specified by the content of LC. Nothing stored in the table if no label is encountered. LC is then incremented by 1 and a new line of code is processed.
- The assembler generates the address symbol table listed in Table 4.15, for the program to Table 4.10. We store each label symbol in two memory locations and terminate it by a comma. If the label contains less than three characters, the memory locations are filled with the code for space. The value found in LC while the line was processed is stored in the next sequential memory location.
- The program has three symbolic addresses: MIN, SUB, and DIF. These symbols represent 12-bit addresses equivalent to hexadecimal 106, 107 and 108, respectively. The address symbol table occupies three words for each label symbol encountered and constitutes the output data that the assembler generates during the first pass.

Table 4.15.

Memory Word	Symbol or (LC)*	Hexadecimal code		Binary representaion			
1	M I	4D	49	0100	1101	0100	1001
2	N,	4E	2C	0100	1110	0010	1100
3	(LC)	01	06	0000	0001	0000	0110
4	S U	53	55	0101	0011	0101	0101
5	B,	42	2C	0100	0010	0010	1100
6	(LC)	01	07	0000	0001	0000	0111
7	D I	44	49	0100	0100	0100	1001
8	F,	46	2C	0100	0110	0010	1100
9	(LC)	01	08	0000	0001	0000	1000

* (LC) designates content of location counter.

4.3.4.3 Second Pass

- With the help of table-lookup procedures, machine instructions are translated during the second pass. A take-lookup procedure is a search of table entries to find whether a specific item matches one of the items stored in the table.
- The assembler uses four tables. Any symbol that is encountered in the program must be available as an entry in one of these tables; otherwise, the symbol cannot be interpreted. We assign the following names to the four tables:
  1. Pseudo-instruction table.
  2. MRI table.
  3. Non-MRI table.
  4. Address symbol table.
- The pseudo-instruction table has the four symbols ORG, END, DEC, and HEX. Each symbol refers the assembler to a subroutine that processes the pseudo-instruction when encountered in the program.
- The MRI table has the seven symbols of the memory-reference instructions and their 3-bit operation code equivalents.
- The non-MRI table has the symbols for the 18 register-reference and input-output instructions and their 16-bit binary code equivalents.

- The address symbol table is created in the first pass. In order to determine its binary value, the assembler searches these tables to find the symbol that it is currently processing.
- The tasks performed by the assembler during the second pass are described in the flowchart of Fig. 4.4. LC is initially set to 0. Lines of code are then analyzed one at a time. Labels are neglected during the second pass, so the assembler goes immediately to the instruction field and proceeds to check the first symbol encountered. It first checks the pseudo-instruction table. A match with ORG sends the assembler to a subroutine that sets LC to an initial value. A match with END terminates the translation process. An operand is placed in the memory location specified by the content of LC. The location counter is then incremented by 1 and the assembler continues to analyze the next line of code.

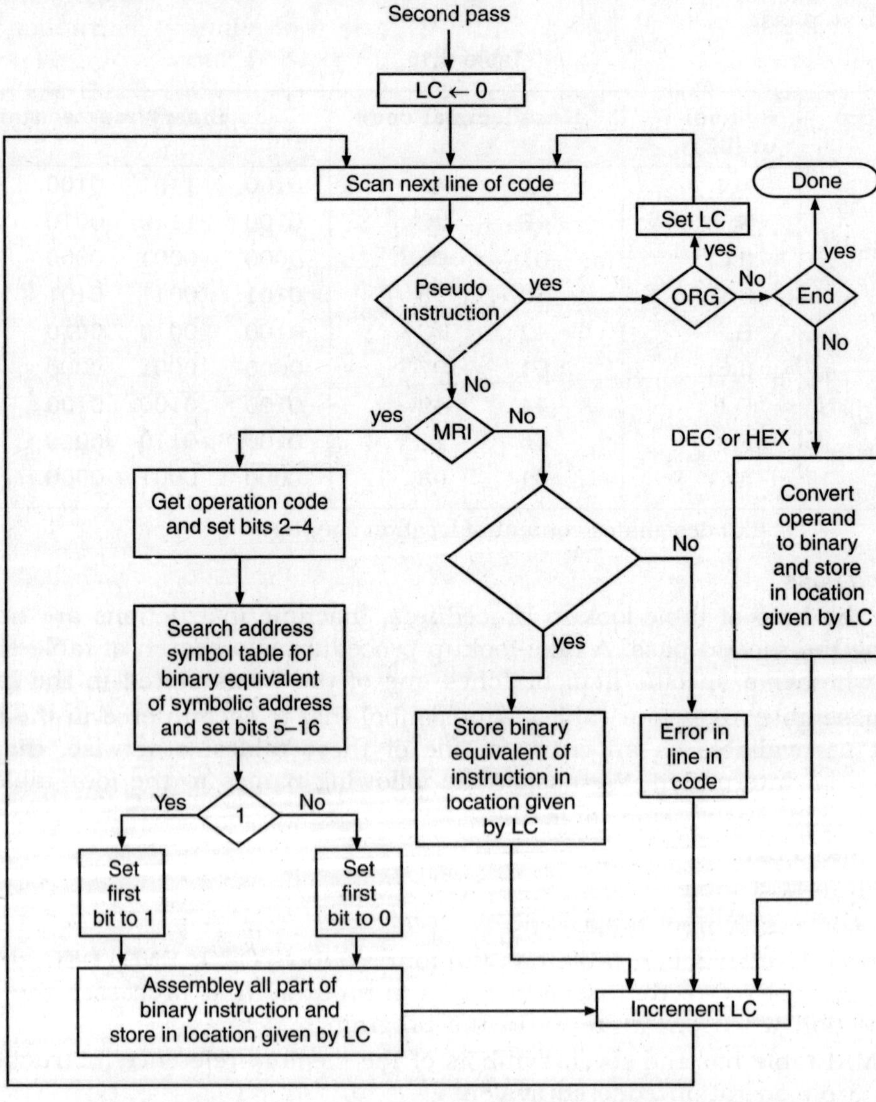

Fig. 4.4. Flowchart for second pass of assembler

- If the symbol obtained is not a pseudo-instruction, the assembler goes to the MRI table. The assembler refers to the non-MRI table if the symbol is not found in MRI table. A symbol found in the non-MRI table corresponds to a register reference or input-output instruction. The assembler stores the 16-bit instruction code into the memory word designated by LC. The location counter is incremented and a new line analyzed.
- When we get a symbol in the MRI table, the assembler extracts its equivalent 3-bit code and inserts it in bits-2 through 4 of a word.
- A memory reference instruction is designated by two or three symbols. The second symbol is a symbolic address and the third, which may or may not be present, is the letter I.
- By searching the address symbol table the symbolic address is converted to binary. The first bit of the instruction is set to 0 or 1, depending on whether the letter I is absent or present. The three parts of the binary instruction code are assembled and then stored in the memory location specified by the content of LC. The location counter is incremented and the assembler continues to process the next line.

An important job of the assembler is to check for possible errors. We can call it "error diagnostics". One example of such an error may be an invalid machine code symbol which is detected by its being absent in the MRI and non-MRI tables. The assembler is unable to translate such a symbol because it does not know its binary equivalent value. In such a case, the assembler prints an error message to inform the programmer that his symbolic program has an error at a specific line of code. Another possible error may occur if the program has a symbolic address that did not appear also as a label. The assembler cannot translate the line of code properly because the binary equivalent of the symbol will not be found in the address symbol table generated during the first pass. Other errors may occur and a practical assembler should detect all such errors and print an error message for each.

SOLVED PROBLEMS

Q.1. Define the following :

(a) Control memory

(b) Control word

(c) Sequencer

(d) Pipe line register.

Solution:

(a) Control Memory : Control units that use dynamic microprogramming employ a writable control memory. This type of memory can be used for writting, but is used mostly for reading. A memory that is part of a control unit is received to as control memory.

(b) Control word : The control variables at any given time can be represented by a string of 1's and 0's called *control word.*

(c) Sequencer : The next address generator is sometimes called a *microprogram sequencer,* as it determines the address sequence that is read from control memory. The address of the next microinstruction can be specified is several ways, depending on the sequencer inputs.

(*d*) **Pipe line register :** The control data register holds the present microinstruction while the next address is computed and read from memory. The data register is some times called *pipe line register.*

Q.2. How many types of control organizations are there ? Which one is most beneficial for computer system and why ?

Solution: There are two major types of control organization :

(*a*) Hardware control

(*b*) Microprogram control.

Hardware control is more advantage as compared to microprogram control.

In the hardware organization, the control logic is implemented with gates, Flip-flops, decoders and other digital circuits. It has the advantage that it can be optimized to produce a fast mode of operation.

Mostly computer based on the reduced instruction set computer (RISC) architecture concept use hardwired control rather than a control memory with a microprogram. A hardwired control, as the name implies, requires changes in the wiring among the various components, if the design has to modified or changed.

In the microprogrammed organization, the control information is stored in a control memory. The control memory is programmed to initiate the required sequence of micro-operation. The main advantage of microprogrammed control is the fact that once the hardwired configuration is established, there should be no need for further hardware or wiring changes.

Q.3. Define the computer instruction format. What is possible reference instructions, describe it ?

Solution: In computer instruction format, there are three fields.

15	14 11	0
I	Opcode	Address

It consists of 2 byte length [(0 – 15) bit].

(*i*) **Indirect addressing (I) :** Its length is 1 bit field,

(*ii*) **Opcode :** It is 4 bit operation code.

(*iii*) **Address :** It is 11 bit address field.

There are 16 possible memory-reference instructions we have describe four of them.

Symbol	***Opcode***	***Description***
ADD	0000	$AC \leftarrow AC + M[EA]$
BRANCH	0001	If $(AC < 0)$ then $(PC \leftarrow EA)$
STORE	0010	$M[EA] \leftarrow AC$
EXCHANGE	0011	$AC \leftarrow M[EA], M[EA] \leftarrow AC$

- The ADD instruction adds the content of the operand found in the effective address to the content OF AC.
- The **BRANCH** instruction causes a branch to the effective address if the operand in AC is negative. The program proceeds with the next consecutive instruction if AC is not negative. The AC is negative if its sign bit is a 1.
- **STORE** instruction transfers the content of AC into memory word specified by the effective address.

- **EXCHANGE.** The exchange instruction swaps the data between AC and the memory word specified by the effective address.

Each computer instruction must be microprogrammed.

Q.4. Describe the micro-instruction format for the control memory.

Solution: Micro-instruction format is 20 bits in length. It divided into four functional parts. The three fields F_1, F_2 and F_3 specify micro-operations for the computer. The CD field selects status bits condition :

- the BR field specifies the type of Branch to be used.
- the AD field contains a branch address. The address field is seven bits wide, since the control memory has 128 = 2^7 words.

3	3	3	2	2	7
F_1	F_2	F_3			AD

F_1, F_2, F_3 : Micro-operation fields
CD : Condition for branching
BR : Franchfield
AD : Address field

- Micro-operations are sub-divided into three fields of three bits each. These bits in each field are encoded to specify seven distinct micro operations. This gives 21 micro-operations. If fewer than three micro operation are used, one or more of the helds will use the binary code 000 for no operation.

Symbols and Binary Code for Micro Instruction Fields

F_1	Micro-operation	Symbol
000	None	NOP
001	AC ← AC + DR	ADD
010	AC ← 0	CLRAC
011	AC ← AC + 1	INCAC
100	AC ← DR	DRTAC
101	AR ← DR(0 – 10)	DRTAR
110	AR ← PC	DCTAR
111	M [AR] ← DR	WRITE

F_2	Micro-operation	Symbol
000	None	NOP
001	AC ← AC – DR	SUB
010	AC ← AC ∪ DR	OR
011	AC ← AC ∩ DR	AND
100	DR ← M[AR]	READ
101	DR ← AC	ACTDR
110	DR ← DR + 1	INCDR
111	DR (0 – 10) ← PC	PCTDR

F_3	Micro-operation	Symbol
000	None	NOP
001	AC ← AC ⊕ DR	XOR

010	AC ← $\overline{AC}$	COM
011	AC ← SNAC	SHL
100	AC ← SHRAC	SHR
101	PC ← PC + 1	INCPC
110	PC ← AR	ARTPC
111	Reserved	

CD	Conditions	Symbol	Comments
00	Always = 1	U	Unconditional branch
01	DR (15)	I	Indirect Address bit
10	AC (15)	S	Sign bit of AC
11	AC = 0	Z	Zero value in AC

CD	Conditions	Symbol Comments
00	JMP	CAR ← AD if condition = 1 CAR ← CAR + 1 if condition = 0
01	CALL	CAR ← AD SBR ← CAR + 1 if condition = 1 CAR ← CAR + 1 if condition = 0
10	RET	CAR ← SBR (Return from sub routine)
11	MAP	CAR (2–5) ← DR (11 – 14), CAR (0, 1, 6) ← 0

Q.5. A logic network has two data inputs *A* and *B*, and two control inputs C0 and C_1. It implements the function *F* according to the following Table :

C_1	C_2	F
0	**0**	$\overline{A + B}$
0	**1**	$A + B$
1	**0**	$A \oplus B$

Implement the circuit using one 4 to AB multiplexer, one 2 input exclusive OR gate, one 2 input AND gate, and 2 input OR gate and one inverter.

Solution:

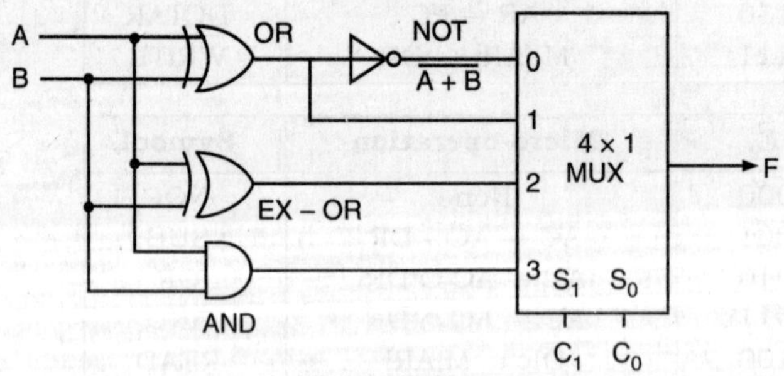

Fig. 4.5.

Q.6. Write the difference between a microprocessor and a microprogram? Is it possible to design a microprocessor without a microprogram? Are all microprogrammed computers also microprocessors?

Solution: A microprocessor is a small size CPU (computer on a chip). Microprogram is a program for a sequence of micro-operations. The control unit of a microprocessor can be hardwired or microprogrammed, depending on the specific design. A microprogrammed computer does not have to be a microprocessor.

Q.7. Explain the difference between hardwired control and microprogrammed control. Is it possible to have a hardwired control associated with a control memory?

Solution: Hardwired control, by definition, does not contain a control memory.

Q.8. Define the following:

(*a*) Micro-operation **(*b*) Microinstruction**

(c) Microprogram **(*d*) Microcode**

Solution: (*a*) Micro-operation: It is an elementary digital computer operation.

(*b*) Microinstruction: It is an instruction stored in control memory.

(c) Microprogram : It is a sequence of microinstruction.

(*d*) Microcode: It is same as microprogram.

Q.9. The microprogrammed control organization shown in figure given below has the following propagation delay times. 40ns to generate the next address, 10 ns to transfer the address into the control address register, 40 ns to across the control memory ROM, 10 ns to transfer the microinstruction into the control data register, and 40 ns to perform the required microoperations specified by the control word. What is the maximum clock frequency that the control can use? What would the clock frequency be if the control data register is not used?

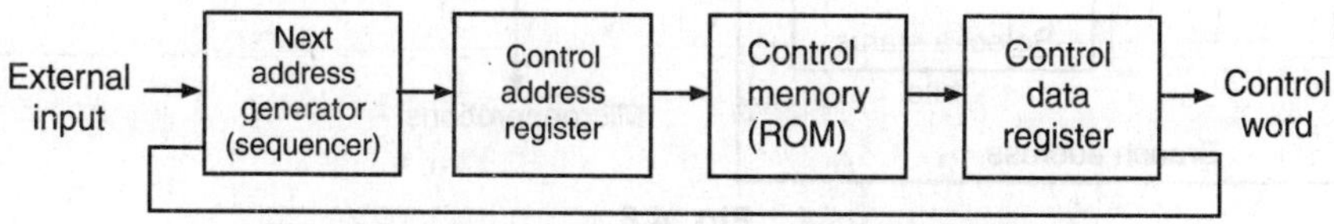

Fig. 4.6. Microprogrammed control organization

Solution:

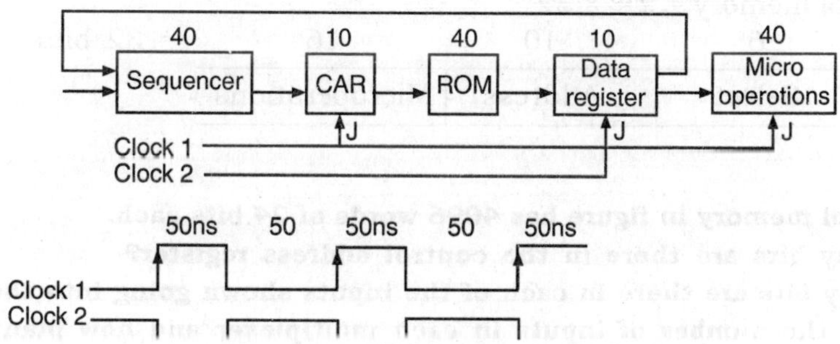

Fig. 4.7.

$$= \frac{1000}{100} \times 10^6 = 10 \text{ MHz}$$

If the data register is removed, we can use a single phase clock with a frequency

$$= \frac{1}{90 \times 10^{-9}} = 11.1 \text{ MHz}$$

Q.10. The system shown in figure given below uses a control memory of 1024 words of 32-bits each. The microinstruction has three fields as shown in the diagram. The microoperations field has 16-bits.

(*a*) How many bits are there in the branch address field and the select field?

(*b*) If there are 16 status bits in the system, how many bits of the branch logic are used to select a status bit?

(c) How may bits are left to select an input for the multiplexers?

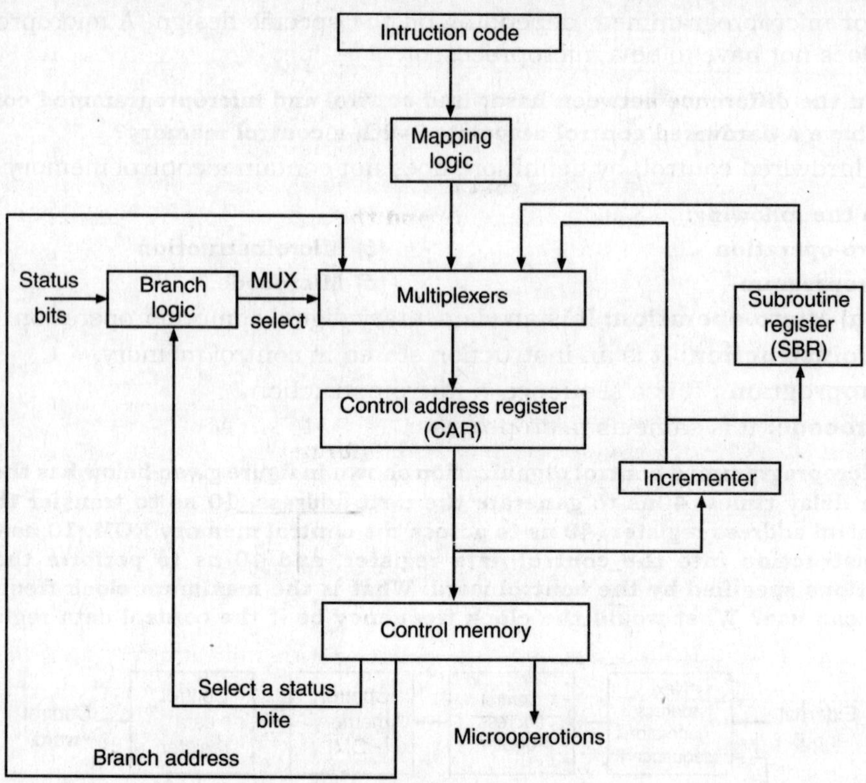

Fig. 4.8.

Solution: Control memory = $2^{10} \times 32$

(*a*)

6	10	16	= 32 bits
Select	Address	Microperations	

(*b*) 4 bite

(*c*) 2 bits

Q.11. The control memory in figure has 4096 words of 24 bits each.

(*a*) How many bits are there in the control address register?

(*b*) How many bits are there in each of the inputs shown going into the multiplexers?

(*c*) What are the number of inputs in each multiplexer and how many multiplexesrs are needed?

Solution: Control memory = $2^{12} \times 24$

(*a*) 12 bits

(*b*) 12 bits

(*c*) 12 multiplexers, each of size 4-to-1 line.

Q.12. Using the mapping procedure described in figure given below give the first micro instruction address for the following operation code :

(*a*) 0010

(*b*) 1011

(*c*) 1111

Computer Instruction :

Mapping bits :

Microinstruction address :

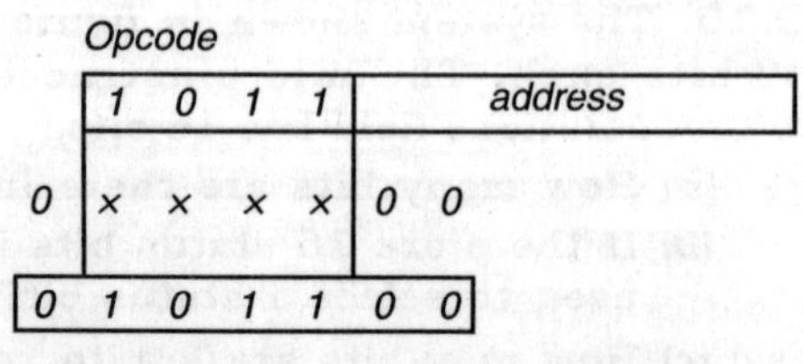

Fig. 4.9. Mapping from instruction code to microinstruction address

Solution:

(*a*) 0001000 = 8

(*b*) 0101100 = 44

(*c*) 0111100 = 00

Q.13. Formulate a mapping procedure that provides eight consecutive microinstructions for each routine. The operation code has six bits and the control memory has 2048 words.

Solution: Opcod = 6 bits

Control memory address = 11 bits

$$00\left|\begin{matrix}\times & \times & \times & \times & \times & \times \\ \times & \times & \times & \times & \times & \times\end{matrix}\right|000$$

Q.14. Why do we need the two multiplexers in the computer hardware configuration shown in figure given below? Is there another way that information from multiple sources can be transferred to a common destination ?

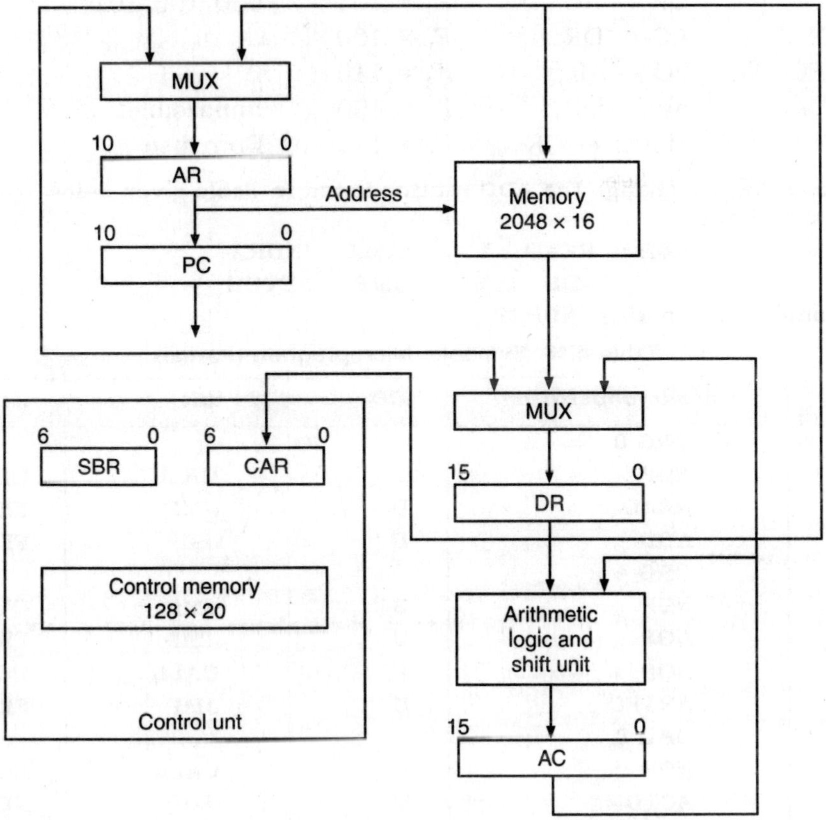

Fig. 4.10. Computer hardware configurtion

Solution: Either multiplexers, three state gates or gate logic (equivalent to a mux) are needed to transfer information from many sources to a common destination.

Q.15. Give the 9-bit Micro-operation field for the following micro-operalion:

(*a*) **AC ← AC + 1** **DR ← DR + 1**

(*b*) **PC ← PC + 1** **DR ← M[AR]**

(*c*) **DR ← AC** **AC ← DR**

Solution:

	F_1	F_2	F_3			
(*a*)	011	110	000	INCAC	INCUR	NOP
(*b*)	000	100	101	NOP	READ	INCPC
(*c*)	100	101	000	DRTAC	ACTDR	NOP

Q.16. Convert the following symbolic micro-operations to register transfer statements and to binary.

(*a*) READ, INCPC

(*b*) ACTDR, DRTAC

(*c*) ARTPC, DRTAC, WRITE

Solution:

				Binary
(*a*)	READ	DR ← M[AR]	F_2 = 100	000 100 101
	DRTAC	AC ← DR	F_3 = 101	
(*b*)	ACTDR	DR ← AC	F_2 = 101	000 100 101
	DRTAC	AC ← DR	F_1 = 100	
(*c*)	ARTPC	PC ← AR	F_3 = 110	
	DRTAC	AC ← DR	F_1 = 100	Impossible,
	WRITE	M[AR] ← DR	F_1 = 111	Both use F_1

Q.17. Suppose that we change the ADD routine listed in Table given below to the following two microinstructions.

ADD : READ I CALL INDR2
ADD U JMP FETCH

What should be subroutine INDR2?

Table 4.16. Symbolic Microprogram (Partial)

Label	*Microoperations*	*CD*	*BR*	*AD*
ADD :	**ORG 0**			
	NOP	**I**	**CALL**	**INDIRECT**
	READ	**U**	**JMP**	**NEXT**
	ADD	**U**	**JMP**	**FETCH**
	ORG 4			
BRANCH :	**NOP**	**S**	**JMP**	**OVER**
	NOP	**U**	**JMP**	**FETCH**
OVER :	**NOP**	**I**	**CALL**	**INDIRECT**
	ARTPC	**U**	**JMP**	**FETCH**
	ORG 8			
STORE :	**NOP**	**I**	**CALL**	**INDIRECT**
	ACTDR	**U**	**JMP**	**NEXT**
	WRITE	**U**	**JMP**	**FETCH**
	ORG 12			
EXHANGE :	**NOP**	**I**	**CALL**	**INDIRECT**
	READ	**U**	**JMP**	**NEXT**
	ACTDR, DRTAC	**U**	**JMP**	**NEXT**
	WRITE	**U**	**JMP**	**FETCH**
	ORG 64			

FETCH :	**PCTAR**	**U**	**JMP**	**NEXT**
	READ, INCPC	**U**	**JMP**	**NEXT**
	DRTAR	**U**	**MAP**	
INDRCT :	**READ**	**U**	**JMP**	**NEXT**
	DRTAR	**U**	**RET**	

Solution : If $I = 0$, the operand is read in the first microinstruction and added to *AC* in the second. If $I = 1$, the effective address is read into *DR* and control goes to INDR2. The subroutine must read the operand into *DR*.

INDR2 :	DRTAR	U	JMP	NEXT
	READ	U	RET	—

Q.18. The following is a symbolic microprogram for an instruction in the computer

ORG40			
NOP	**S**	**JMP**	**FETCH**
NOP	**Z**	**JMP**	**FETCH**
NOP	**I**	**CALL**	**INDRCT**
ARTPC	**U**	**JMP**	**FETCH**

(a) Specify the operation performed when the instruction is executed.

(b) Convert the four microinstructions into their equivalent binary form.

Solution:

(a) Branch if $S = 0$ and $Z = 0$ (Positive and non-zero *AC*) – See last instruction in problem 7 – 16.

(b)

40 :	000	000	000	10	00	1000000
41 :	000	000	000	11	00	1000000
42 :	000	000	000	01	01	1000011
43 :	000	000	110	00	00	1000000

Q.19. Add the following instructions to the computer (*EA* is the effective address). Write the symbolic microprogram for each routine as in Table given in Q. 17. (Note that *AC* must not change in value unless the instruction specifies a change in *AC*).

Symbol	*Opcode*	*Symbolic Function*	*Description*
AND	**0100**	**AC ← AC M[EA]**	**AND**
SUB	**0101**	**AC ← AC – M[EA]**	**Subtract**
ADM	**0110**	**M[EA] ← M[EA] + AC**	**Add to memory**
BTCL	**0111**	**AC ← AC M[EA]**	**Bit clear**
BZ	**1000**	**If (AC – O) then (PC ← EA)**	**Branch if AC zero**
SEQ	**1001**	**If (AC = M[EA]) then (PC ← PC + 1)**	**Skip if positive**
BPNZ	**1010**	**If (AC > O) then (PC ← EA)**	**Branch if positive and non - zero**

Solution:

	ORG16			
AND :	NOP	1	CALL	INDRCT
READ	U	JMP	NEXT	
ANDOP :	AND	U	JMP	FETCH

	ORG 20			
SUB :	NOP	I	CALL	INDRCT
	READ	U	JMP	NEXT
	SUB	U	JMP	FETCH
	ORG 24			
ADM :	NOP	I	CALL	INDRCT
	READ	U	JMP	NEXT
	DRTACACTDR	U	JMP	NEXT
	ADD	U	JMP	EXCHANGE + 2
	ORG 28			
BTCL :	NOP	I	CALL	INDRCT
	READ	U	JMP	NEXT
	DRTACACTDR	U	JMP	NEXT
	COM	U	JMP	ANDOP
	ORG 32			
BZ :	NOP	Z	JMP	ZERO
	NOP	U	JMP	FETCH
ZERO :	NOP	I	CALL	INDRCT
	ARTPC	U	JMP	FETCH
	ORG 36			
SEQ :	NOP	I	CALL	INDRCT
	READ	U	JMP	NEXT
	DRTAC, ACTDR	U	JMP	NEXT
	XOR (OR SUB)	U	JMP	BEQ 1
	ORG 69			
BEQ1 :	DRTAC, ACTDR	Z	JMP	EQUAL
	NOP	U	JMP	FETCH
EQUAL :	INCPC	I	JMP	FETCH
	ORG40			
BPNZ :	NOP	S	JMP	FETCH
	NOP	Z	JMP	FETCH
	NOP	I	CALL	INDRCT
	ARTPC	U	JMP	FETCH

Q.20. Write a symbolic microprogram routine for the ISZ (increment and skip if zero) instruction (Table given below). Note that DR = 0 status condition is not available in the CD field of the computer. However can exchange AC and DR and check if AC = 0 with the Z bit.

Table 4.17. Memory-Reference Instructions

Symbol	*Operation decoder*	*Symbolic description*
AND	D_0	**AC ← AC M[AR]**
ADD	D_1	**AC ← AC + M[AR], EC_{out}**
LDA	D_2	**AC ← M[AR]**
STA	D_3	**M[AR] ← AC**
BUN	D_4	**PC ← AR**
BSA	D_5	**M[AR] ← PC, PC ← AR + 1**
ISZ	D_6	**M[AR] ← M[AR] + 1**
		If M[AR] + 1 = 0 then PC ← PC + 1

Solution:

ISZ :	NOP	I	CALL	INDRECT
	READ	U	JMP	NEXT
	INCDR	U	JMP	NEXT
	DRTAC, ACTOR	U	JMP	NEXT (or past Indrct)
	DRTAC, ACTOR	Z	JMP	ZERO
	WRITE	U	JMP	FETCH
ZERO :	WRITE, INCPC	U	JMP	FETCH

Q.21. Write the symbolic microprogram routines for the BSA (branch and save address) instructions defined in (Table give in Q.20). Minimize the number of microinstructions.

Solution:

BSA :	NOP	I	Call	INDRCT
	PCTDR, ART PC	U	JMP	NEXT

Q.22. Show how outputs 5 and 6 of decoder F3 in figure given below are to be connected to the progran counter PC.

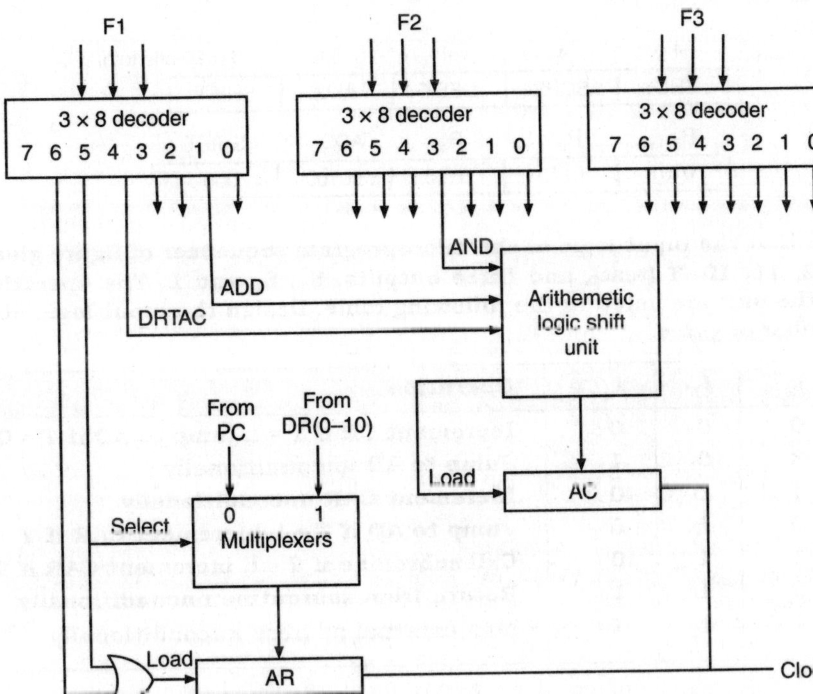

Fig. 4.11. Decoding of micro-operation field

Solution:

F = 101 (5) PC ← PC + 1

From

F_3 output 6

F_3 output 5

Q.23. Show how a 9-bit micro-operation field in a microinstruction can be divided into sub fields to specify 46 micro-operations. How many micro-operations can specified in one microinstrucion?

Solution: A field of 5 bits can specify $2^5 - 1 = 31$ micro-operations.

A field of 4 bits can specify $2^4 - 1 = 15$ micro-operations

9 bits 46 micro-operations

Q.24. A computer has 16 registers, an ALU (arithmetic logic unit) with 32 operations, and a shifter with eight operations, all connected to a common bus system.

(a) Formulate a control word for a microoperation.

(b) Specify the number of bits in each field of the control word and give a general encoding scheme.

(c) Show the bits of the control word that specify the micro-operation $R_4 \leftarrow R_5 + R_6$.

Solution :

(a)

3	3	3	5
SELA	SELB	SELD	OPR

(b) 16 registers need 4-bits, ALU need 5-bits, and the shifter need 3-bits to encode all operations.

4	4	4	5	3 = 20-bits total
SRC1	SCR2	DEST	ALU	SHIFT
R_5	R_6	R_4	ADD	SHIFT
0101	0110	0100	00100	000

Q.25. Assume that the input logic of the microprogram sequencer of figure given below has four inputs 12, 11, 10, T (test), and three outputs, S_1, S_0 and *L*. The operations that are performed in the unit are listed in the following table. Design the input logic circuit using a minimum number of gates.

I_2	I_1	I_o	*Operation*
0	0	0	Increment CR if $T = 1$, jump to AD if $T = 0$
×	0	1	Jump to AD unconditionally
1	0	0	Increment CAR unconditionally
0	1	0	Jump to AD if $T = 1$, increment CAR if $T - 0$
1	1	0	Call subrouine if $T - 1$, increment CAR if $T = 0$
0	1	1	Return from subroutine unconditionally
1	1	1	Map external address unconditionally

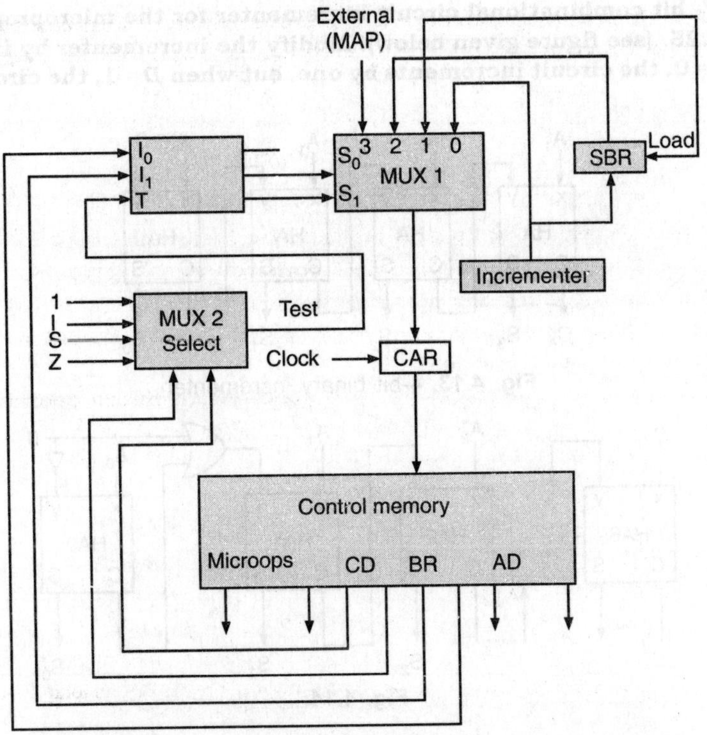

Fig. 4.12. Microprogram sequencer for a control memory

Solution :

I_2	I_1	I_0	T	S_1	S_0	L	
0	0	0	0	0	1	0	AD (1)
0	0	0	1	0	0	0	INC (0)
0	0	1	0	0	1	0	AD (1)
0	0	1	1	0	1	0	AD (1)
0	1	0	0	0	0	0	INC (0)
0	1	0	1	0	1	0	AD (1)
0	1	1	0	1	0	0	RET (2)
0	1	1	1	1	0	0	RET (2)
1	0	0	0	0	0	0	INC (0)
1	0	0	1	0	0	0	INC (0)
1	0	1	0	0	1	0	AD (1)
1	0	1	1	0	1	0	AD (1)
1	1	0	0	0	0	0	INC (0)
1	1	0	1	0	1	1	CALL (1)
1	1	1	0	1	1	0	MAP (3)
1	1	1	1	1	1	0	MAP (3)

I_2T_1 \ I_0T	00	01	11	10
00	1		1	1
01		1		
11		1	1	1
10			1	1

I_2I_1 \ I_0T	00	01	11	10
00				
01			1	1
11			1	1
10				

$$S_0 = I_2 I_1 + I_1 I_0 T + I_2 I_1 T$$

$$L = I_2 I_1 I_0 T$$

Q.26. Design a 7- bit combinational circuit incrementer for the microprogram sequencer of figure given in Q.25. (see figure given below). Modify the incrementer by including a control input *D*. When *D* = 0, the circuit increments by one, but when *D* = 1, the circuit increments by two.

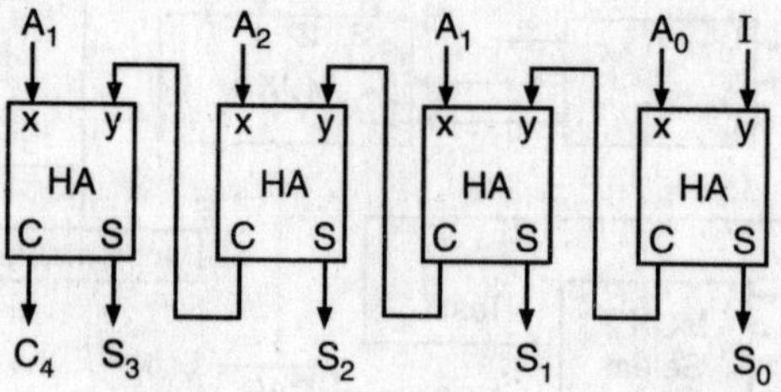

Fig. 4.13. 4-bit binary incrementer

Solution:

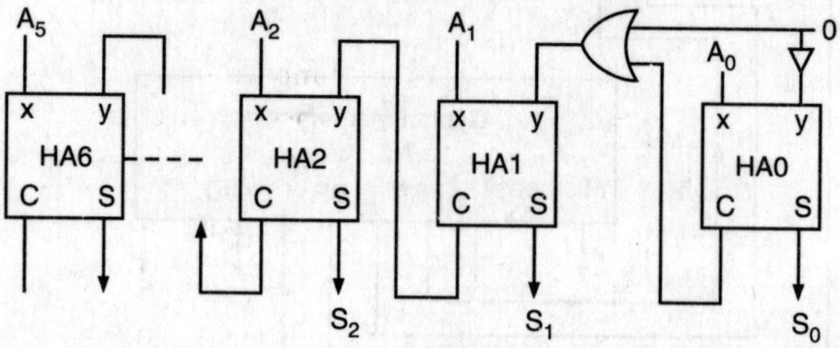

Fig. 4.14.

Q.27. Insesrt an exclusive–OR gate between MUX 2 and the input logic of figure given in Q.25. One input to the gate comes from the test output of the multiplexer. The other input to the gate comes from a bit labeled *P* (for polarity) in the microinstruction from control memory. The output of the gate goes to the input *T* of the input logic. What does the polarity control *P* accomplish?

Solution: *P* is used to determine the polarity of the selected status bit.

When $P = 0$, $T = G$ because $G \oplus = G$.

When $P = 1$, $T = G$ because $G \oplus 1 = G$.

When *G* is the value of the selected bit in MUX2.

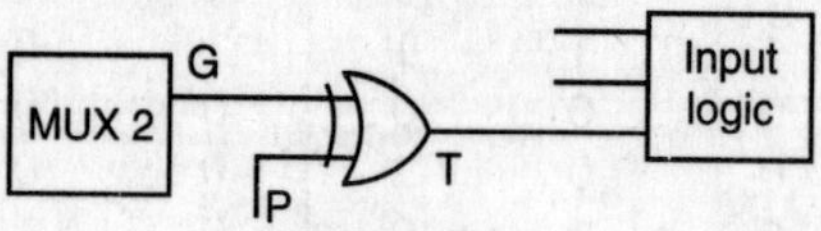

Fig. 4.15.

SUMMARY

1. Computer instructions are normally stored in consecutive memory locations and are executed sequentially one at a time.
2. The timing got all registers in the basic computer controlled by a clock.
3. An instruction is a group of bits that instruct the computer to program a specific operation.
4. A label consists of one, two or three, but not more than three alpha numeric characters.
5. A control memory is the unit which handles micro instructions.
6. Program resides in control memory is called micro program, instruction is called microinstruction and operation is called micro-operation.
7. Microinstruction format is of 20 bits.

8. Micro programmed control unit will have two sperate memories— a main memory and a control memory.
9. The timing for all registers in the basic computer is controlled by master clock generator.
10. Macros in an assembly language are used to repeat a sequence of instructions several times in a program. The instructions are stored in a macro body and whenever a macro name is encountered by the assembler it is replaced by the instruction.

EXERCISE

(I) Fill in the Blanks (Solved)

1. Computer .... are normally stored in consecutive memory locations and are executed sequentially one at a time.
2. The timing for all registers in the basic computer is controlled by a ....

(II) True or False (Solved)

1. A two phase consists of one, two, or three, but not more than three alphanumeric characters.
2. An interrupt cycle is a hardware implementation of branch and save return address operation.

(III) Fill in the Blanks (Unsolved)

1. An ....is a group of bits that instruct the computer to program a specific operation.
2. A .... consists of one, two, or three, but not more than three alphanumeric characters.

(IV) True or False (Unsolved)

1. The memory that is a part of a control unit is referred to as an accumulator.
2. The data register is sometimes called a buffer.

(V) Detailed Questions

1. What is the difference between a direct and an indirect address instruction? How many references to memory are needed for such type of instruction to bring an operand into a processor register?
2. Explain why each of the following micro-operations cannot be executed during a single clock pulse.
   (*a*) IR M[PC] (*b*) AC AC + TR (*c*) DR DR + AC (AC does not change)
3. Describe the control gates associated with the program counter PC in the basic computer.
4. Can the letter I be used as a symbolic address in the assembly language program defined for the basic computers? Justify the answer.
5. Write a program loop, using a pointer and a counter that clears to 0 the contents of hexadecimal location 500 through 5FF.
6. Is it possible to have a hardwired command associated with a control memory?
7. Define the following :
   (*a*) Micro-operation (*b*) Micro instruction (*c*) Micro program (*d*) Micro code
8. What is the difference between a microprocessor and a microprogram? Is it possible to design a microprocessor without a microprogram? Are all microprogrammed computers also microprocessors?

Answers

(I) Fill in the Blanks (Solved)

1. instructions 2. clock

(II) True or False (Solved)

1. False 2. True

Chapter

5

Data Path Design

Chapter Outline

- Introduction
- Fixed Point Arithmetic
- Floating Point Arithmetic
- Pipelining
- Addition and Subtraction of Floating Point Number
- Multiplying Floating Point Numbers
- Floating Point Division

5.0 INTRODUCTION

In many high-level programming languages we have a facility for specifying floating-point numbers. The most common way is by a real declaration statement. High level programming languages must have a provision for handling floating-point arithmetic operations. The operations are generally built in the internal hardware. If no hardware is available, the compiler must be designed with a package of floating-point software subroutine. Although the hardware method is more expensive, it is much more efficient than the software method. Therefore, floating- point hardware is included in most computers and is omitted only in very small ones.

5.1 FIXED POINT ARITHMETIC

Registers and adder used for fixed-point arithmetic are used for processing the mantissas. The difference lies in the way the exponents are handled.

The register organization for floating-point operations is shown in Fig. 5.1. Three registers are there, BR, AC, and QR. Each register is subdivided into two parts. The mantissa part has the same uppercase letter symbols as in fixed-point representation. The exponent part may use corresponding lower-case letter symbol.

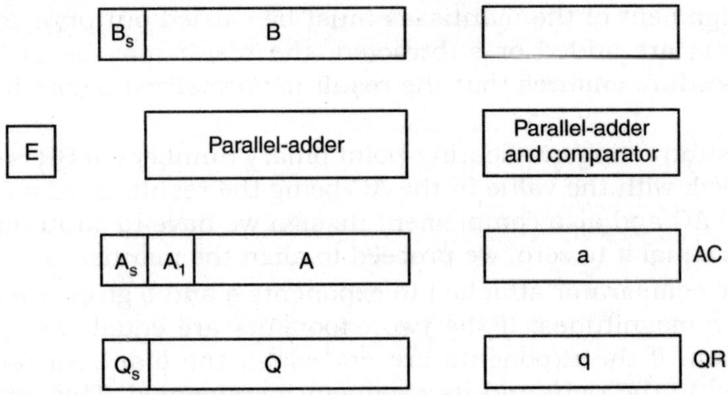

Fig. 5.1. Registers for Floating Point arithmetic operations

Assuming that each floating-point number has a mantissa in signed-magnitude representation and a biased exponent. Thus the AC has a mantissa whose sign is in As, and a magnitude that is in A. The diagram shows the most significant bit of A, labeled by A1. The bit in his position must be a 1 to normalize the number. Note that the symbol AC represents the entire register, that is, the concatenation of As, A and a.

In the similar way, register BR is subdivided into Bs, B, and b and QR into Qs, Q and q. A parallel-adder adds the two mantissas and loads the sum into A and the carry into E. A separate parallel adder can be used for the exponents. The exponents do not have a district sign bit because they are biased but are represented as a biased positive quantity. It is assumed that the floating-point number are so large that the chance of an exponent overflow is very remote and so the exponent overflow will be neglected. The exponents are also connected to a magnitude comparator that provides three binary outputs to indicate their relative magnitude.

The number in the mantissa will be taken as a fraction, so they binary point is assumed to reside to the left of the magnitude part. Integer representation for floating point causes certain scaling problems during multiplication and division. To avoid these problems, we adopt a fraction representation.

The numbers in the registers should initially be normalized. After each arithmetic operation, the result will be normalized. Thus all floating-point operands are always normalized.

5.2 FLOATING POINT ARITHMETIC

5.2.1 Addition and Subtraction of Floating Point Numbers

During addition or subtraction, the two floating-point operands are kept in AC and BR. The sum or difference is formed in the AC. The algorithm can be divided into four consecutive parts:

1. Check for zeros.
2. Align the mantissas.
3. Add or subtract the mantissas
4. Normalize the result

A floating-point number cannot be normalized, if it is 0. If this number is used for computation, the result may also be zero. Instead of checking for zeros during the normalization process we check for zeros at the beginning and terminate the process if necessary. The alignment of the mantissas must be carried out prior to their operation. After the mantissas are added or subtracted, the result may be un-normalized. The normalization procedure ensures that the result is normalized before it is transferred to memory.

For adding or subtracting two floating-point binary numbers, if BR is equal to zero, the operation is stopped, with the value in the AC being the result. If AC = 0, we transfer the content of BR into AC and also complement its sign we have to subtract the numbers. If neither number is equal it to zero, we proceed to align the mantissas.

The magnitude comparator attached to exponents a and b gives three outputs, which show their relative magnitudes. If the two exponents are equal, we go to perform the arithmetic operation. If the exponents are not equal, the mantissa having the smaller exponent is shifted to the right and its exponent incremented. This process is repeated until two exponents are equal.

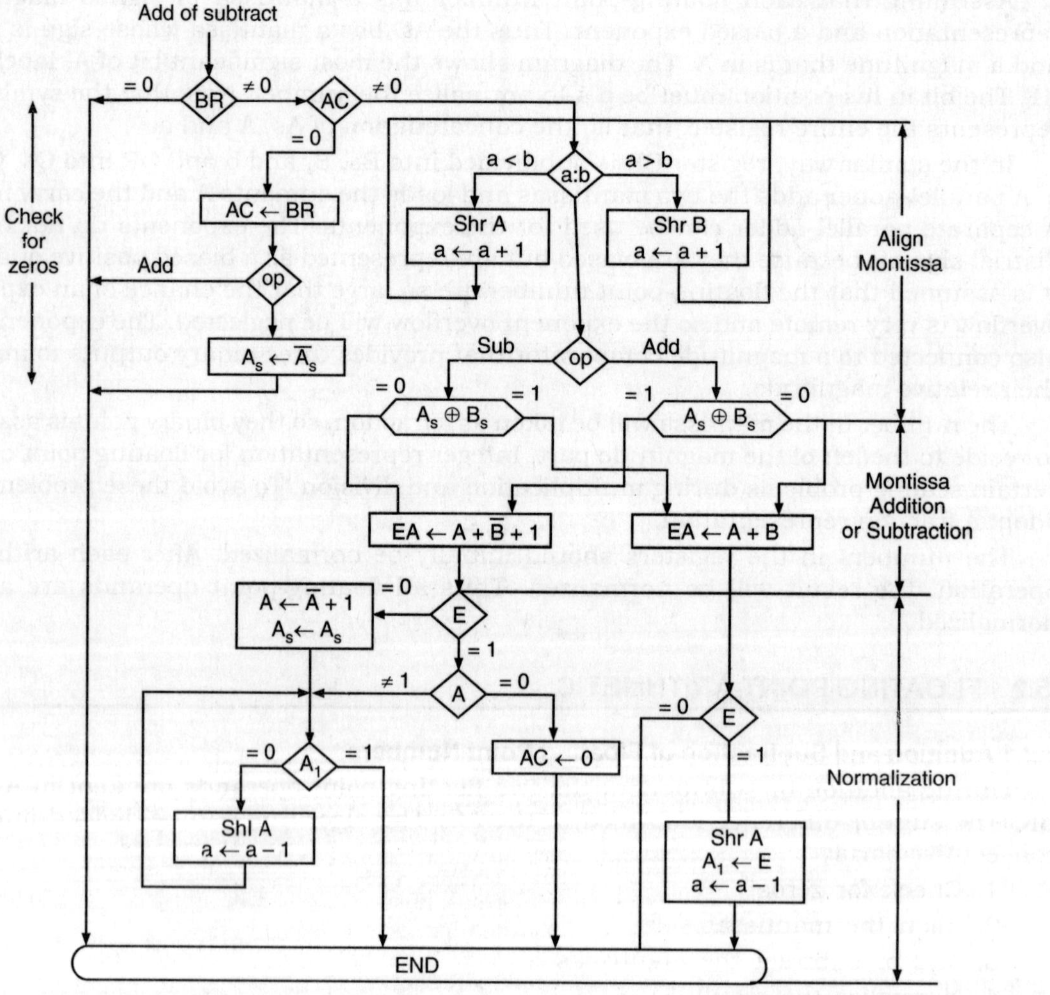

Fig. 5.2. Addition and Subtraction of Floating-Point Numbers

The addition and subtraction of the two mantissas is similar to the fixed-point addition and subtraction algorithm presented in Fig. 5.2. The magnitude part is added or subtracted depends on the operation and the signs of the two mantissas. If an overflow occurs when the magnitudes are added, it is transferred into flip-flop E. If E = 1, the bit is transferred into A1 and all other bits of A are shifted right. The exponent must be incremented so that it can maintain the correct number. No underflow may occur in this case this is because the original mantissa that was not shifted during the alignment was already in a normalized position.

If the magnitudes were subtracted, there may be zero or may have an underflow in the result. If the mantissa is equal to zero the entire floating-point number in the AC is cleared to zero. Otherwise, the mantissa must have at least one bit that is equal to 1. The mantissa has an underflow if the most significant bit in position A1, is 0. In that case, the mantissa is shifted left and the exponent decremented. The bit in A1 is checked again and the process is repeated until A1 = 1. When A1 = 1, the mantissa is normalized and the operation is completed.

5.2.2 Multiplying Float Numbers

The multiplication of two floating-point numbers requires multiplying the mantissas and adding the exponents. There is no need of comparison of exponents or alignment of mantissas. The multiplication of the mantissas is done in the same way as in fixed-point to provide a double-precision product. The double-precision answer is used in fixed-point numbers to increase the accuracy of the product. In floating-point, the range of a single-precision combined with the exponent is usually accurate enough. So the half most significant bits of the mantissa product and the exponent are taken together to form a single-precision floating-point product.

We divide the multiplication algorithm into four parts:

1. Check for zeros.
2. Addition of the exponents
3. Multiplication of the mantissas
4. Normalization of the product

We can carry out steps 2 and 3 simultaneously if we have separate adders available for the mantissas and exponents respectively.

Floating point multiplication is given in Fig. 5.3. The two operands are checked whether they contain a zero. If either operand is zero, then product is zero and AC is cleared to 0. If neither of the operands is equal to zero, the process continues with the exponent addition.

The exponent of the multiplier is in q and the adder is between exponents a and b. We need to transfer the exponents from q and a, add the two exponents and transfer the sum into a. Since both exponents are biased by the addition of a

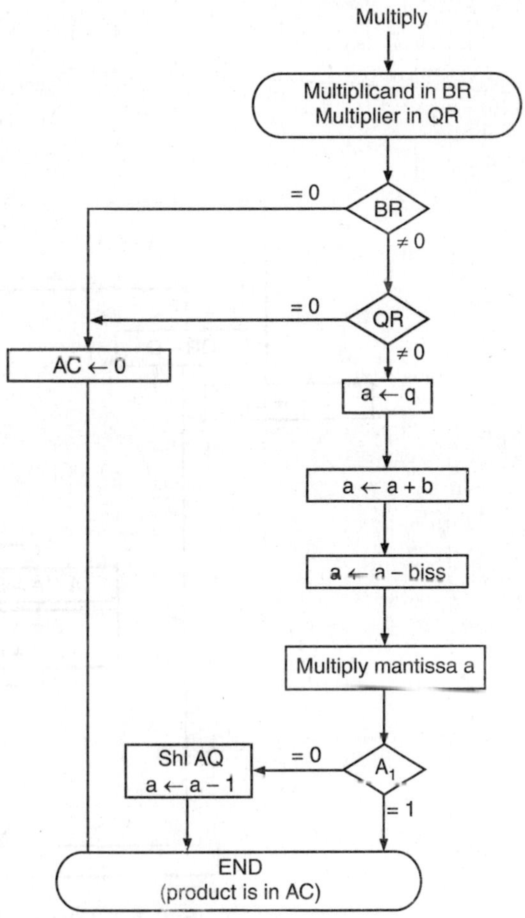

Fig. 5.3. Multiplication of Floating-point Numbers

contract, the exponents' sum will have double this bias. We obtain the correct biased exponents for the product by subtracting the bias number from the sum.

5.2.3 Floating Point Division

The division algorithm has five parts:

1. Check for zero.
2. Initialize registers and evaluate the sign
3. Align the dividend
4. Subtract the exponents
5. Divide the mantissas

The flowchart for floating-point division is given in Fig. 5.4. The two operands are checked for zero. If the division is zero, it shows an attempt to divide by zero, which is illegal. So the operation is terminated with an error message. An alternative procedure would be set the quotient in QR to the most positive (if the divided is negative). If the dividend in AC is zero, the quotation in QR is made zero and the operation terminates.

If the operands are not zero, we determine the sign of the quotient and store it in Qs. The sign of the dividend in As, is left unchanged to be the sign of the remainder. The Q register is cleared and the sequence counter SC is set to a number equal to the number of bits in the quotient.

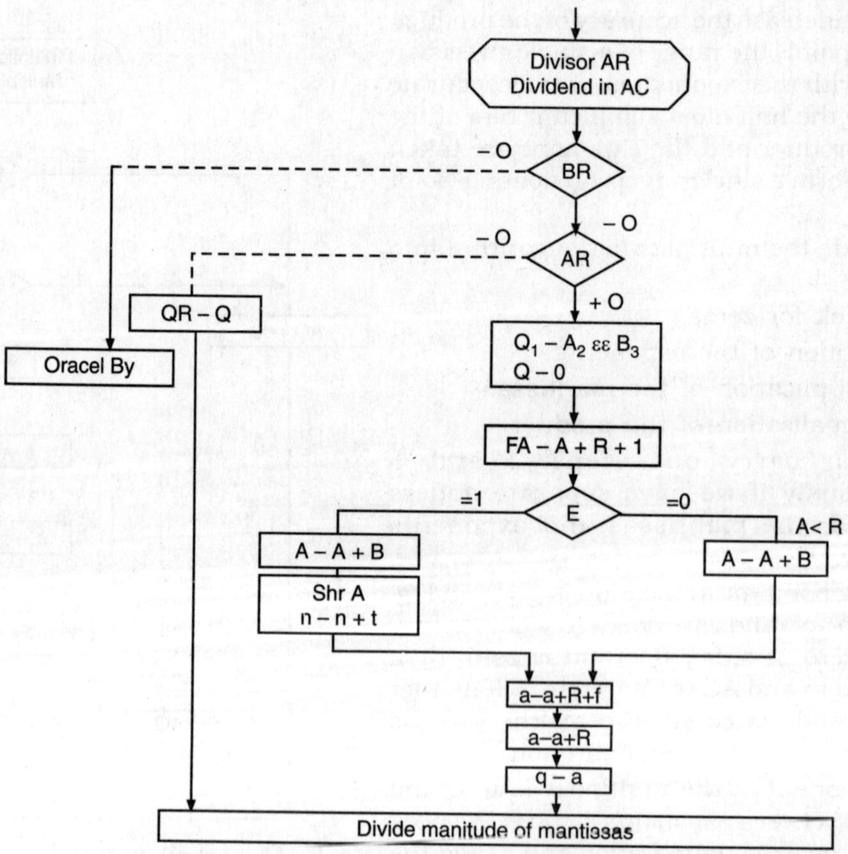

Fig. 5.4. Division O Floating-point Numbers

The proper dividend alignment requires that the fraction division test. The carry in E tells their relative magnitude. The dividend fraction is restored to its original value by adding the divisor. If A < B, we need to shift A once to the right and increment the dividend exponent. This alignments ensure that A < B, since both operands are normalized.

Now, the divisor exponent is subtracted from the dividend exponent. Since both exponents were originally biased, the subtraction operation gives the difference without the bias. The bias is often added and the result transferred into q because the quotient is stored in QR.

5.3 PIPELINING

The most commonly used technique for parallel processing is pipelining. It is a sequential process disintegrated into sub operations, with each sub process being executed in a special dedicated segment that operates concurrently with all other segments. A pipeline can be visualized as a collection of processing segments acting as a channel for binary information flow. Each segment performs partial processing dictated by the way the task is segregated. The result obtained from the computation in each segment is transferred to the next segment in the pipeline. The first result is obtained after the data have passed through all segments. It is characteristic of pipelines that several computations can be in progress in distinct segments at the same time. The overlapping of computation is made possible by associating a register with each segment in the pipeline. The registers provide isolation between each segment so that each can operate on distinct clock at the same time simultaneously. For example, suppose that we want to perform the combined multiply and add operations with a stream of numbers:

Ai + Bi*Ci for i = 1, 2, 3......7

Each segment has one or two registers and a combinational circuit as shown in Fig. 5.5.

The sub operations performed in each segment of the pipeline are as follows:

R1	Ai R2	Bi ← Input Ai and Bi
R3	R1 *R2	R4 ← Ci Multiply and input Ci
R5	R3 + R4	Add Ci to product

Pipeline and vector processing

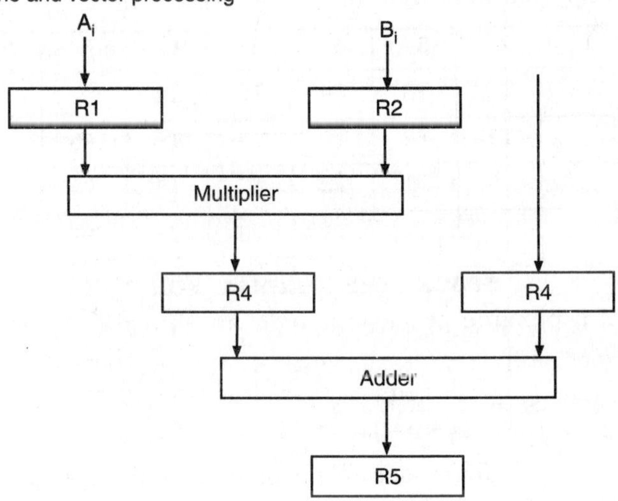

Fig. 5.5. Example of Pipelining process

The five registers are loaded with new data every clock pulse. The effect of each clock is shown in Table 5.11.

Table 5.1. Content of Registers in Pipeline

	Segment 1		Segment 2		Segment 3
Number	R_1	R_2	R_3	R_4	R_5
1	A_1	B_1	–	–	–
2	A_2	B_2	$A_1{*}B_1$	C_1	
3	A_3	B_3	$A_2{*}B_2$	C_s	$A_1{*}B_1 + C_1$
4	A_4	B_4	$A_3{*}B_3$	C_3	$A_2{*}B_2 + C_2$
5	A_5	B_5	$A_4{*}B_4$	C_4	$A_3{*}B_3 + C_3$
6	A_6	B_6	$A_5{*}B_5$	C_5	$A_4{*}B_4 + C_4$
7	A_7	B_7	$A_6{*}B_6$	C_6	$A_5{*}B_6 + C_5$
8	–	–	$A_7{*}B_7$	C_7	$A_6{*}B_6 + C_6$
9	–	–	–	–	$A_7{*}B_7 + C_7$

It takes three clock pulses to fill up the pipe and retrieve the first output from R5. From there on, each clock produces a new output and moves the data one step down the pipeline. Any operation that can be decomposed into a sequence of sub operations of about the same complexity can be implemented by a pipeline processor. The technique is efficient for those applications that need to repeat the same task many times with different sets of data. The behavior of a pipeline can be illustrated with a space time diagram. This is a diagram that shows the segment utilization as a function of time. The space time diagram of a four segment pipeline is demonstrated in

Fig. 5.6. The horizontal axis displays the time in clock cycles and the vertical axis gives the segment number. The diagram shows six tasks T1 through T6 executed in four segments. Initially task T1 is handled by segment 1. After the first clock, segment 2 is busy with T1, while segment 1 is busy with task T2. Continuing in this manner, the first task T1 is completed after the fourth clock cycle. From then on, the pipe completes a task every clock cycle. No matter how many segments there are in the system, once the pipeline is full, it takes only one clock period to obtain an output.

		1	2	3	4	5	6	7	8	9	→ Clock cycle
Segment:	1	T1	T2	T3	T4	T5	T6				
	2		T1	T2	T3	T4	T5	T6			
	3			T1	T2	T3	T4	T5	T6		
	4				T1	T2	T3	T4	T5	T6	

SPACE-TIME DIAGRAM FOR PIPELINE

The speed up of a pipeline processing over an equivalent non-pipeline processing is defined by the ratio,

$$S = \frac{nt}{(k+n-1)t_p}$$

As the number of tasks increases, n becomes much larger than k - 1, and k + n- 1 approaches the value of n. Under this condition, the speedup becomes,

$$S = \frac{t_n}{t_p}$$

If we assume that the time it takes to process a task is the same in the pipeline and non-pipeline circuits, we will have tn = ktp. Including this assumption, the speedup reduces to;

$$S = \frac{kt_p}{t_p} = k$$

To clarify the meaning of the speed up ratio, consider the following numerical example. Let the time it takes to process a sub operation in each segment be equal to tp = 20 ns. Assume that the pipeline has K = 4 segments and executes n =100 tasks in sequence. The pipeline system will take (k + n – 1) tp = (4 +9 9) × 20 = 2060 ns to complete. Assuming that tn = ktp = 4 × 20 = 80 ns, a non pipelining system requires nktp = 100 × 80 = 8000 ns to complete 100 tasks. The speed up ratio is equal to 8000/2060 = 388. As the number of tasks increases, the speed up will approach to 4. There are two areas of computer design where the pipeline divides an arithmetic operation into sub operations for execution in the pipeline segments.

An instruction pipeline operates on a stream of instruction by overlapping the fetch decode and execute phases of the instruction cycle. Pipeline arithmetic units are an essential part in very high speed computers. They are employed to implement floating point operations, multiplication of fixed point numbers and similar computations encountered in scientific problems. A pipeline multiplier is essentially an array multiplier with special address designed to minimize the carry propagation time through the partial products. Floating point operations are easily decomposed into sub operations. An instruction pipeline reads consecutive instructions from the memory while previous instructions are being executed in other segments. This causes the instruction and execute phases to overlap and perform simultaneous operations. One possible digression associated with such a scheme is that an instruction may cause a branch out of sequence. In that case the pipeline must be emptied and all the instructions that have been read from memory after the branch instruction must be discarded. The instruction fetch segment can be implemented by means of a first-in first-out (FIFO) buffer. This is a type of unit that forms a queue rather than a stack. The design of an instruction pipeline will be most efficient if the instruction cycle is divided into segments of equal duration. The time that each step takes to ful-fill its function depends on the instruction and the way it is executed.

SOLVED PROBLEMS

Q.1. What is the use of booth multiplication algorithm ? Describe hardware for booth diagram.

Solution: Booth algorithm gives a procedure for multiplying binary integers in signed 2's complement representation. It operates, when strings of 0's in the multiplier require no addition but just shifting and a string of 1's in the multiplier from bit weight 2^k to weight 2^m can be treated as $2^{k+1} - 2^m$.

e.g. Binary number 001110 (+14) has a string of 1's from 2^3 to 2^1. $K = 3$, $m = 1$.

The number can be represented as $2^{KH} - 2^m = 2^4 - 2^1 = 14$.

Therefore multiplication $M \times 14$, where M is multiplication and 14, the multiplier can be done as $M \times 2^4 - M \times 2^1$.

Thus the product can be obtained by shifting the binary multiplication *M* four times to the left and subtracting *M* shifted left once.

Booth algorithm requires examination of the multiplier bits and shifting of the partial product. Prior to the shifting, the multiplication may be added to the partial product, subtract from the partial product or left unchanged according to the following rules.

Hardware Ror Both algorithrn

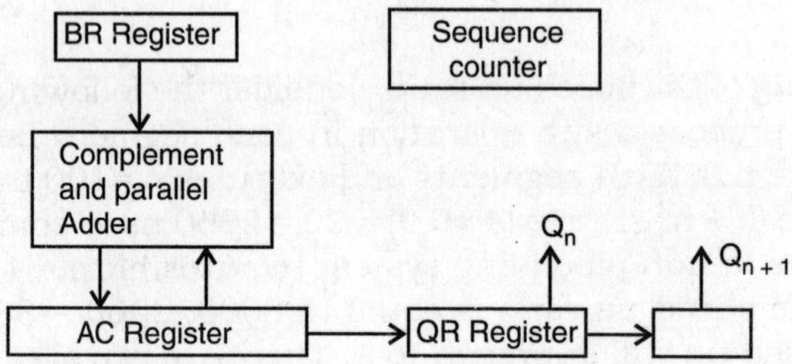

Fig. 5.6.

Q.2. Show the addition and subtraction with signed 2's complement.

Solution: The left most bit of a binary number represents the sign bit : 0 for positive and 1 for negative. The sign bit is 1, the entire number is represented in 2's complement form.

e.g.

+33 represented as : 00100001

–33 represented as : 1101111

The addition of two numbers in signed 2's complement form consists of adding the numbers with the sign bit treated same as the other bits of the number. The carry out of the sign bit position is discarded.

*The subtraction consists of first taking the 2's complement of the subtrahend and then adding it to the minuend.

Hardware for signed 2's complement addition and subtraction :

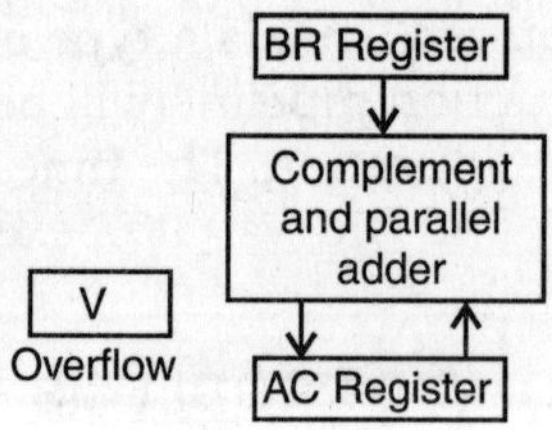

Fig. 5.7.

Q.3. Define over flow. How we detected over flow ?

Solution: When two numbers of *n* digits each are added and the sum occupies *n* + 1 digits, we say that an over flow occured. The effect of an over flow on the sum of two signed 2's complement number.

An over flow can be detected by inspecting the last two carries out of the addition. When two carries are applied to an exclusive *OR* gate, the over flow is detected when the output of the gate is equal to 1.

Q.4. Show adding and subtracting numbers is signed 2's complement representation.

Solution:

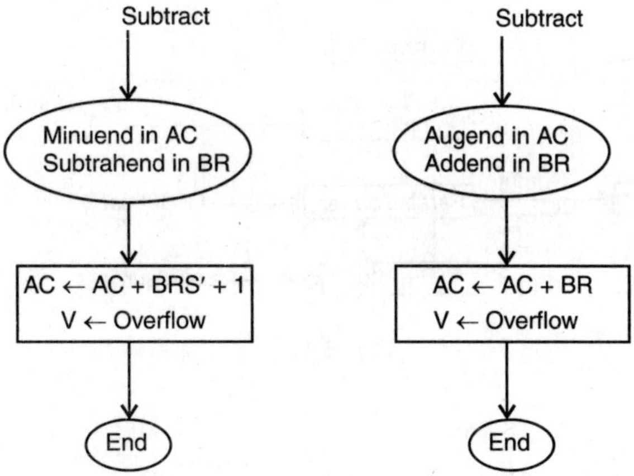

Fig. 5.8.

Q.5. Describe the register configuration for floating point operation.

Solution: The register configuration for floating point operations is quite similar to the layout for fixed point operations. The register for floating point operation is shown below.

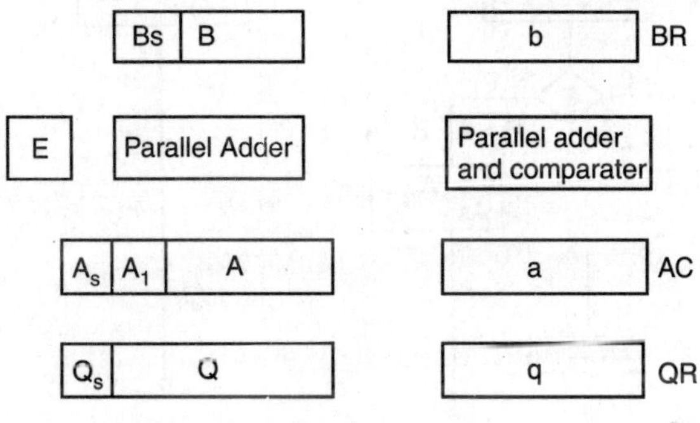

Fig. 5.9.

There are three registers *BR, AC* and *QR*. Each register is sub-divided into two parto the mantissa part has the same upper case letter symbols as in fixed point representation. The exponent part uses the corresponding lower case letter symbol. It is assumed that each floating point number has a mantissa in a signed magnitude representation and a biased exponent. Thus the *AC* has a mantissa whose sign is in A_s and magnitude that is in *A*. The exponent is in the part of the register denoted in the lower case letter symbol *a*. The diagram shows explicitly the most significant of *A*. lebeled by A_1. The bit in this position must be a_1 for the number to be normalized. Symbol *AC* represent entire register that is contentation of A_S, *A* and *a*.

Similarly, register *BR* is sub-divided into B_S, *B* and *b* and Q_R into Q_S, *a* and *q*. A parallel adder adds the two mantissa and transfers the sum into *A* and carry into *E*. A separate parallel adder is used for exponents.

Q.6. The complement shown in figure below is not needed if instead of performing $A + B^1 + 1$ we perform $B + A^1$. Derive an algorithm in flow chart form for addition and subtraction of

fixed point binary numbers in signed magnitude representation with the magnitudes subtracted by the two micro-operations $A \leftarrow \overline{A}$ and $EA \leftarrow A + B$.

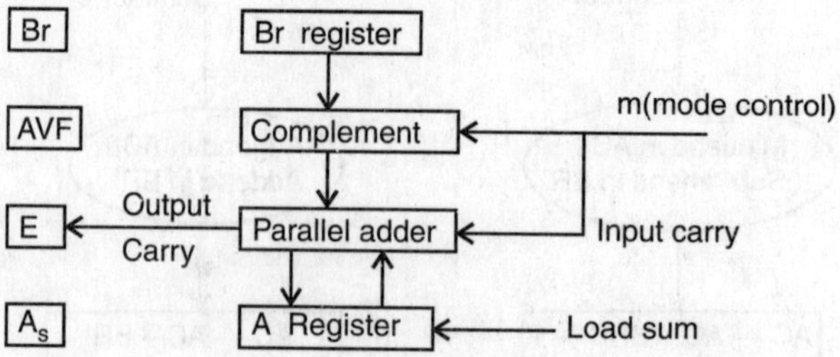

Fig. 5.10.

Solution:

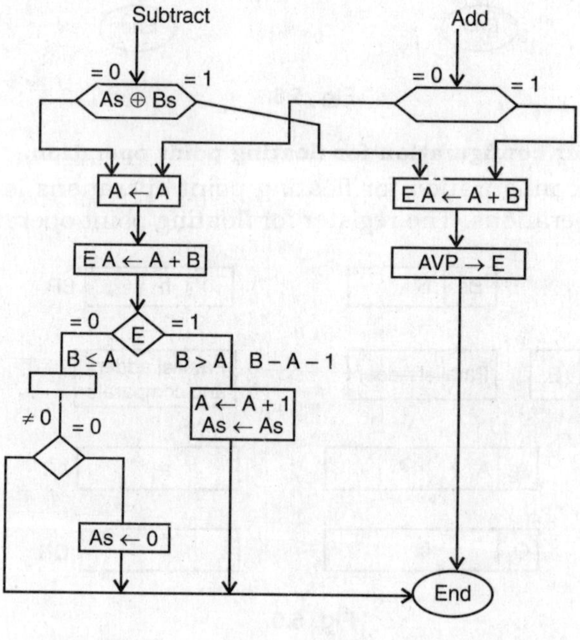

Fig. 5.11.

Q.7. Perform the arithmetic operations below with binary numbers and with negative numbers in signed 2's complement representation. Use seven bits to accommodate each number together with its sign. In each case, determine if there is an overflow by checking the carries into and out of the sign bit position.

(*a*) (+35) + (+40)

(*b*) (–35) + (–40)

Solution: (*a*) +35 = 0100011

→ 75 = 1001011

+40 = 0101000

Here, $F = 0$ and $E = 1$

$F \oplus E = 1$ overflow.

(*b*) +351011101
-401011000

-700110101

Here, $F = 1$ and $E = 0$

$$F \oplus E = 1 \text{ overflow.}$$

Q.8. Formulate a hardware procedure for detecting an overflow by comparing the sign of the sum with the signs of the augend and addend. The numbers are in signed 2's complement representation.

Solution: Transfer augend sign into T_s

Then add : $AC \leftarrow AC + BR$

As will hare sign of sum. Truth table for combine circuit.

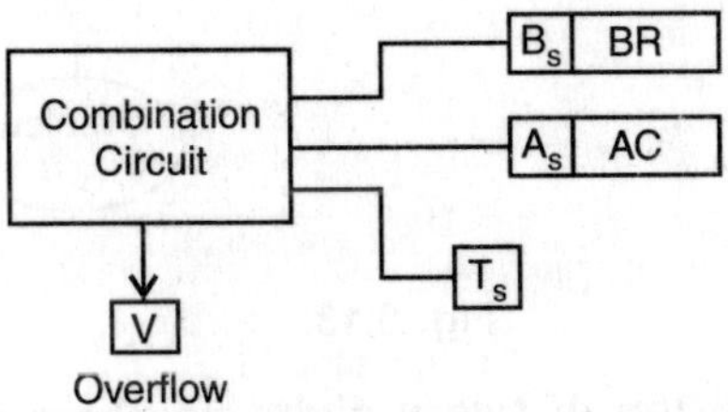

Fig. 5.12.

Boolean function for circuit :

$$V = T_S{}^1 \,.\, B_S{}^1 A_S + T_S B_S A_S{}^1.$$

Q.9. (*a*) Perform the operation (–9) + (+6) = –15 with binary numbers is signed –1's complement representation using only five bits to represent each number (including the sign). Show that the overflow detection procedure of checking the inequality of the last two carries fails in this case.

(*b*) Suggest a modified procedure for detecting an overflow when signed –1's complement numbers are used.

Solution: (*a*) –9 = 10110
+6 = 01001

–15 = 11111
01111
+1

1000 = –15

$F = 1, E = 0 \leftarrow$ Carries

$E \oplus F = 1$ but there should be no overflow since result is –15.

(*b*) The procedure $V \leftarrow E \oplus F$ is valid for 1's complement numbers provided we check the result.

$$\frac{0}{A_s} \quad \frac{1111\ldots11}{A} \text{ when } V = 1$$

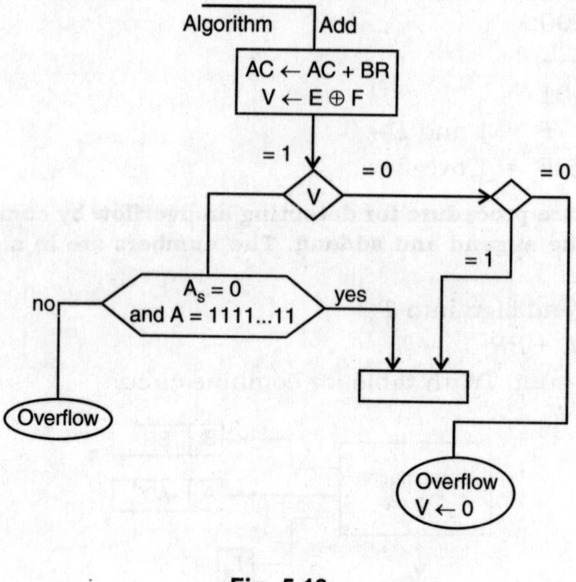

Fig. 5.13.

Q.10. Prove that the multiplication of two *n*-digits numbers in base *r* gives a product no more than 2*n* digits in length. Show that this statement implies that no overflow can occur in the multiplication operation.

Solution: Maximum value of numbers = $r^n - 1$

It is necessary to show that maximum product is less than or equal r^{2n-1}

$$\text{Maximum product} = (r^n - 1)(r^n - 1)$$
$$= r^{2n} - 2r^n + 1$$
$$\leq z^{2n} - 1$$

which gives $\quad 2 \leq 2\, r^n$

or $\quad 1 \leq r^n$

This is always true since $r \geq z$ and $n \geq 1$.

Q.11. Show that adding *B* after the operation *A* + *B* + 1 restores the original value of *A*. What should be done with the end carry ?

Solution: $A + B + 1$ performs $A + 2^n - B = 2^n + A - B$

Adding B : $(2^K + A - B) + B = 2^n + A$

Remove end-carry 2^n to obtain A.

Q.12. Design an array multiplier that multiplies two 4-bit numbers. Use *AND* gates and binary address.

Solution:

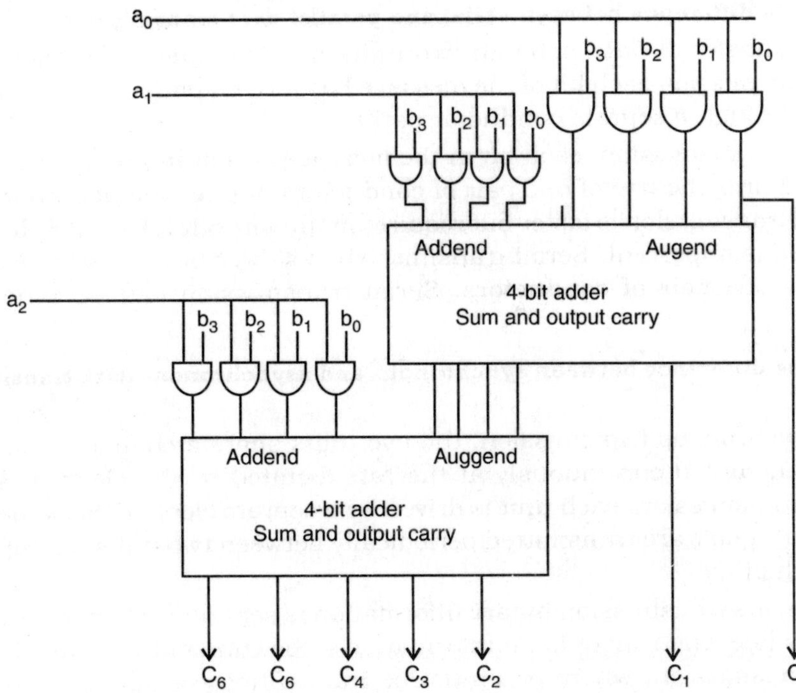

Fig. 5.14.

Q.13. What do you mean by isolated ? Define memory mapped input/output ?

Solution: Many computers use common bus to transfer information between memory and input/output and the CPU. The distinction between a memory transfer and input/output transfer is made through seprate read and write lines. The CPU specifies whether the address on the address line is for memory word or for an interface register by enabling one of two possible read or write lines. The input/output read input/output write control lines are enabled during an input/output transfer.

The memory read and memory write control lines are enabled during a memory transfer. This configuration isolates all input/output interface addresses from the addresses assigned to memory and is referred to as the isolated input/outputmethod for assigning addresses in a common bus.

In the isolated input/output configuration, the CPU has distinct input and output instructions and each of these instructions is associated with the address of an interface register. The isolated input/output method isolates memory and input/output address, so that memory address values are not affected by interface address assignment since each has its own address space.

In memory mapped input/output, it use the same address space for both memory and input/output. This is the case in computers that employ only one set of read and write signals and do not distinguish between memory and input/output address.

Computers with memory mapped input/output can use memory type instructions to access input/output data with memory mapped input/output. All instructions that refer to memory are also available for input/output.

Q.14. What is the difference between serial and parallel data transmission.

Solution: The transfer of data between two units may be done in parallel or serial. In parallel data transmission each bit of the message has its own path and the total messages is transmitted through *n* seprate conductor paths.

In serial data transmission, each bit in the message is sent in sequence one at a time. This method requires the use of one pair of conductors or one conductor and a common ground. Parallel transmission is faster but requires many wires. It is used for short distances and where speed is important. Serial transmission is slower but is less expensive. Since it requires only one pair of conductors. Serial transmission can be synchronous or asynchronous.

Q.15. What is the difference between synchronous and asynchronous data transfer in case of serial transfer?

Solution: In synchronous transmission, the two units share a common clock frequency and bits are transmitted continuously at the rate dictated by the clock pulses. In long distance serial transmission, each unit is driven by a seprate clock of the same frequency. Synchronization signals are transmitted periodically between two units to keep their clock in step with each other.

In asynchronous transmission binary information is sent only when it is available and the line remains idle when there is no information to be transmitted. This is contrast to synchronous transmission, where bits must be transmitted continuously to keep clock frequency in both units synchronized with each other.

A serial asynchronous data transmission technique used in many interactive terminals employs special bits that are inserted at both ends of the character code with this technique. Each character consists of three parts:

(*a*) Start bit

(*b*) Character bit

(*c*) Stop bits

The convention is that the transmitted resets at the 1 state when no characters are transmitted. The first bit called the start bit, is always 0 and is used to indicate the beginning of a character. The last bit called the stop bit is always a 1.

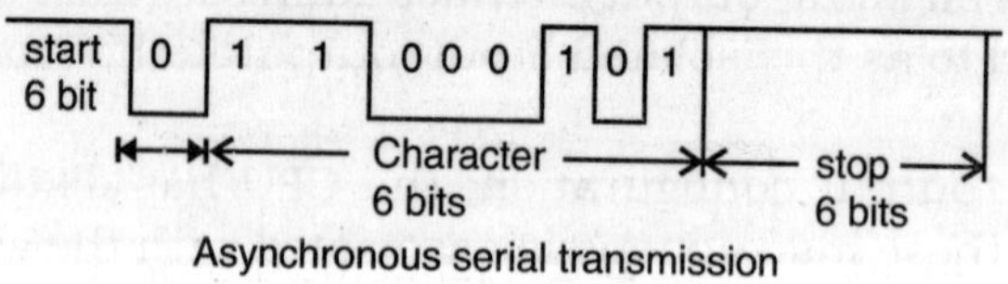

Fig. 5.15.

Q.16. Describe different types of modes of transfer ?

Solution: Data transfer between the central computer and input/output devices may be handled in a variety of modes. Some modes use the CPU as an intermediate path, others transfer the data directly to and from the memory unit. Data transfer to and from peripherals may be handled in one of the possible modes.

(*i*) Programmed input/output

(*ii*) Interrupt initiated input/output

(*iii*) Direct memory access (DMA).

***(i)* Programmed input/output :** Programmed input/output operations are the result of *I/O* instructions written in the computer program. Each data item transfer is initiated by an instruction in the program usually, the transfer is to and from a CPU register and peripheral. Other instructions are needed to transfer the data to and from CPU and memory. In the programmed input/output method the CPU stage in program loop until the input/output method the CPU stage in program loop until the input/output unit indicates that it is ready for data transfer. Transfer of data under programmed input/output between CPU and peripheral.

***(ii)* Interrupt Initiated input/output :** When an interface determine that the device is ready for the data transfer, it generates an interrupt request to the computer upon detecting the external interrupt signal. The CPU, momentarily stops the task, it is processing, branches to a servic program to process the input/output transfer and then returns to the task it was originally performing.

***(iii)* Direct Memory Access :** In direct memory access (DMA), the interface transfers data into and out of the memory unit through the memory bus. The CPU initiates the transfer by supplying the interface with the starting address and number of words needed to be transferred and then proceeds to execute other tasks when the transfer is made. The DNA requests memory cycles through the memory bus when the request is granted by the memory controller, the DMA transfers the data directly into memory.

Q.17. Define the following :

(*a*) Priority interrupt

(*b*) Polling

(*c*) Vector interrupt.

Solution: (*a*) Priority interrupt : A priority interrupt is a system that establishes a priority over the various sources to determine which condition is to be serviced first when two or more requests arrive simultaneously. The system may also determine which conditions are permitted to interrupt the computer while another interrupt is being serviced. Higher priority interrupt levels are assigned to requests which, if delayed or interrupted could have serious consequences.

(*b*) Polling : Establishing the priority of simultaneous interrupt can be done by software or hardware. A polling procedure is used to identify the highest priority sources by software means. In this method there is one common branch address or all interrupts. The program that takes care of interrupts beginning at the branch address and poll the interrupt sources in sequences. The order in which they are tested first and if its interrupt signal is on, control branches to service routine for this source. Otherwise the next lower-priority sources is tested and so on.

(*c*) Vector Interrupt : The way that the processor chooses the branch address of the service routine varies from one unit to another. In principle there are two methods for accomplishing this.

- **(*i*) Non-vector interrupt :** In a non-vector interrupt, the branch address is assigned to fixed location in memory.
- **(*ii*) Vector interrupt :** In a vector interrupt, the source that interrupts supplies the branch information to the computer. This information is called the interrupt vector. In some computers the interrupt vector is the first address of the I/O service routine.

Q.18. Give peripheral devices that produce an acceptable output for a person to understand.

Solution:

(*i*) Character printer
(*ii*) Line printer
(*iii*) Laser printer
(*iv*) Digital printer
(*v*) Graphics display
(*vi*) Voice output
(*vii*) Digital to analog converter
(*viii*) Instrument indicator.

Q.19. Indicate whether the following constitute a control status or data transfer commands

(*a*) Skip next instruction if flag is set

(*b*) Seek a given record on a magnetic disk

(*c*) Check if input/output device is ready

(*d*) Move printer paper to beginning of next page

(*e*) Read interface status register.

Solution: (*a*) Status command — Checks status of flag bit

(*b*) Control command — Moves magnetic head in disk

(*c*) Status command — Check if device power is on

(*d*) Control command — Moves paper position

(*e*) Data input command — Reads value of a register.

Q.20. A CPU with a 20 MHz clock is connected to a memory unit whose access time is 40 *ns*. Formulate a read and write timing diagrams using a READ strobe and a WRITE strobe. Include the address in the timing diagram.

Solution: 20 MHz = 20×10^6 Hz

$$T = \frac{10^{-6}}{20} = 50\ nx$$

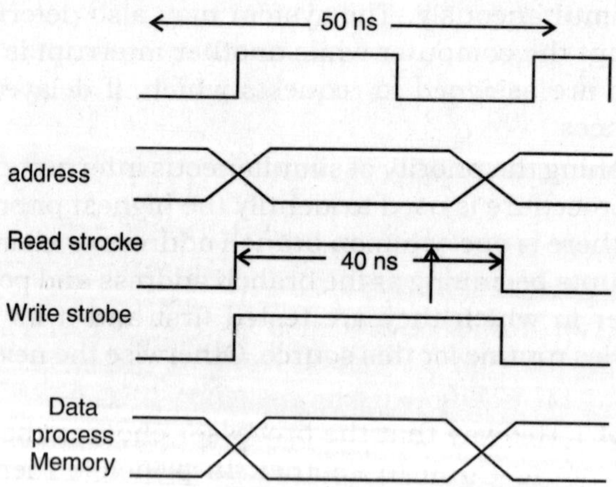

Fig. 5.16.

Q. 21. Give six status conditions for the setting of individual bits in the status register of an asynchronous communication interface.

Solution: Status Conditions :

(*i*) Output flag to indicate when transmitter register is empty.

(*ii*) Input flag to indicate when receiver register is full.

(*iii*) Enable interrupt if any flag is set.

(*iv*) Parity error.

(*v*) Framing error.

(*vi*) Overrun error.

Q.22. What is use of booth multiplication algorithm ? Describe Hardware for booth algorithm.

Solution: Booth algorithm gives a procedure for multiplying binary integers in signed 2's complement representation. It operates, when strings of 0's in the multiplier require no addition but just shifting and a string of 1's in the multiplier from bit weight 2^k weight 2^m can be treated as $2^{k+1} - 2^m$.

e.g. Binary number **001110 (+ 14)** has a string of 1's from 2^3 to 2^1, $k = 3$, $m = 1$.

The number can be represented as $2^{k+1} - 2^m - 2^m = 2^4 - 2^1 = 14$

Therefore multiplication $M \times 14$, where M is multiplication and 14 the multiplier,can be done as $M \times 2^4 - M \times 2$.

Thus the product can be obtained by shifting the binary multiplication M four times to the left and subtracting M shifted left one.

Booth algorithm requires examination of the multiplier bits and shifting of the partial product. Prior to the shifting, the multiplication may be added to the partial product, subtracted from the partial product or left unchanged according to the following rules.

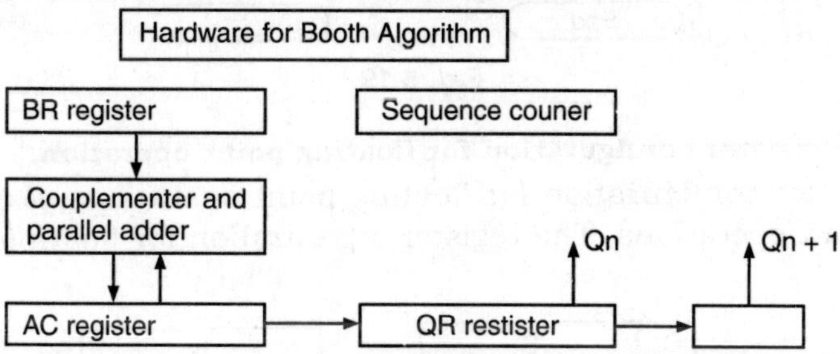

Fig. 5.17.

Q.23. Show addition and subtraction with signed 2's complement.

Solution: The leftmost bit of a binary number represents the sign bit : 0 for positive and 1 for negative.

The sign bit is 1, the entire number is represented in 2's complement form.

e.g.

+ 33 represented as : 00100001

– 33 represented as : 1101111

* The addition of two numbers in signed 2's complement from consists of adding the numbers with the sign bit treated same as the other bits of the number. The carry out of sign bit position is discarded.
* The subtraction consists of first taking the 2's complement of the subtrahend and then adding it to the minuend.

Hardware for signed 2's complement addition and subtraction

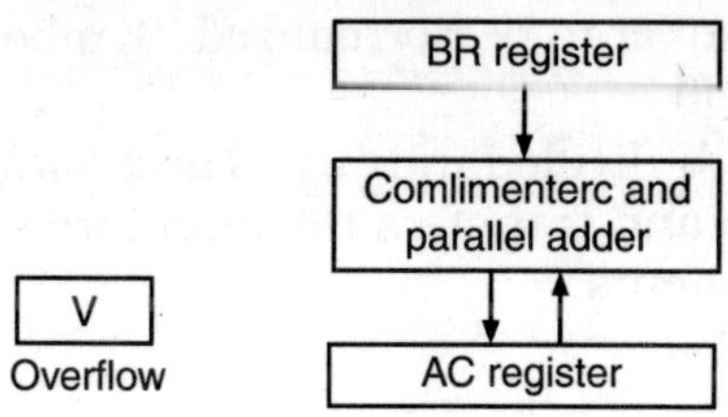

Fig. 5.18.

Q.24. Define over flow. How we detected over flow ?

Solution: When two numbers of n digits each are added and the sum occupies $n + 1$ digits, we say that an overflow occurred. The effect of an overflow on the sum of two signed 2's complement number.

An overflow can be detected by inspecting the last two carries out of the addition. When two carries are applied to an exclusive — OR gate, the over flow is detected when the output of the gate is equal to 1.

Q.25. Show adding and subtracting numbers in signed 2's complement representation.

Solution:

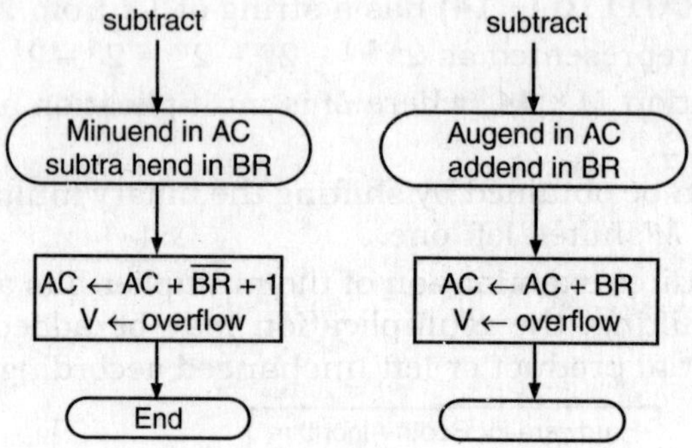

Fig. 5.19.

Q.26. Describe the register configuration for floating point operation.

Solution: The register configuration for floating point operations is quite similar to the layout for fixed point operations. The register organization for floating-point operation is shown below :

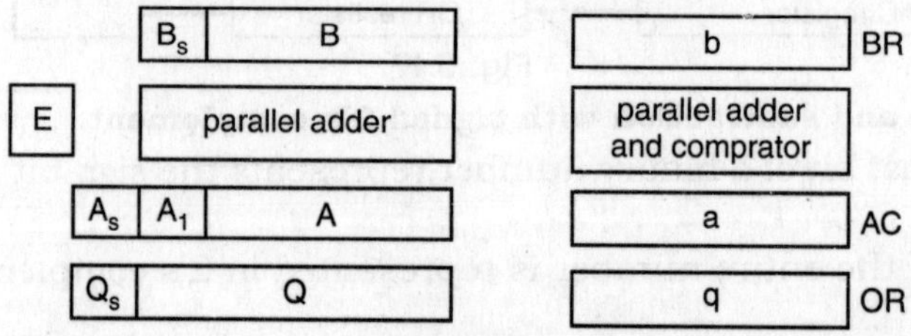

Fig. 5.20.

There are three registers *BR*, *AC* and *QR*. Each register is subdended into two parts. The mantissa part has the same upper case letter symbols as in fixed point representation. The exponent part uses the corresponding lower case letter symbols. It is assumed that each floating-point number has a mantissa in signed magnitude representation and a biased exponent. Thus the *AC* has a mantissa whose sign is in A_S and a magnitude that is in *A*. The exponent is in the part of the register denoted by the lower case letter symbols *a*. The diagram shows explicitly the most significant bit of *A*, labeled by A_1. The bit in this position must be a_1 for the number to be normalized. Symbol *AC* represent entire register, that is contestation of A_S, *A* and *a*.

Similarly, register *BR* is sub-divided into B_S, *B* and *b* and *QR* into *Q* and *q*. A parallel adder adds the two mantissas and transfers the sum into *A* and carry into *E*. A separate parallel adder is used for exponents.

Q.27. The complement shown in figure below is not needed if instead of performing $A + \overline{B} + 1$ we perform $B + \overline{A}$ (B plus the 1's complement of A). Derive an algorithm in flow chart form for addition and subtraction of fixed-point binary numbers in signed-magnitude representation with the magnitudes subtracted by the two micro-operations $A \leftarrow \overline{A}$ and $EA \leftarrow A + B$.

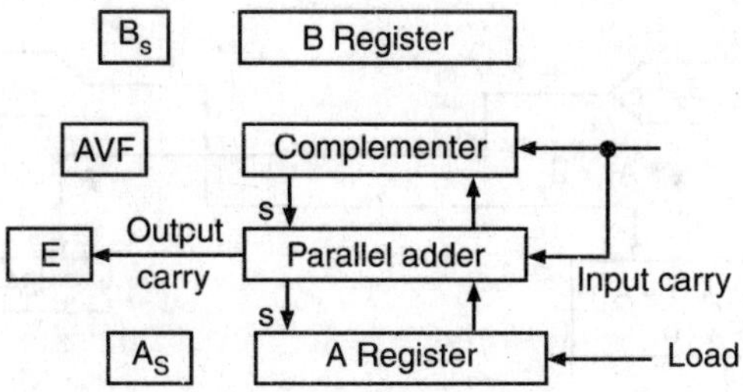

Fig. 5.21. Hardware for signed-magnitude addition and subtraction

Solution:

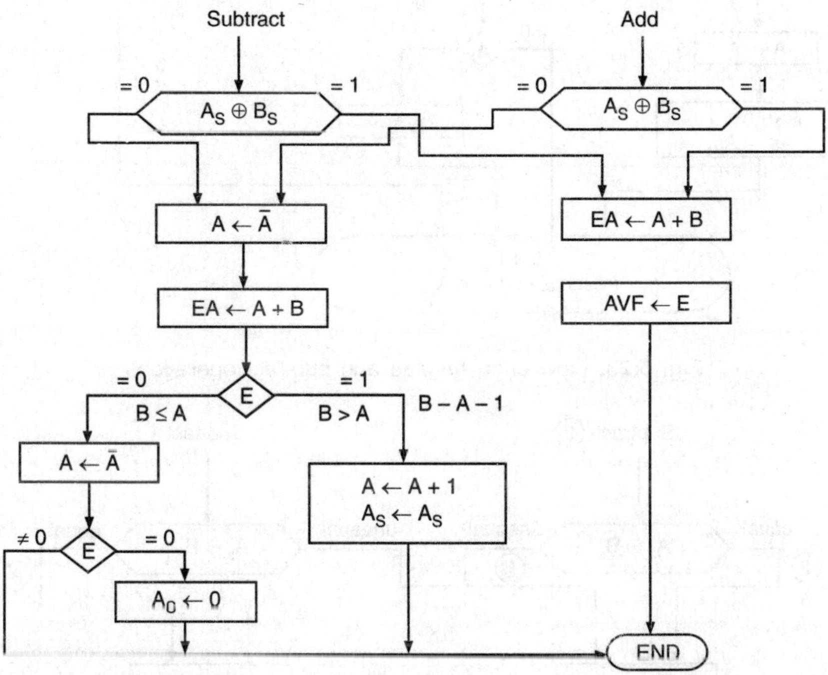

Fig. 5.22.

Q.28. Mark each individual path in the flow chart of figure given below by a number and then indicate the overall path the algorithm takes when the following signed magnitude numbers are computed. In each case give the value of AVF. The left most bit in the following numbers represents the sign bit.

(a) 0 101101 + 0 011111

(b) 1 011111 + 1 101101

(c) 0 101101 + 0 011111

(d) 0 101101 + 0 101101

(e) 1 011111 + 0 101101

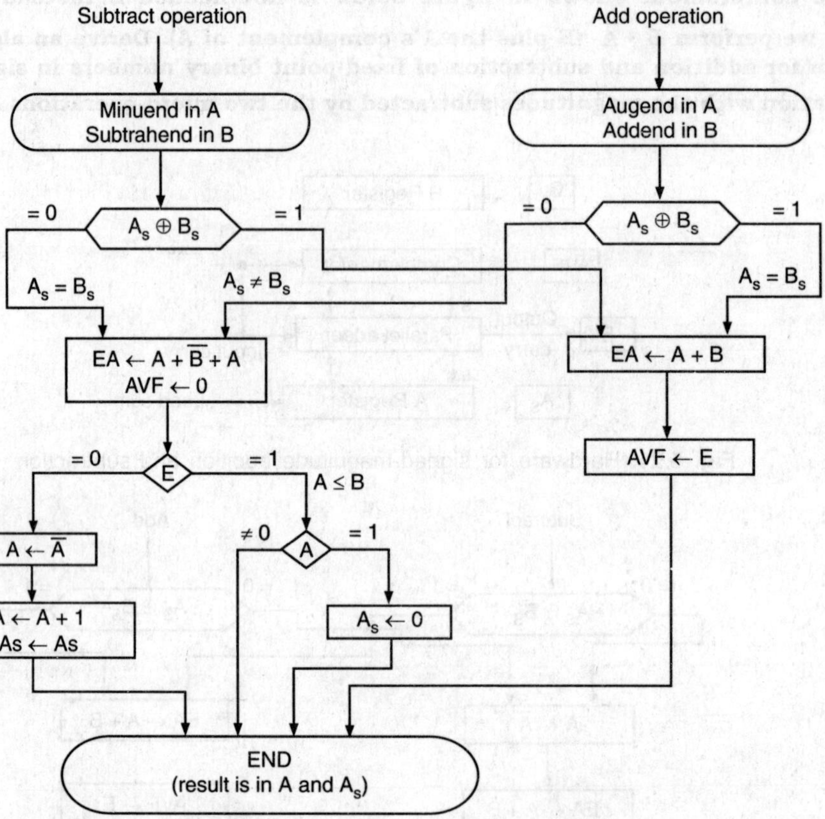

Fig. 5.23. Flow chart for add and subtract operations

Solution:

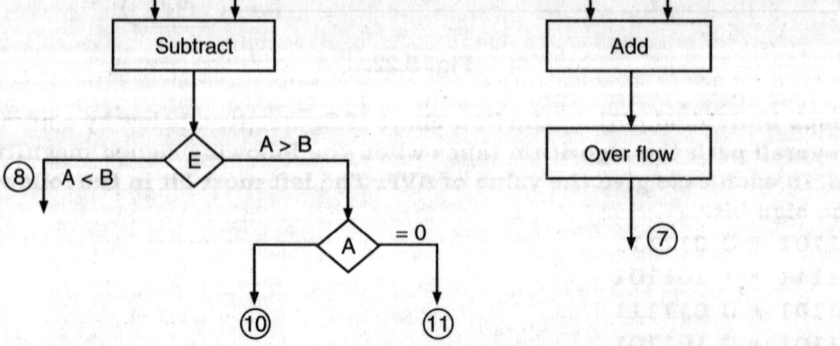

Fig. 5.24.

$2^6 - 1 = 63$, over flow if sum greater than |63|

(a) (+ 45) + (+ 31) = 76	Path (1) (3) (7)	AVF = 1
(b) (– 31) + (– 45) = – 76	Path (1) (3) (7)	AVF = 1
(c) (+ 45) – (+ 31) = 14	Path (2) (6) (9) (10)	AVF = 0
(d) (+ 45) – (+ 45) = 0	Path (2) (6) (9) (11)	AVF = 0
(e) (– 31) – (+ 45) = – 76	Path (2) (5) (7)	AVF = 1

Q.29. Perform the arithmetic operations below with binary numbers and with negative numbers in signed 2's complement representation. Use seven bits to accommodate each number together with its sign. In each case, determine if there is an overflow by checking the carries into and out of the sign bit position.

(a) (+ 35) + (+ 40)

(b) (– 35) + (– 40)

Solution:

(a) + 350 100011
+ 400 101000

+ 75 1 001011

Here $F = 0$ and $E = 1$

$F \oplus E = 1$; overflow

(b) + 35 1011101
– 40 1011000

– 70 0110101

Here $F = 1$ and $E = 0$

$F \oplus E = 1$; overflow

Q.30. Consider the binary numbers when they are in signed 2's complement representation. Each number has *n* bits : one for the and $k = n - 1$ for the magnitude. A negative number – *X* is represented as $2^k + (2^k - X)$, where the first 2^k designates the sign bit and $(2^k - X)$ is the 2's complement of *X*. A positive number is represented as $0 + X$, where the 0 designates the sign bit, and *X*, the *k*-bit magnitude. Using these generalized symbols, prove that the sum $(\pm X) + (\pm Y)$ can be formed by adding the numbers including their sign bits and discarding the carry-out of the sign bit position. In other words, prove the algorithm for adding two binary numbers in signed – 2's complement representation.

Solution:

Case	Operation in sign-magnitude	Operation in complement	Required result in sign 2's complement
1.	$(+X) + (+Y)$	$(0 + X) + (0 + Y)$	$0 + (X + Y)$
2.	$(+X) + (-Y)$	$(0 + X) + 2^k + (2^k - Y)$	$0 + (X - Y)$ if $X ³ Y$ $2^k + 2^k - (Y - X)$ if $X < Y$
3.	$(-X) + (+Y)$	$2^k + (2^k - X) + (0 + Y)$	$0 + (Y - X)$ if $Y ³ X$ $2^k + 2^k - (X - Y)$ if $Y < X$
4.	$(-X) + (-Y)$	$(2^k + 2^k - X) + (2^k + 2^k - Y)$	$2^k + 2^k - (X + Y)$

It is necessary to show that the operations in column (*b*) produce the results listed in.

Case 1. Column (*b*) = column (*c*)

Case 2. If $X ³ Y$ then $(X - Y) ³ 0$ and consists of *k* bits.

Operation in column (*b*) gives : $2^{2k} + (X - Y)$.

Discard carry $2^k = 2^n$ to get $0 + (X - Y)$ as in column (c)

If $X < Y$, then $(Y - X) > 0$,

Operation gives $2^k = 2^k - (Y - X)$ as in column (c)

Case 3. Is the same azs case 2 with X and Y reversed.

Case 4. Operation in column (b) gives : $2^{2k} + 2^k - 2^k - (X - Y)$.

Discard carry $2^{2k} = 2^n$ to obtain result of (c) :

$$2^k + (2^k - X - Y)$$

Q.31. Formulate a hardware procedure for detecting an overflow by comparing the sign of the sum with the signs of the augend and addend. The numbers are in signed – 2's complement representation.

Solution: Transfer augend sign into T_s

Then add : $AC \neg AC + BR$

A_s will have sign of sum. Truth Table for combine circuit.

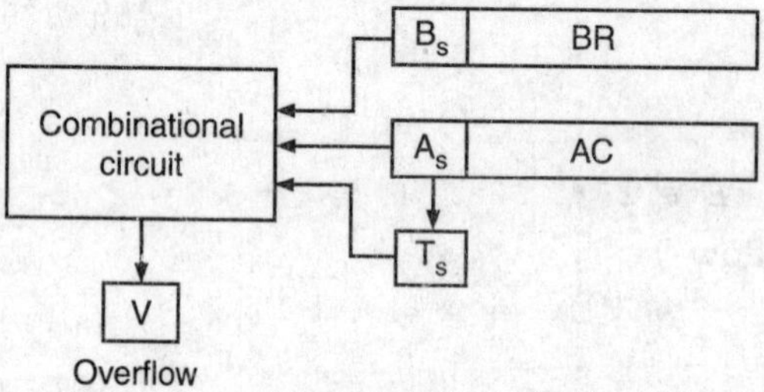

Fig. 5.25.

Boolean function for circuit :

$$V = T_s¢\ B_s¢\ A_s + T_s\ B_s\ A_s¢$$

Q.32. (a) Perform the operation (– 9) + (+ 6) = – 15 with binary numbers is signed – 1's complement representation using only five bits to represent each number (including the sign). Show that the overflow detection procedure of checking the inequality of the last two carries fails in this case.

(b) Suggest a modified procedure for detecting an overflow when signed – 1's complement numbers are used.

Solution: (a) – 910110 Add end around carry F as needed in signed – 1's complement addition

– 611001

– 1501111

01111

+ 1 = – 15

10000

F = 1 E = 0 ¬ carries

$E\oplus$ F = 1 but there should be no overflow since result is – 15.

(b) The procedure $V \leftarrow E\oplus F$ is valid for 1's complement numbers provided we check the result.

0 1111...11

As A When $V = 1$

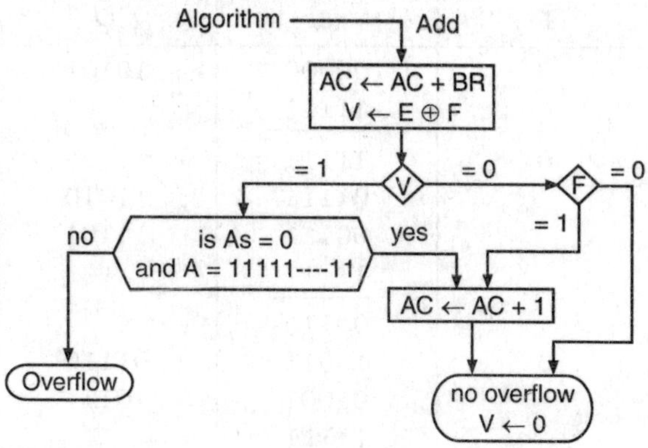

Fig. 5.26.

Q.33. Prove that the multiplication of two n-digit numbers in base r gives a product no more than 2n digits in length. Show that this statement implies that no overflow can occur in the multiplication operation.

Solution: Maximum value of numbers = $r^n - 1$

It is necessary to show that maximum product is less than or equal $r^{2n} - 1$.

$$\text{Maximum product} = (r^n - 1)(r^n - 1) = r^{2n} - 2r^n + 1$$
$$\leq r^{2n} - 1$$

which gives, $2 \leq 2r^n$

or $1 \leq r^n$

This is always true since $r \geq 2$ and $n \geq 1$.

Q.34. Show the contents of registers *E*, *A*, *Q* and *SC* (as in Table given below) during the process of multiplication of two binary numbers, 11111 (multiplication) and 10101 (multiplier). The signs are not included.

Table 5.2. Numerical Example for Binary Multiplier

Multiplicand B = 10111	*E*	*A*	*Q*	*SC*
Multiplier in Q	**0**	**0000**	**1011**	**101**
Q_n = 1; add *B*	**10111**			
First partial product	**0**	**10111**		
Shift right *EAQ*	**0**	**01011**	**11001**	**100**
Q_n − 1; add *B*		**10111**		
Second partial product	**1**	**00010**		
Shift right *EAQ*	**0**	**10001**	**01100**	**011**
Q_n = 0; shift right *EAQ*	**0**	**01000**	**10110**	**010**
Q_n = 0; shift right *EAQ*		**00100**	**01011**	**001**
Q_n = 1; add *B*		**10111**		
Fifth partial product	**0**	**11011**		
Shift right *EAQ*	**0**	**01101**	**10101**	**000**

Solution: Multiplication $B = 11111 = (31)_{10}$ $\quad 31 \times 21 = 651$

Multiplier in Q	*E*	*A*	*Q*	*SC* $Q = (21)_{10}$
	0	00000	10101	101
Q_n = 1, add B		11111		
	0	11111		
Shr EAQ		01111	11010	100
Q_n = 0, shr EAQ		00111	11101	011
Q_n = 1, add B		11111		
	1	00110		
Shr EAQ	0	10011	01110	010
Q_n = 0 shr EAQ		01001	10111	001
Q_n = 1, add B		11111		
	1	01000		
Shr EAQ		10100	01011	000
		(651)	10	

Q.35. Show the contents of registers *E, A, Q* and *SC* (as in figure given below) during the process of division of

(*a*) 10100011 by 1011;

(*b*) 00001111 by 0011. (Use of dividend of eight bits)

Divisor B = 10001, $\overline{B}$ + 1 = 01111

	E	***A***	***Q***	***SC***
Dividend :		**01110**	**00000**	**5**
shl *EAQ*	**0**	**11100**	**0000**	
add $\overline{B}$ + 1		**01111**		
***E* = 1**	**1**	**01011**		
Set Q_n = 1	**0**	**01011**	**00001**	**4**
shl *EAQ*		**10110**	**00010**	
Add $\overline{B}$ + 1		**01111**		
***E* = 1**	**1**	**00101**		
Set Q_n = 1	**1**	**00101**	**00011**	**3**
shl *EAQ*	**0**	**01010**	**00110**	
Add $\overline{B}$ + 1		**01111**		
***E* = 0; leave Q_n = 0**	**0**	**11001**	**00110**	
Add *B*		**10001**		
Restore remainder		**01010**		**2**
shl *EAQ*	**0**	**10100**	**01100**	
Add *B* + 1		**01111**		
***E* = 1**	**1**	**00011**		
Set Q_n = 1	**1**	**00011**	**01101**	**1**
shl *EAQ*	**0**	**00110**	**11010**	
Add $\overline{B}$ + 1		**01111**		

$E = 0$; leave $Q_n = 0$	**0**	**10101**	**11010**	
Add B		**10001**		
Restore remainder	**1**	**00110**	**11010**	**0**
Neglect E				
Remainder in A :		**00110**		
Quotient in Q :			**11010**	

Solution: (*a*) $\frac{10100011}{1011} = 1110 + \frac{1001}{1011} \quad \frac{163}{11} = 14 + \frac{9}{11}$

$$B = 1011 \quad \overline{B} + 1 = 0101 \quad DVF = 0$$

	E	*A*	*Q*	*SC*
Dividend in AQ	0	1010	0011	100
shl EAQ	1	0100	0110	
add $\overline{B}$ + 1, suppress carry		0101		
$E = 1$, set Q_n to 1	1	1001	0111	011
shl EAQ	1	0010	1110	
add $\overline{B}$ + 1, suppress carry		0101		
$E = 1$, set Q_n to 1	1	0111	1111	010
shl EAQ	0	1111	1110	
add $\overline{B}$ + 1, carry to E		0101		
$E = 1$, set Q_n to 1	1	0100	1111	001
shl EAQ	0	1001	1110	
add $\overline{B}$ + 1, carry to E		0101		
$E = 0$, leave $Q_n = 0$	0	1110	1110	
add B		1011		
restore remainder	1	1001	1110	000
	remainder		quotient	

(*b*) $\frac{1111}{0011} = 0101 \quad B = 0011 \quad \overline{B} + 1 = 1101$

	E	*A*	*Q*	*SC*
Dividend in Q, $A = 0$		0000	1111	100
shl EAQ	0	0001	1110	
add $\overline{B}$ + 1, suppress carry		1101	1110	
$E = 0$, leave $Q_n = 0$	0	1110		
add B		0011		
restore partial remainder	1	0001		011
shl EAQ	0	0011	1100	
add $\overline{B}$ + 1		1101		

$E = 1$, set Q_n to 1	1	0000	1101	010
shl EAQ	0	0001	1010	
add $\overline{B}+1$		<u>1101</u>		
$E = 1$, set Q_n to 1	1	0000	1101	010
shl EAQ	0	0001	1010	
add $\overline{B}+1$		<u>1101</u>		
$E = 0$, leave $Q_n = 0$	0	1110	1010	
add B		<u>0011</u>		
restore partial remainder	1	0001		001
shl EAQ	0	0011	0100	
and $\overline{B}+1$		<u>1101</u>		
$E = 1$, set Q_n to 1	1	0000	0101	000
	remainder		quotient	

Q.36. Show that adding *B* after the operation *A* + *B* + 1 restores the original value of *A*. What should be done with the end carry ?

Solution: $A + B + 1$ performs : $A + 2^n - B = 2^n + A - B$

adding B : $(2^k + A - B) + B = 2^n + A$

remove end – carry 2^n to obtain A.

Q.37. Why should the sign of the remainder after a division be the same as the sign of the dividend?

Solution: To correspond with correct result. In general $\frac{A}{B} = Q + \frac{R}{B}$

where A is dividend, Q the quotient and R the remainder. Four possible signs for A and B:

$$\frac{+52}{+5} = +10 + \frac{+2}{+5} = +10.4 \qquad \frac{-52}{+5} = -10 + \frac{-2}{+5} = -10.4$$

$$\frac{+52}{-5} = -10 + \frac{+2}{-5} = -10.4 \qquad \frac{-52}{-5} = +10 + \frac{-2}{-5} = +10.4$$

The sign of the remainder (2) must be same as sign of dividend (52).

Q.38. Design an array multiplier that multiplies two 4-bit numbers. Use AND gates and binary address.

Solution: Add one more stage of figure given below with 4 AND gates and a 4-bit adder.

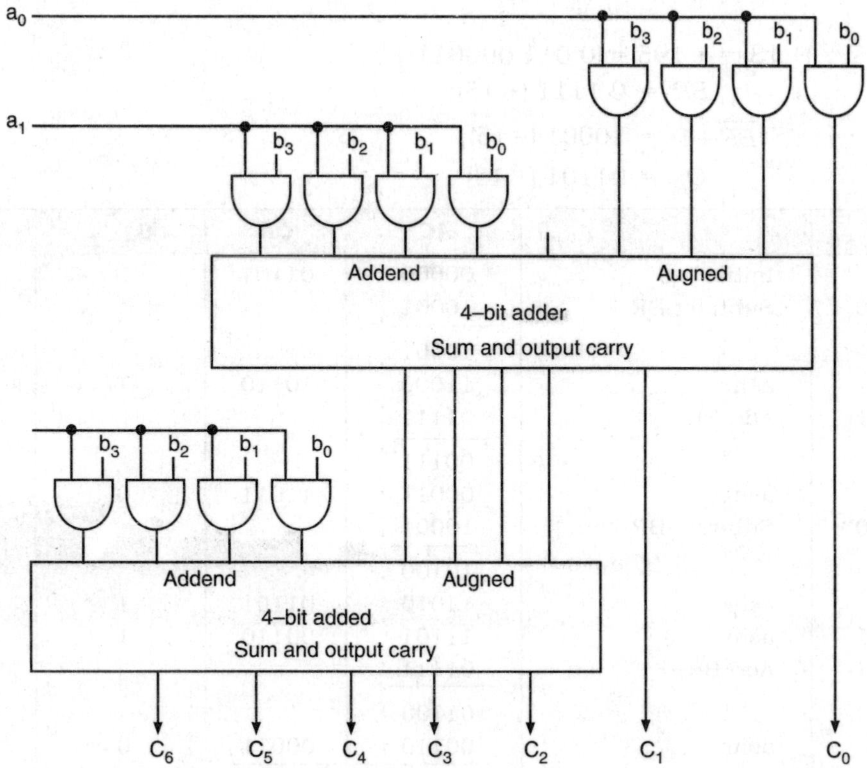

Fig. 5.27. 4-bit by 3 bit array multiplier

Q.39. Show the step by step multiplication process using Booth algorithm (as in Table given below) when the following binary numbers are multiplied. Assume 5-bit registers that hold signed numbers. The multiplicand in both cases is + 15.

(*a*) (+15) × (+ 13)

(*b*) (+15) × (– 13)

Table 5.3. Example of Multiplication with Booth Algorithm

Q_n	Q_{n+1}	BR = 10111 $\overline{BR}$ + 1 = 01001	AC	QR	Q_{n+1}	SC
		Initial	00000	10011	0	101
1	0	Subtract BR	01001			
			01001			
		ashr	00100	11001	1	100
1	1	ashr	00010	01100	1	011
0	1	Add BR	10111			
			11001			
		ashr	11100	10110	0	010
0	0	ashr	11110	01011	0	001
1	0	Subtract BR	01001			
			00111			
		ashr	00011	10101	1	000

Solution:

(*a*) (+ 15) × (+ 13) = + 195 = $(0\ 011\ 000011)_2$

BR = 0 1111 (+ 15);

$\overline{BR}+1$ = 10001 (– 15);

Q_R = 01101 (+ 13)

Q_n	Q_{n+1}		AC	QR	Q_{n+1}	SC
		Initial	00000	01101	0	101
1	0	Subtract BR	10001			
			10001			
		ashr	11000	10110	1	100
0	1	Add BR	01111			
			00111			
		ashr	00011	11011	0	011
1	0	Subtract BR	10001			
			10100			
		ashr	11010	01101	1	010
1	1	ashr	11101	00110	1	001
0	1	Add BR	01111			
			01100			
		ashr	00110	00011	0	000
			+ 195			

(*b*) (+ 15) × (– 13) = – 195 = (11 00111101)2¢ Complement

BR = 011111 (+ 15);

($\overline{BR}+1$ = 10001 (–15);)

QR = 10011 (– 13)

Q_n	Q_{n+1}		AC	QR	Q_{n+1}	SC
		Initial	00000	10011	0	101
1	0	Subtract BR	10001			
			10001			
		ashr	11000	11001	1	100
1	1	ashr	11100	01100	1	011
0	1	Add BR	01111			
			01011			
		ashr	00101	10110	0	010
0	0	ashr	00010	11011	0	001
1	0	Subtract BR	10001			
			10011			
		ashr	110001	11101	1	000
			–195			

Q.40. Drive an algorithm in flowchar form the nonrestoring method of fixed-point binary division:

Solution:

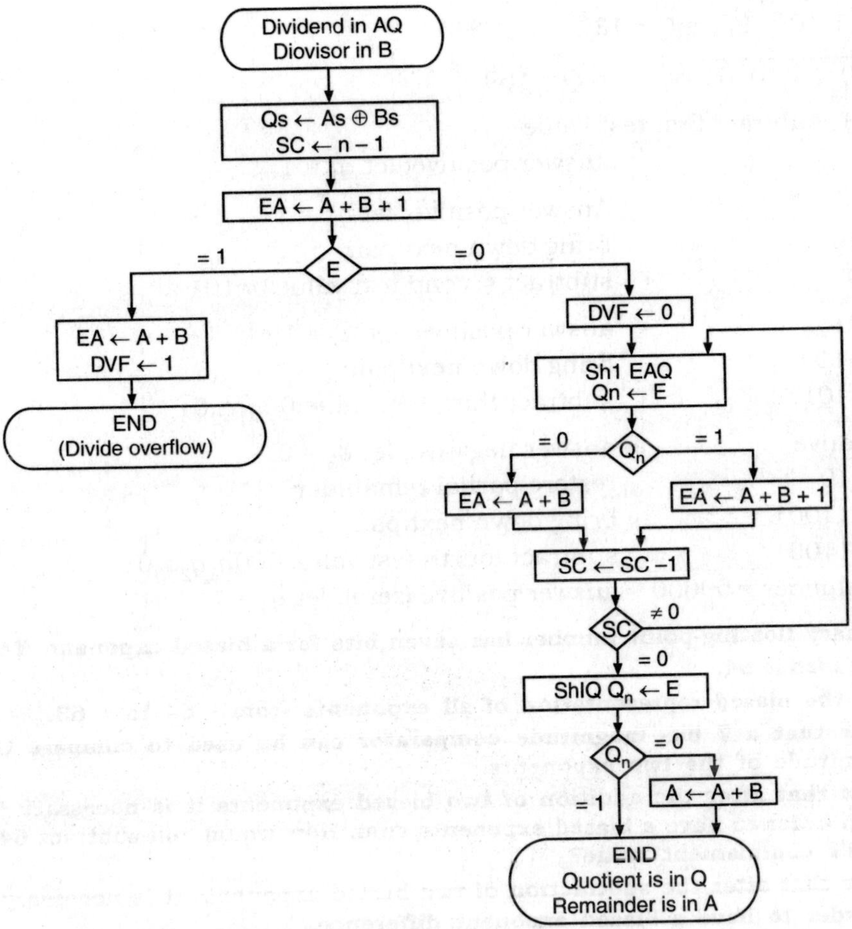

Fig. 5.28.

Q.41. Derive an algorithm for evaluating the square root of a binary fixed-point number.

Solution: The algorithm for square-root is similar to division with the radicant being equivalent to the divident and a "test value" being equivalent to the divisor.

Let *A* be the radicant, *Q* the square-root; and *R* the remainder such that $Q^2 + R = A$ or $\sqrt{A} = Q$ and a remainder.

General comments :

(*i*) For *k* bits in *A* (*k* even), *Q* will have $\frac{k}{2}$ bits : $Q = q_1\ q_2\ q_3 \cdots \frac{qk}{2}$

(*ii*) The first test value if 01

The second test value is $0_{q1}01$

The third test value is $00_{q1\ q2}01$

The fourth test value is $000_{q1q2q3}01$ etc.

(*iii*) Mark the bits of A in groups of two starting from left.

(*iv*) The procedure is similar to the division restoring method as shown in the following example :

$q_1\ q_2\ q_3\ q_4$

1 1 0 1 = Q = 13

$\sqrt{10\ 10\ 10\ 01}$ = Q = 169

01 subtract first test value	
01	answer positive; let q_1 = 1
01	Answer positive; let $q1$ = 1
0110	bring down next pair
0101	subtract second test value $0q101$
0001	answer positive; let q_2 = 1
000110	bring down next pair
001101	subtract third test value $00q_1q_201$
negative	answer negative; let q_3 = 0
00110	restore partial remainder
00011001	bring down next pair
00011001	subtract fourth test value $000q_1q_2q_301$
Remainder = 00000	answer positive (zero); let q_4 = 1

Q.42. A binary floating-point number has seven bits for a biased exponent. The constant used for the bias is 64.

(*a*) List the biased representation of all exponents from – 64 to + 63.

(*b*) Show that a 7 bit- magnitude comparator can be used to compare the relative magnitude of the two exponents.

(*c*) Show that after the addition of two biased exponents it is necessary to subtract 64 in order to have a biased exponents sum. How would you subtract 64 by adding its 2's complement value?

(*d*) Show that after the subtraction of two biased exponents it is necessary to add 64 in order to have a biased exponent difference.

Solution: (*a*)

e* = exponent *e	***e* + 64**	***e* + 64 = biased exponent biased exponent**
– 64	– 64 + 64 = 0	0 000 000
– 63	– 63 + 64 = 1	0 000 001
– 62	– 62 + 64 = 2	0 000 010
– 1	– 1 + 64 = 63	0 111 111
– 0	0 + 64 = 64	1 000 000
+ 1	1 + 64 = 65	1 000 001
+ 62	62 + 64 = 126	1 111 110
+ 63	63 + 64 = 127	1 111 111

(*b*) The biased exponent follows the same algorithm as a magnitude comparator

(*c*) $(e_1 + 64) + (e_2 + 64) = (e_1 + e_2 + 64) + 64$

subtract 64 to obtain biased exponent sum.

(*d*) $(e_1 + 64) - (e_2 - 64) = e_1 + e_2$

Add 64 to obtain biased exponend difference.

Q.43. Device an algorithm in flowchart from the comparison of two signed binary numbers when negative numbers are in signed – 2's complement representation:

(*a*) By means of a subtraction operation with the signed – 2's complement numbers.

(*b*) By scanning and comparing pairs of bits from left to right.

Solution: (*a*)
$$AC = As\ A_1\ A_2\ A_3 \ldots A_n$$
$$BS = B_s\ B_1\ B_2\ B_3 \ldots B_n$$

If signs are unlike – The one with a 0 (plus) is larger.

If signs are alike – both numbers are either positive or negative.

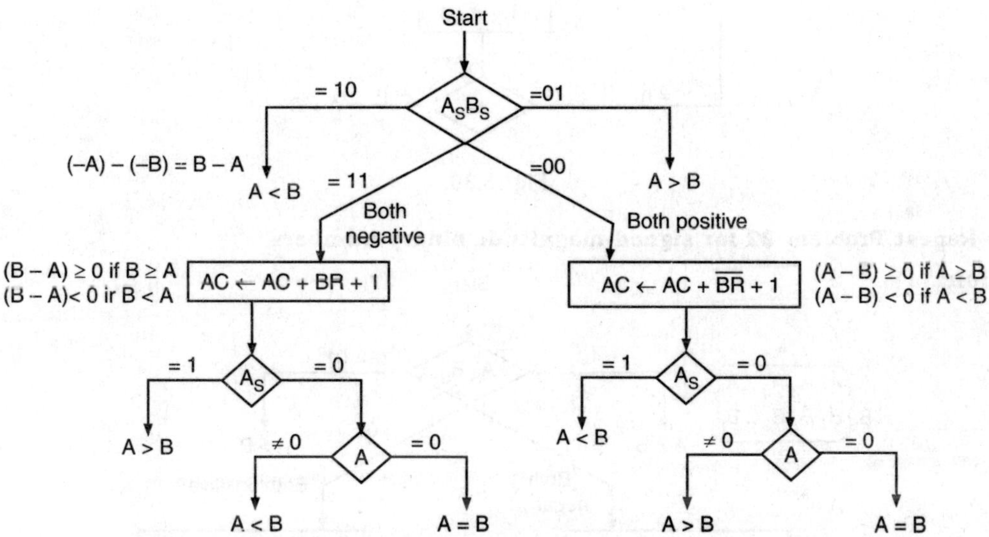

Fig. 5.29.

(*b*)

	A_s	A_1	$A_2 \ldots A_n$
+ 2	0	0	0 0 0 1 0
+ 1	0	0	0 0 0 0 1
0	0	0	0 0 0 0 0
– 1	1	1	1 1 1 1 1
– 2	1	1	1 1 1 1 0
– 3	1	1	1 1 1 0 1

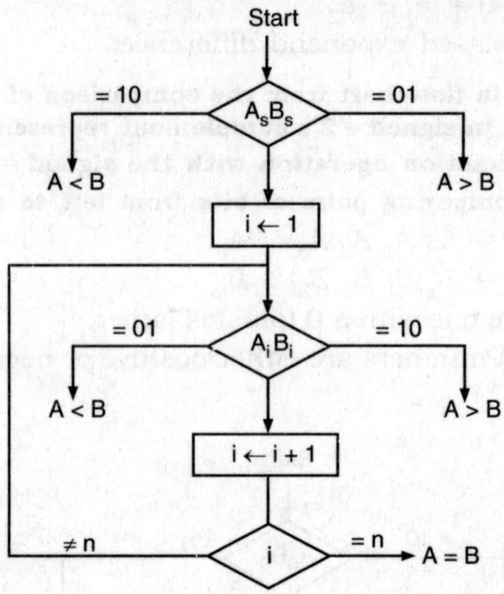

Fig. 5.30.

Q.44. Repeat Problem 22 for signed-magnitude binary numbers.

Solution:

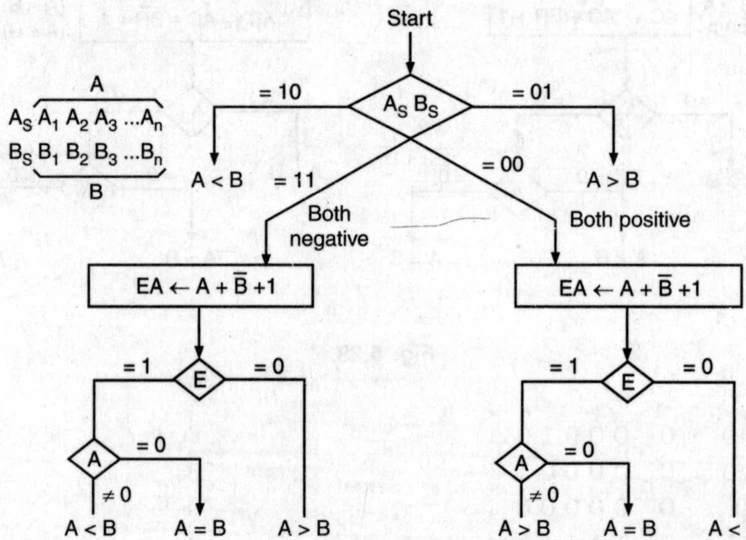

Fig. 5.31.

Q.45. Let *n* be the number of bits of the mantissa in binary floating point number. When the mantissas are aligned during the addition or subtraction the exponent difference may be greater than *n* – 1. If this occurs, the mantissa with the smaller exponent is shifted entirely out of the registor. Modify the mantissa alignment in figure given below by including a sequence counter. SC that count the number of shifts. If the number of shifts is greater than *n* – 1,the large number is then used to determine the results.

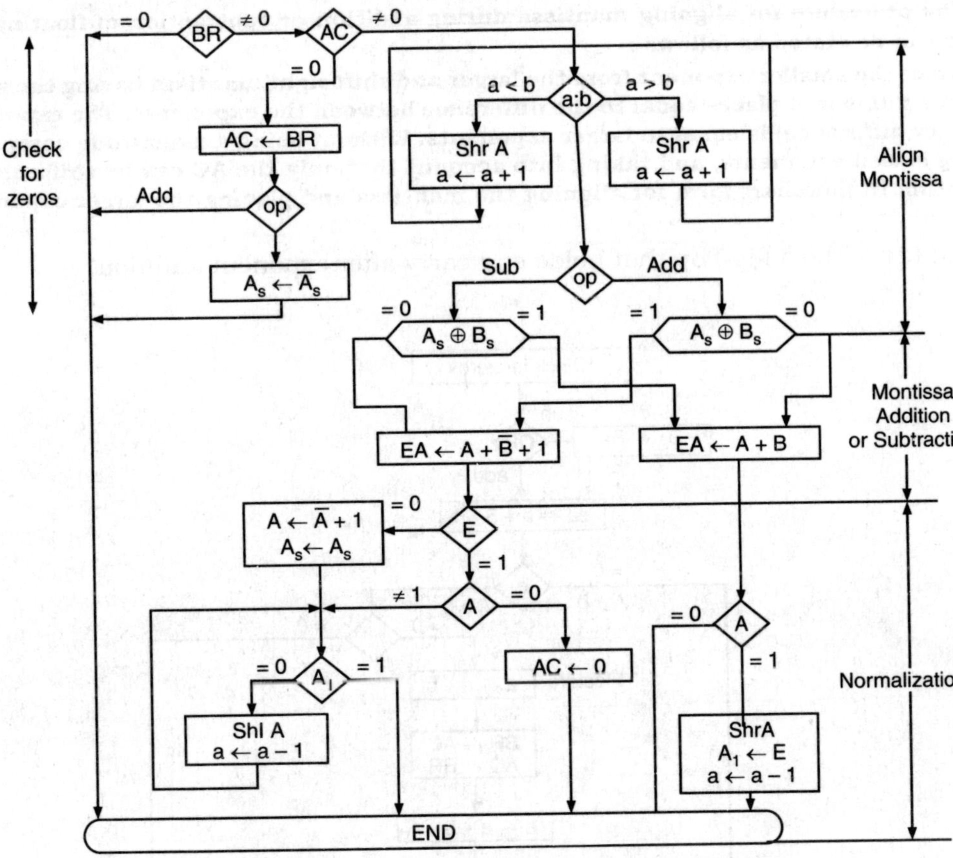

Fig. 5.32. Addition and subtraction of floating-point numbers

Solution: Figure below shows mantissa alignment

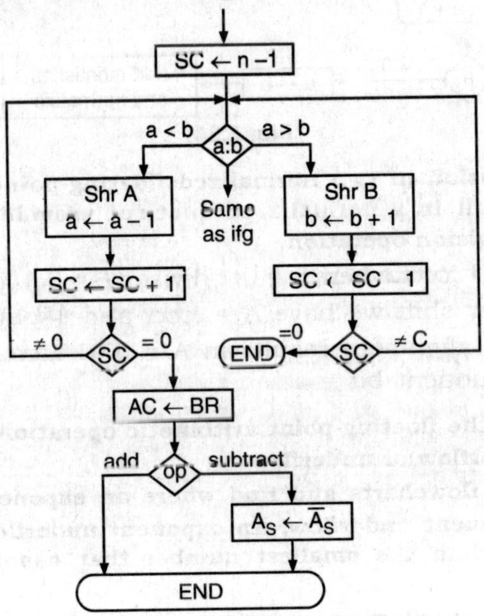

Fig. 5.33.

Q.46. The procedure for aligning mantissa during addition or subtraction of floating point numbers can be stated as follows :

Subtract the smaller exponent from the larger and shift right mantissa having the smaller exponent a number of places equal to the difference between the exponents. The exponent of the sum (or difference) is equal to larger exponents. Without using a magnitude comparator, assuming biased exponents, and taking into account that only the AC can be shifted, derive an algorithm in flowchart form for aligning the mantissa and placing the larger exponent in the AC.

Solution: Let "*e*" be a flip-flop that holds end carry after exponent addition.

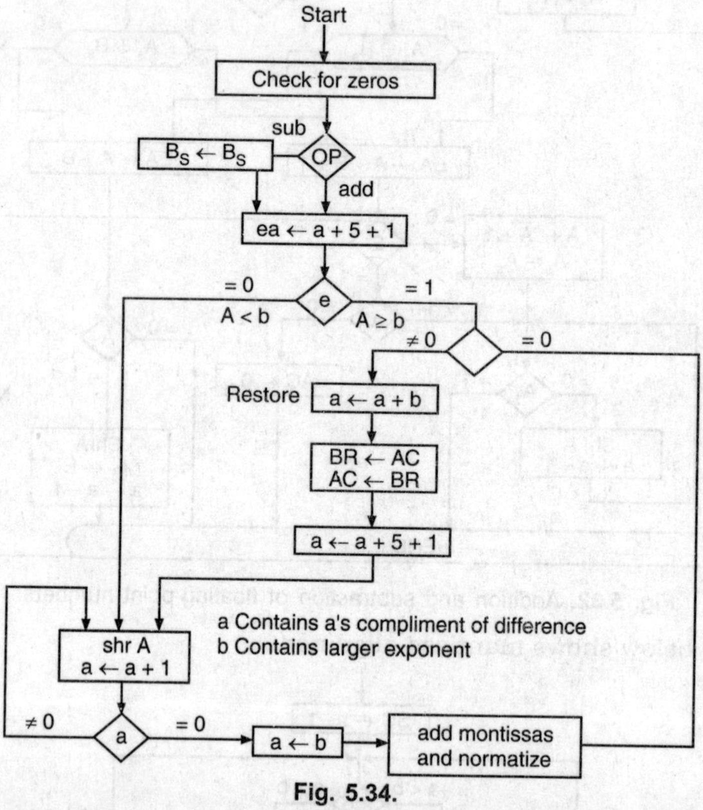

Fig. 5.34.

Q.47. Show that the division of two normalized floating-point numbers with fractional mantissas will always result in a normaliazed quotient provided a dividend alignment is carried out prior to the division operation.

Solution: Dividend A = 0.1 *xxxx* where x = 0.1 divisor *B* = 0.1 *xxxx*

(*a*) If $A < B$ then after shift we have A = *xxxx* and 1st quotient bit is 1.

(*b*) If $A \geq B$, dividend alignment results in A = 0.01 *xxxx*. Then after the left shift $A \geq B$ and first quotient bit = 1.

Q.48. The algorithms for the floating-point arithmetic operations in Sec. 10-5 neglect the possibility of exponent overflow or underflow.

(*a*) Go over the three flowcharts and find where an exponent overflow may occur.

(*b*) Repeat (*a*) for exponent underflow. An exponent underflow occurs if the exponent is more negative than the smallest number that can be accommodated in the register.

(c) Show an exponent overflow or underflow can be detected by the hardware.

Solution:

(*a*) When the exponents are added or incremented.

(*b*) When the exponents are subtracted or decremented.

(*c*) Check end-carry after addition and carry after increment or decrement.

Q.49. If we assume integer representation for the mantissa of floating-point numbers, we encounter certain scaling problems during multiplication and division. Let the number of bits in the magnitude part of the mantissa be (*n* – 1). For integer representation.

(*a*) Show that if a single-precision product is used, (*n* – 1) must be added to the exponent product in the *AC*.

(*b*) Show that if a single-precision mantissa dividend is used, (*n* – 1) must be subtracted from the exponent dividend when Q is cleared.

Solution: Assume integer mantissa of $n - 1 = 5$ bits (excluding sign)

(*a*) Product : *A* *Q*

x x x x x *x x x x x.* $*2^z$

binary-point for integer

Product in AC : *x x x x x* . $* 2^{Z+5}$

(*b*) Single precision normalized divident : *x x x x x.* $* 2^z$

Divident in AQ : *A* *Q*

x x x x x 00000.$* 2^{z-5}$

Q.50. Show the hardware to be used for the addition and subtraction of two decimal numbers in signed-magnitude representation. Indicate how an overflow is detected.

Solution: Neglect B_e and A_e from

Apply carry directly to *E*.

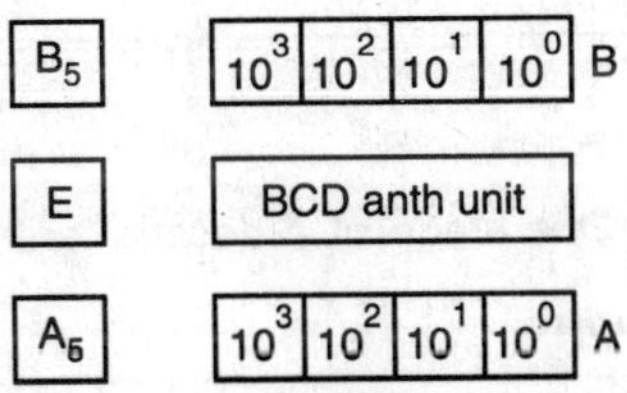

Fig. 5.35.

Q.51. Show that 673 – 356 can be computed by adding 673 to the 10's complement of 356 and discarding the end carry. Draw the block diagram of three-stage decimal arithmetic unit and show how this operation is implemented. List all input bits and output of the unit.

Solution: 673 673

–356 +

317 10 s complement of 356 = 644

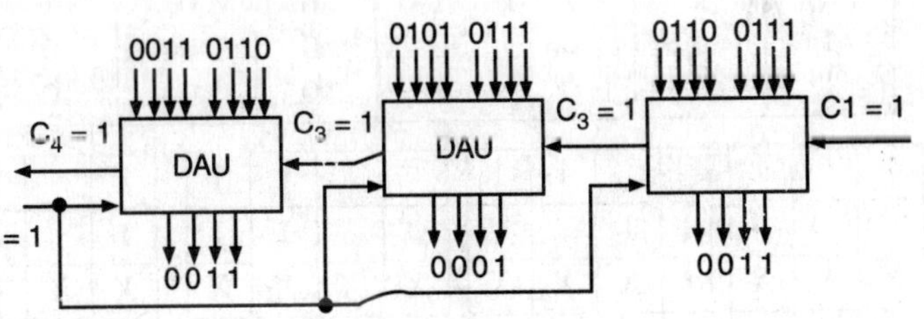

Fig. 5.36.

Q.52. Show that the lower 4 bit binary adder in figure given in the figure below can be replaced by one full-adder and two half-adders.

Solution:

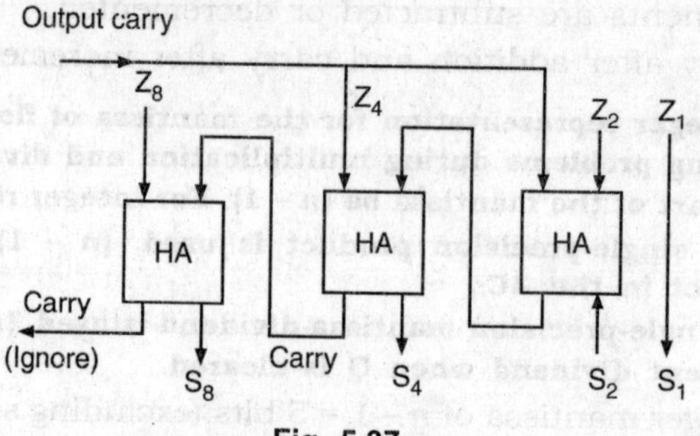

Fig. 5.37.

Q.53. Using combinational circuit design techniques, derivet he Boolean functions for the BCD 9's complementer of figure given below. Drasw the logic diagram.

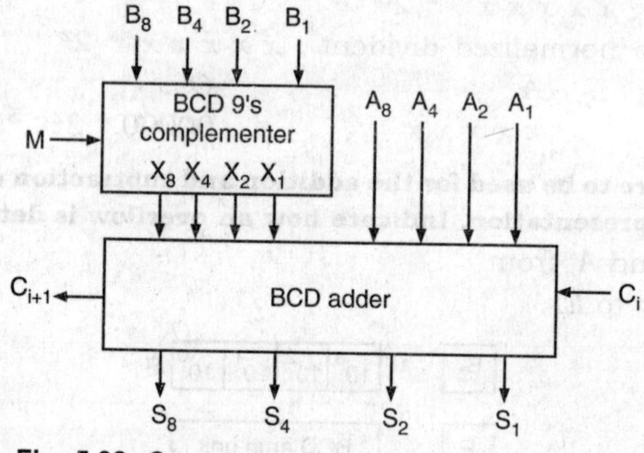

Fig. 5.38. One stage of a decimal arithmetic unit

Solution:

	Input				Output				
	B_8	B_4	B_2	B_1	X_8	X_4	X_2	X_1	
0	0	0	0	0	1	0	0	1	9
1	0	0	0	1	1	0	0	0	8
2	0	0	1	0	0	1	1	1	7
3	0	0	1	1	0	1	1	0	6
4	0	1	0	0	0	1	0	1	5
5	0	1	0	1	0	1	0	0	4
6	0	1	1	0	0	0	1	1	3
7	0	1	1	1	0	0	1	0	2
8	1	0	0	0	0	0	0	1	1
9	1	0	0	1	0	0	0	0	0

1	1		
X	X	X	X
		X	X

		1	1
1	1		
1	X	X	X
		X	X

		1	1
		1	1
X	X	X	X
		X	X

1			1
1			1
X	X	X	X
1		X	X

$d(B_8\, B_2\, B_1) = \Sigma\,(10, 11, 12, 13, 14, 15)$ are don't-care conditions

$$X_8 = B'_8\, B'_4\, B'_2$$
$$X_4 = B_4\, B'_2 + B'_4\, B'_2$$
$$X_2 = B_2$$
$$X_1 = B'_2$$

Q.54. It is necessary to design an adder for two decimal digits represented in the excess –3 code. Show that the correction after adding two digits with a 4-bit binary adder is as follows:

(*a*) The output carry is equal to the uncorrected carry.

(*b*) If output carry = 1, add 0011.

(*c*) If out put carry = 0, add 1101 and ignore the carry from this addition. Show that the excess – 3 adder can be constructed with seven full-adders and two inverters.

Solution:

	Z	*Y*			*Z*		*Y*
dec	***uncorrected***	***corrected***	***dec***		***uncorrected***		***corrected***
0	0110	0011	10	1	0000	1	0011
1	0111	0100	11	1	0001	1	0100
2	1000	0101	12	1	0010	1	0101
3	1001	0110	13	1	0011	1	0110
4	1010	0110	14	1	0100	1	0111
5	1011	1000	15	1	0101	1	1000
6	1100	1001	16	1	0110	1	1000
7	1101	1010	17	1	0111	1	1010
8	1110	1011	18	1	1001	1	1011
9	1111	1100	19	1	1001	1	1100

No output carry

$Y = Z - 3 = Z + 13 - 16$

ignore carry

Uncorrected carry = output carry

$Y - Z + 3$

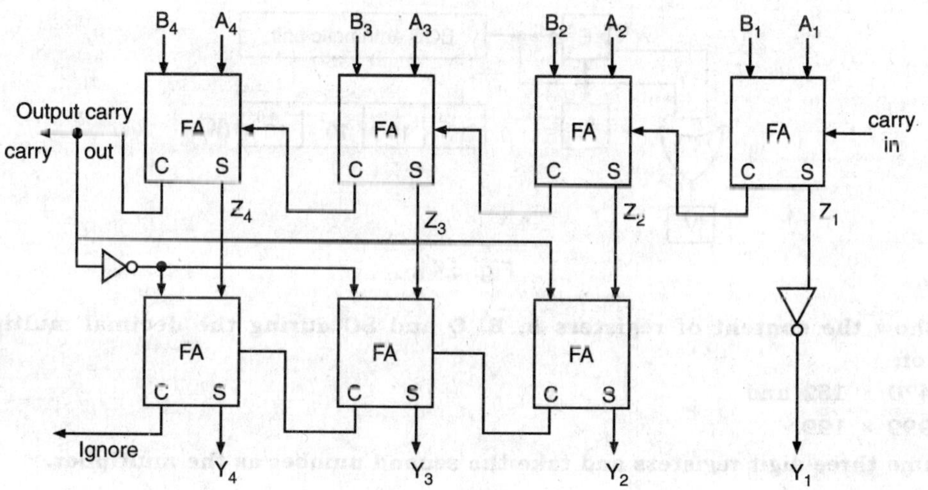

Fig. 5.39.

Q.55. Derive the circuit of a 9's complement when decimal digits are represented in the excess-3 code. A mode control input determines whether the digit is complemented or not. What is the advantage of using this code over BCD?

Solution: The excess-3 code is self complementing code. Therefore, to get 9's complement we need to complement each bit.

$M = 0$ for $x = B$

$M = 0$ for $x = 9$'s complement of B

M	B_i	$X_i = B_i \oplus M$
0	0	0 $X_i = B_i$
0	1	1
1	0	1 $X¢_i = B¢_i$
1	1	0

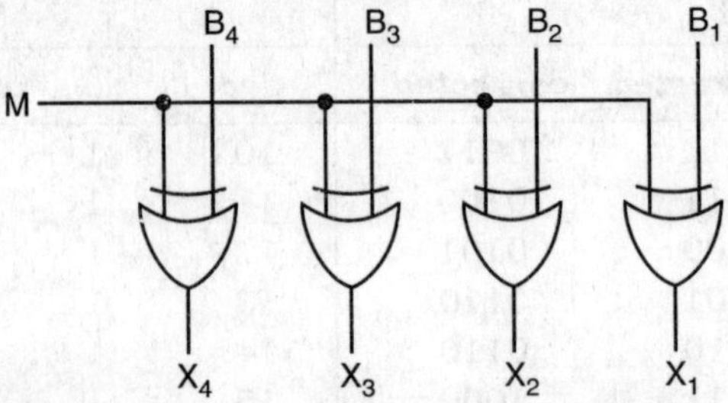

Fig. 5.40.

Q.56. Show the hardware to be used for the addition and subtraction of two decimal numbers with negative numbers insigned – 10's complement representation. Indicate how an overflow is detected. Derive the flowchart algorithm and try a few numbers to convince yourself that the algorithm produces correct results.

Solution:

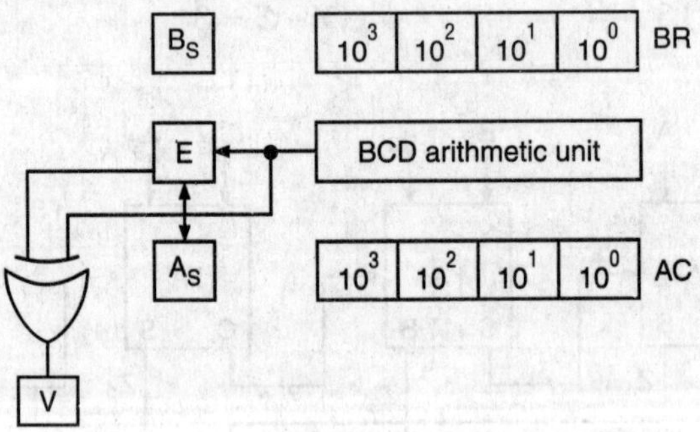

Fig. 5.41.

Q.57. Show the content of registers A, B, Q and SC during the decimal multiplication (Figure) of:

(*a*) 470 × 152 and

(*b*) 999 × 199.

Assume three-digit registers and take the second number as the multiplier.

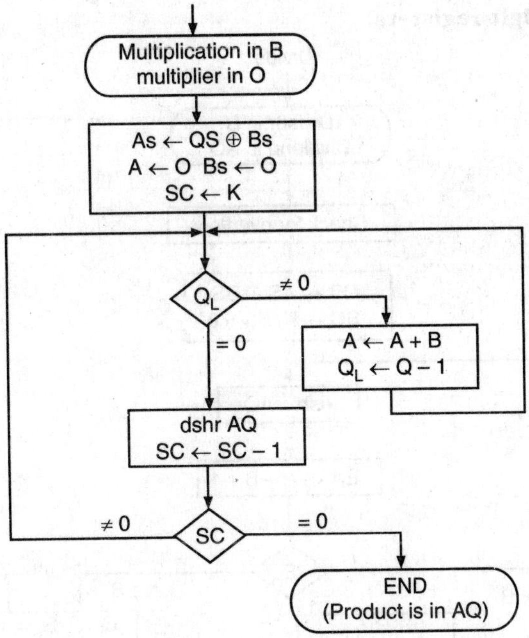

Fig. 5.42. Flowchart for decimal multiplication

Solution: (*a*) $B = 740$

	Ae	*A*	$Q = Q_1$	*SC*
Initial	0	000	152	3
Q_L ¹ 0	0	470	151	
Q_L ¹ 0	0	940	150	
$Q_L = 0$, dshr	0	094	015	2
	0	564	014	
	1	034	013	
Q_L ¹ 0	1	504	012	
	1	974	011	
	2	444	010	
Q_L ¹ 0, dshr	0	244	401	1
Q_L ¹ 0	0	714	400	
$Q_L = 0$, dshr	0	071	440	0
		Product		

(*b*)

999
× 199

8991 first partial product $A_a = 8$
+ 89910

98901 second partial product $A_e = 9$
+ 99900

198801 final product $A_e = 1$

Q.58. Show the content of registers A, E, Q and SC during the decimal division (figure) of 1680/32. Assume two digit registers.

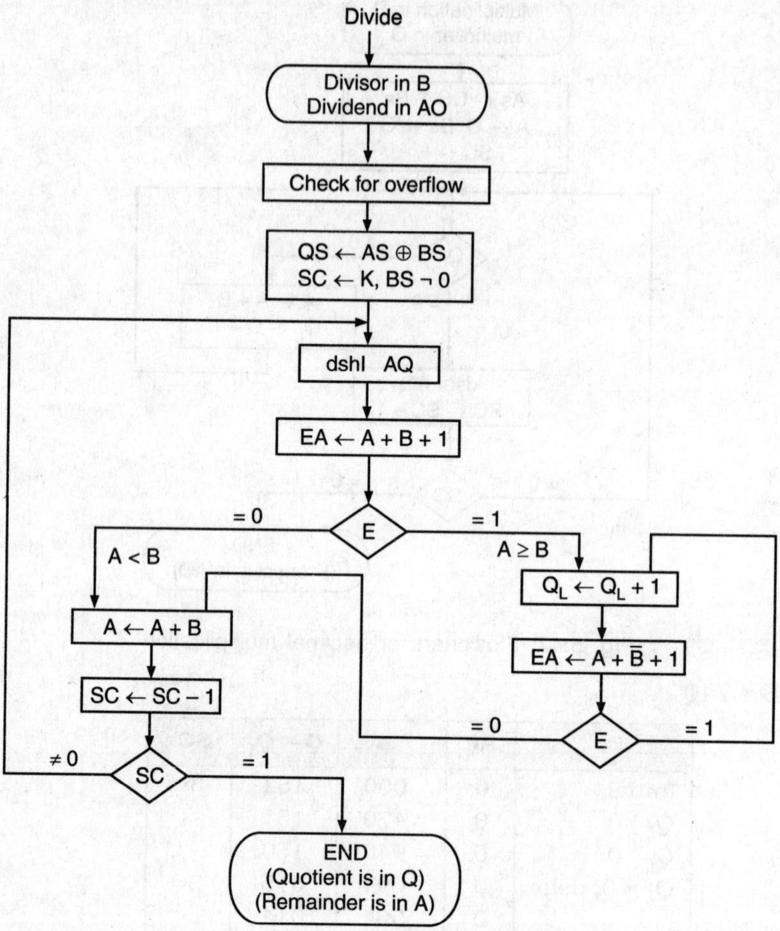

Fig. 5.43. Flowchart for decimal division

SUMMARY

1. The four basic arithmetic operations are addition, subtraction, multiplication and division.
2. Multiplication of two fixed-point binary numbers in signed magnitude representation is done with paper and pencil by a process of successive shift and add operations.
3. Booth algorithm gives a procedure for multiplying binary integers in signed-2's complement representation.
4. Division of two fixed-point binary numbers in signed magnitude representation is done with paper and pencil by a process of successive compare, shift and subtract operations.
5. The division operation may result in a quotient with an overflow.
6. Representation for floating point numbers is known as the IEEE 754 floating point standard.
7. The IEEE 754 standard has two main representations :
   (*i*) single precision, (*ii*) double precision
8. Arithmetic processing unit in central processing unit is used to execute all these operations. The arithmetic instructions are performed generally on binary or decimal data. Fixed-point

numbers are used to represent integers or fractions. We can have signed or unsigned negative numbers. Fixed-point addition is the simplest arithmetic operation.

9. A processor has an arithmetic processor as a sub part of it) that executes arithmetic operations. The data type, assumed to reside in processor, registers during the execution of an arithmetic instruction.

10. Negative numbers may be in a signed magnitude or signed complement representation. There are three ways of representing negative fixed point - binary numbers signed magnitude, signed 1's complement or signed 2's complement. Most computers use the signed magnitude representation for the mantissa.

EXERCISE

(I) Fill in the Blanks (Solved)

1. A .... system is an inter connection of two or more CPUs with memory and input-output equipment.
2. The advantage of biased exponents is that they contain only .... numbers.

(II) True or False (Solved)

1. There are many different topologies for multistorage switching networks to control processor communication. One of them is omega switching network.
2. A priority interrupt system is a communication of branch and loop techniques.

(III) Fill in the Blanks (Unsolved)

1. There are some similarities between multi processor and .... system since both support concurrent operations.
2. The .... needs the usual circuits of an interface to communicate with the CPU and input/output device.

(IV) Detailed Questions

1. Show the step-by-step multiplication process using booth algorithm, when the following binary numbers are multiplied. Assume 5 - bit registers to hold signed numbers.

   (*a*) + 10 x + 14 (*b*) + 10 x - 14
2. Derive an algorithm for evaluating the square root of a binary fixed-point number.
3. Why should the sign of the remainder after a division be the same as the sign of the dividend?
4. Show that there can be no mantissa overflow after a multiplication operation.
5. Show the hardware to be used for the addition and subtraction.
6. List four peripheral devices that produce an acceptable output for a person to understand.
7. What is the difference between isolated I/O and memory mapped I/O? What are the advantages and disadvantages of each?
8. Give at least six status conditions for the setting of individual bits in the status register of an asynchronous communication interface.
9. What programming steps are required to check when a source interrupts the computer while it is still being serviced by a previous interrupt request from the same source?
10. Why does DMA have priority over the CPU when both request a memory transfer?
11. What is the advantage of using interrupt - initiated data transfer over transfer under program control without an interrupt?
12. The following transfer statements specify a memory. Explain the memory operation in each case

    (*i*) $R_2 \leftarrow M[AR]$ (*ii*) $M(AR) \leftarrow R_3$

    (*iii* $R_5 \leftarrow M(R_5)$.

13. Register *A* holds the 8 bit binary 11011001. Determine the *B*-operand and the logic micro-operation to be performed in order to change the value in *A* to :
    (*i*) 01101101
    (*ii*) 11111101.
14. An 8 bit register contains binary value 11001010. What is the register value after a circular shift right.
15. Starting from an initial value of *R* = 11011101, determine the sequence of binary values in *R* after a logical shift-left, followed by a circular shift-right followed by a logical shift right and a circular shift left.
16. Draw the diagram of bus system using 4 × 1 multiplex.

ANSWERS

(I) Fill in the Blanks (Solved)

1. computer
2. positive

(II) True or False (Solved)

1. True
2. False

❑❑❑

Chapter

6

Control Design

Chapter Outline

6.0 INTRODUCTION

Until this chapter, the control unit has been considered as a black box, with the Instruction Register, the Condition Codes Register and the interrupt lines as inputs and command lines towards all the different parts of the computer as outputs. In chapter 5, the existence of a clock in the control unit was also mentioned. In this chapter the operation of the control unit will be analyzed in somewhat more detail. It will be shown that a control unit is a quite complex device, that can best be build as a programmable automaton.

6.1. THE CONTROL UNIT

The control unit, plays a central role in the organization of a computer. This is clearly visible in Fig. 6.1, borrowed from the introduction to chapter 1.

Its role consists in generating the signals to transfer an instruction from central memory to the Instruction Register, in identifying this instruction and in generating all the signals needed to execute it. In addition, the control unit should also, before fetching an instruction, check whether there is an interrupt pending, and, in case there is, generate all signals needed to react to it. As all the signals generated by a control unit are sequences of binary signals, a timing generator, usually called "clock" is always required (Fig. 6.2). Physically, this clock can be part of the control unit or it can be a separate unit, but logically it should always be considered as part of the control unit.

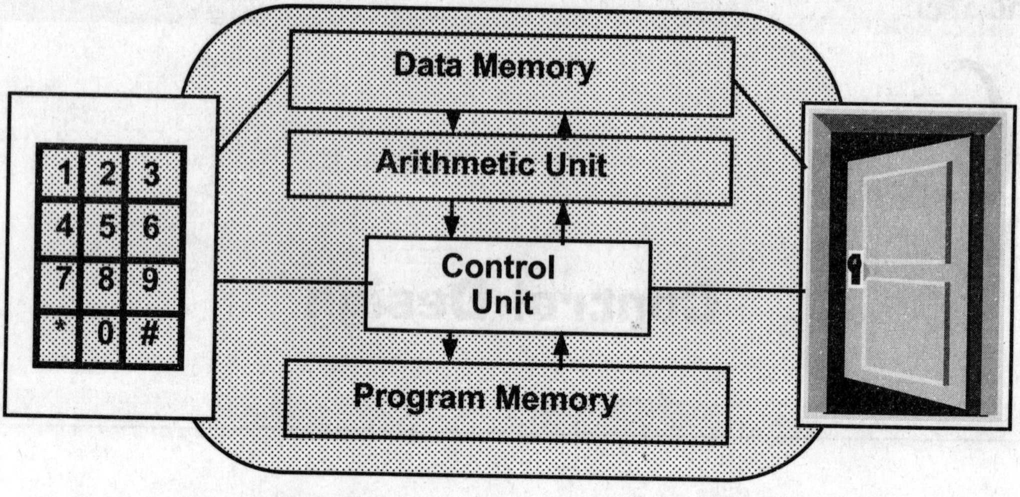

Fig.6.1. The central role of the control unit.

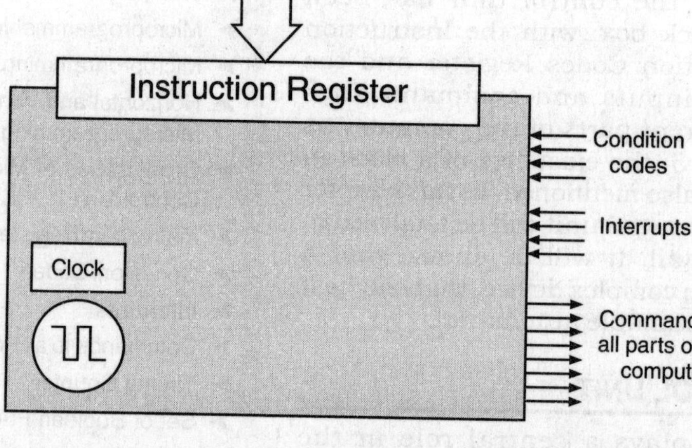

Fig. 6.2. The Control Ubnit at the heart of the CPU

Early control units consisted in a counter, reinitialized at the end of the execution of each instruction, counting the clock ticks, and a large set of Boolean functions combining the states of the timing counter, the instruction register, the condition code register and the interrupt lines to generate the specific control signals needed for instruction fetching and execution and for interrupt handling (Fig. 6.3) The signal resetting the timing counter at the end of the execution of each instruction.

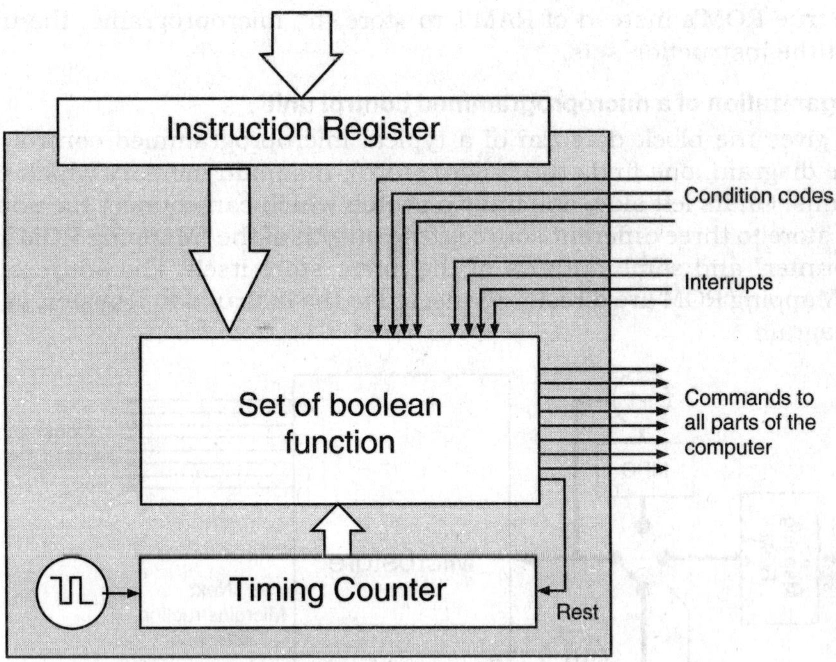

Fig. 6.3. The block diagram of a control unit

6.2. MICROPROGRAMMABLE CONTROL UNITS

6.2.1. Introduction

Fairly early in the history of computers, Maurice WILKES found out that the most economical and versatile method of implementing the large set of complex Boolean functions of a control unit consists in using a Read Only Memory. Implementing a combinatorial Boolean function of k Boolean variables by means of a ROM consists in using the input variables as addresses for the ROM and in storing in each of the 2^k cells the desired value of the Boolean function. When a set of l Boolean functions of the same k variables is needed, it suffices to use a ROM with 2^k words with l bits each. The major advantage of this approach resides in its versatility: whenever a Boolean function needs to be modified, for instance to add a new instruction to the instruction set of the computer, it suffices to replace the ROM rather than to modify the wiring of the control unit. Even if Wilkes published these ideas in the early fifties, the lack of cost effective and fast ROM's delayed their widespread use till the late sixties. At that time, cost effective fast ROM's were not yet available but RAM's were and therefore many control units were build around a very fast RAM which was loaded at power-on time from a special disk unit which kept a copy of all the Boolean functions used in the control unit. The instruction set of such computers could very easily be modified: by replacing the Boolean functions on that disk. On some machines the user could even choose dynamically among several sets of Boolean functions implementing the instructions of different popular computer brands. On other machines, the users could add instructions which would be particularly useful for specific applications. The Boolean functions implementing instruction sets were called microprograms and computers with a RAM based control unit were called "microprogrammable" computers. Modern monolithic processors still use microprograms to define their instruction set, but as they use true ROM's instead of RAM's to store the microprograms, the users can no longer adapt the instruction sets.

6.2.2. The organization of a microprogrammed control unit

Fig. 6.4 gives the block diagram of a typical microprogrammed control unit. In the center of the diagram, one finds the "micro store", the main memory which contains the microprograms. On its left side, one finds a switch which can connect the address inputs of the micro store to three different sources: the output of the "Mapping ROM", the "Micro-Program Counter" and some outputs of the micro store itself. The address lines of the Instruction Mapping ROM are directly connected to the Instruction Register, at the extreme left of the diagram.

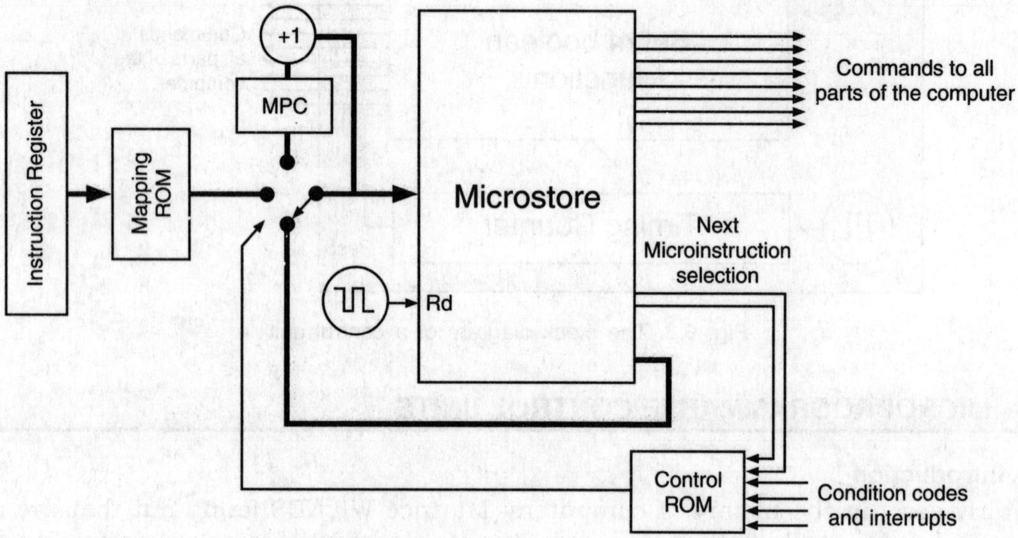

Fig. 6.4. A microprogrammed control unit

To explain the operation of this control unit, it is best to start at the moment an instruction has been fetched from memory and loaded into the Instruction Register.

At that moment, the switch connects the address lines of the microstore to the output of the Mapping ROM, which gives, for every possible opcode loaded in the Instruction Register, the begin address in the Micro Store of the corresponding microprogram. This address is send to the microstore and, simultaneously, incremented by 1 and loaded into to Microprogram Counter (MPC). When the first word of the selected microprogram has been read, the address switch is moved to the MPC and consecutive words of the microprogram are read at each clock tick. At the end of the execution of each instruction, the address switch is moved to the third position and the address in the micro store of the begin of the "fetch" microprogram is obtained from the microstore itself.

It should be clear from the preceding explanations that the position of the address switch is itself controlled by the microprogram. However, condition codes and interrupt lines should also be able to influence the sequencing of the microprograms. Therefore, the control lines of the address switch are, in fact, Boolean functions of some control bits in the micro store, the condition codes and the interrupt lines. These Boolean functions are classically implemented by means of a third ROM, the Control ROM.

From all this, one can deduct the format of a word in the Micro Store. It has three fields, one, The "Next Microinstruction Control" field, with the control bits to select the address switch position, one, the "Destination Address" field, for the address in the Micro Store should the next address have to be fetched from the micro store and one containing all the bits controlling the entire computer (Fig.6.5).

Next µinstr. control	Destination Address	Bits controlling all parts of the computer system

Micro Instruction Format

6.3. MICROPROGRAMMING

Writing microprograms is quite similar to writing programs in machine language for a very primitive computer. It is as time consuming and error prone and requires in depth knowledge of the hardware. Just as for machine language programming, symbolic assembler languages have been developed to facilitate the writing of microprograms. In the early seventies some attempts to develop high level languages and compilers for microcode have even been made, but the evolution towards fully integrated processors with all their microcode in ROM has considerably reduced the interest in such high level languages, as the cost of highly optimized microcode, written by outstanding experts, can be shared by the millions of users of the processor

Table 6.1. Gives, written in a pseudo assembler language with self explaining mnemonics, some typical examples of microcode

label	next µister.	destination	bits controlling all parts of the computer system
fetch	IF inter.		P > mem. add. ; P + 1 > P
	THEN jump	intseq	
	FLSE next		
	inst.map		mem. dat. > IR
	...		
inst.n	next		Wheather is needed to execute instrcution n.
	...		
	jump	fetch	
	...		
intseq	next		Whatever is needed to identify the interrupt, to to save the processor state and to fetch the first instruction of the interrupl handler
	...		
	inst. map		

6.4. HORIZONTAL AND VERTICAL MICROPROGRAMMING

The microinstructions described in section 6.2 had specific bits for each function. This is often called "horizontal microprogramming". It allows to control separately every function in the computer and to perform as many things as possible simultaneously. It also results in extremely wide microinstructions (up to several hundreds of bits) which are difficult to master, even for very good computer designers. Horizontal microprogramming is mainly used in very high performance computers, where simultaneity is the key towards performance.

If one observes Table 6.1, it is obvious that the "Destination Address" field is used in very few microinstructions. As this field is very wide as it contains an address in the microstore, having it in every microinstruction is quite a waste of very expensive microstore space. In addition, one can imagine that many of the bits controlling parts of the computer are not really independent: for instance, it would make no sense to put. The contents of two different registers simultaneously on the same bus. Based on these observations, it would make sense to define several different formats of microinstructions in which each bit can have several different functions, depending on the format. Fig. 6.6 gives an example of such encoded microinstructions and Fig. 6.6 shows the block diagram of a control unit adapted to such encoded microinstructions. Encoding the microinstructions reduces, of course, the freedom of the instruction set designer, so that fewer operations can be performed simultaneously, but it also reduces considerably the size of the microstore and simplifies somewhat the task of writing microprograms. When the microinstructions are encoded, one uses the expression "vertical microprogramming". Vertical microprogramming is mostly used when cost is more important than performance.

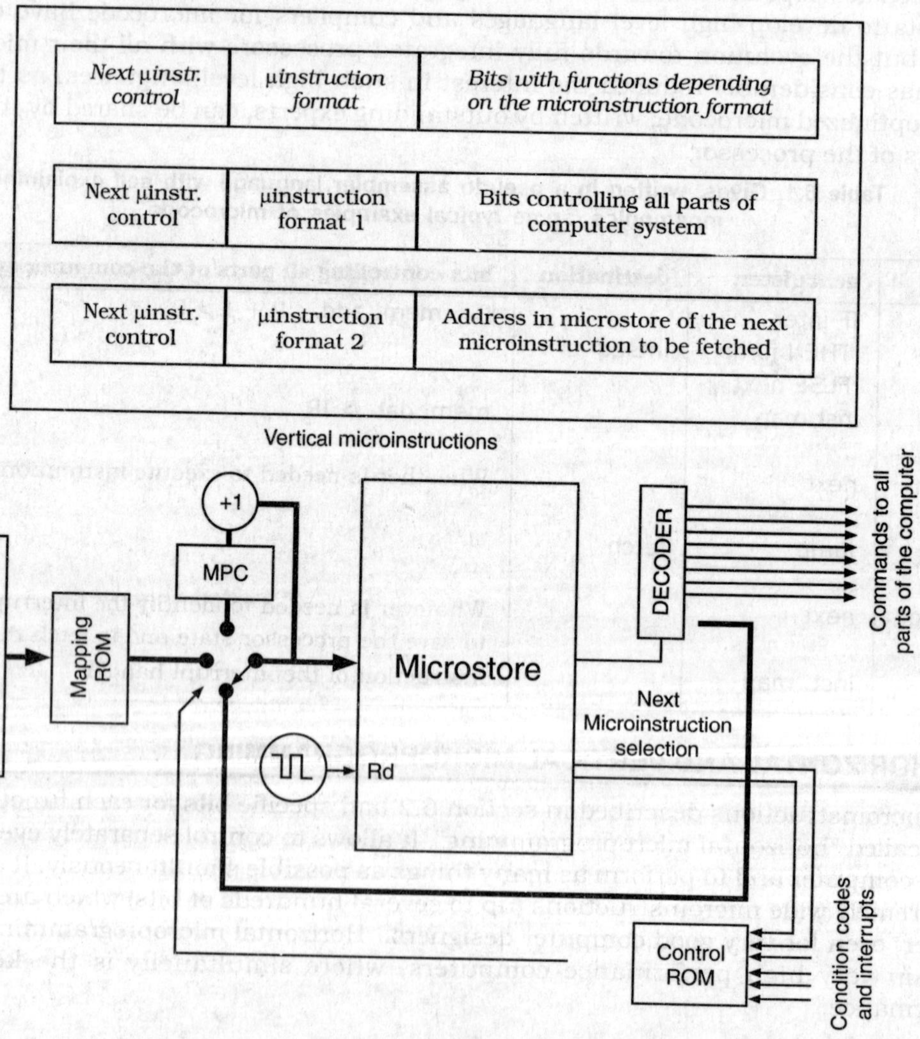

Fig. 6.5. Control unit with vertical microprogramming

Hardwired vs. Micro-programmed Computers

It should be mentioned that most computers today are micro-programmed. The reason is basically one of flexibility. Once the control unit of a hard-wired computer is designed and built, it is virtually impossible to alter its architecture and instruction set. In the case of a micro-programmed computer, however, we can change the computer's instruction set simply by altering the microprogram stored in its control memory. In fact, taking our basic computer as an example, we notice that its four-bit op-code permits up to 16 instructions. Therefore, we could add seven more instructions to the instruction set by simply expanding its microprogram. To do this with the hard-wired version of our computer would require a complete redesign of the controller circuit hardware.

Another advantage to using micro-programmed control is the fact that the task of designing the computer in the first place is simplified. The process of specifying the architecture and instruction set is now one of software (micro-programming) as opposed to hardware design. Nevertheless, for certain applications hard-wired computers are still used. If speed is a consideration, hard-wiring may be required since it is faster to have the hardware issue the required control signals than to have a "program" do it.

SOLVED PROBLEMS

Q.1. What is the difference between a microprocessor and a microprogram? Is it possible to design a microprocessor without a microprogram? Are all microprogrammed computers also microprocessors?

Solution: A microprocessor is a small size CPU (computer on a chip) microprogram is a program for a sequence of microoperations. The control unit of a microprocessor can be hardwired or microprogrammed, depending on the specific design. A microprogram computer does not have to be a microprocessor.

Q.2. Explain the difference between hardwired control and micpprogrammed control. Is it possible to have a hardwired control associated with a control memory?

Solution: Hardwired control, by definition, does not contain a control memory.

Q.3. Define the following : (*a*) microoperation, (*b*) microinstruction, (*c*) microprogram, (*d*) microcode.

Solution: Microoperation – an elemantary digital computer operation.

Microinstruction an instruction stored in control memory.

Microprogram – a sequence of microinstructions.

Microcode – same an microprogram.

Q.4. The microprogrammed control organization shown in Fig. 6.1. has the following propagation delay times. 40 ns to generate the next address, 10 ns to transfer the address into the control address register, 40 ns to access the control memory POM< 10 ns to transfer the microinstruction into the control data register and 40 ns to perform the required microoperations specified by the controword. What is the maximum clock frequency that the control can use? What would the clock frequency be if the control data register is not used?

Solution:

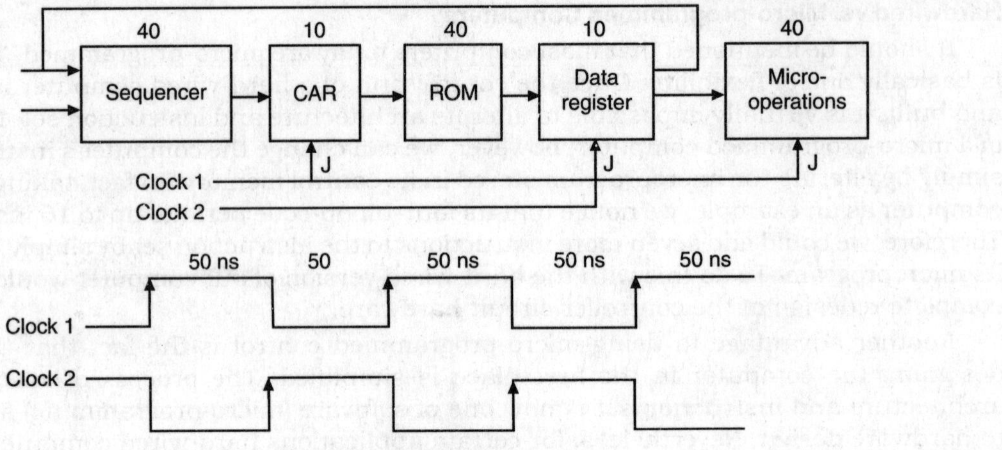

Fig. 6.6.

$$\text{Frequency of each clock} = \frac{1}{100\times10^{-9}} = \frac{1000}{100} \times 106 = 10 \text{ MHz}$$

If the data register is removed, we can ise a single phase clock with a frequency of

$$\frac{1}{90\times10^{-9}} = 11.1 \text{ MHz}$$

Q.5. The system shown in Fig. 7.2. uses a control memory of 1024 words of 32 bits each. The microinstruction has three fields as shown in the diagram. The microoperations field has 16 bits.

(*a*) How many bits are there in the branch address field and the select field?

(*b*) If there are 16 status bits in the system, how many bits of the branch logic are used to select a status bit?

(*c*) How many bits are left to select an input for the multiplexers?

Solution: Control memory = $2^{10} \times 32$

(*a*)

6	10	16	= 32 bits
Select	Address	Microoperations	

(*b*) 4 bits

(*c*) 2 bits

Q.6. The control memory in Fig. 7.2 has 4096 words of 24 bits each.

(*a*) How many bits are there in the control address register?

(*b*) How many bits are there in each of the four inputs shown going into the multiplexers?

(*c*) What are the number of inputs in each multiplexer and how many multiplexers are needed?

Solution: Control memory = $2^{12} \times 24$

(*a*) 12 bits

(*b*) 12 bits

(*c*) 12 multiplexers, each of size 4-to-1 line.

Q.7. Using the mapping procedure described in Fig. 7.3, give the first microinstruction address for the following operation code : (*a*) 0010, (*b*) 1011, (*c*) 1111.

Solution: (*a*) 0001000 = 8

(*b*) 0101100 = 44

(*c*) 0111100 = 60

Q.8. Formuiate a mapping procedure that provides eight consecutive microinstructions for each routine. The operation code has six bits and the control memory has 204° words.

Solution: Opcode = 6 bits

Cont memory

Address = 11 bits

	xxxxxx	
00	xxxxxx	000

Q.9. Explain how the mapping from an instruction code to a microinstruction on address can be done by means of a read-only memory. What is the advantage of this memthod compared to the one in Fig. 7.3?

Solution:

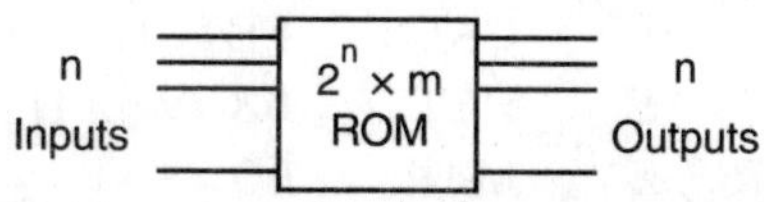

The ROM can be programmed to provide any desired address for a given inputs from the instructions.

Q.10. Why do we need the two multiplexers in the computer hardware configuration shown in Fig. 7.4? Is there another way that information from multiple sources can be transferred to a common destination?

Solution: Either multiplexers, three-state gates, orgate logic (equivalent to a MUX) are needed to transfer information from many sources to a common destination.

Q.11. Using Table 7.1, give the 9-bit microoperation field for the following microoperations :

(*a*) $AC \leftarrow AC + 1$, $DR \leftarrow DR + 1$

(*b*) $PC \leftarrow PC + 1$, $DR \leftarrow M[AR]$

(*c*) $DC \leftarrow AC$, $AC \leftarrow DR$

Solution:	F1	F2	F3	
(*a*)	011	100	000	INCAC INCDR NOP
(*b*)	000	100	101	NOP READ INCPC
(*c*)	100	101	000	DRTACACTDR NOP

Q.12. Using Table 7.1, convert the following symbolic microoperations to register ???? statements and to binary.

(*a*) READ, INCPC

(*b*) ACTDR, DRTAC

(*c*) ARTPC, DRTAC, WRITE

Solution:

				Binary
(*a*)	READ	$DR \leftarrow M[AR]$	F2 = 100	000 100 101
	DRTAC	$AC \leftarrow DR$	F3 = 101	
(*b*)	ACTDR	$DR \leftarrow AC$	F2 = 101	000 100 101
	DRTAC	$AC \leftarrow DR$	F1 = 100	
(*c*)	ARTPC	$DC \leftarrow AR$	F3 = 110	
	DRTAC	$AC \leftarrow DR$	F1 = 100	Impossible
	WRITE	$H[AR] \leftarrow DR$	F1 = 111	Both use F1

Q.13. The computer of Sec. 7.3 has the following binary microprogram.

Address	Binary Microprogram
60	0 1 0 0 0 0 0 1 0 0 0 0 0 1 0 0 0 0 1 1
61	1 1 1 1 0 0 0 0 0 0 1 0 1 1 0 0 0 0 0 0
62	0 0 1 0 0 1 0 0 0 1 0 1 0 0 1 1 1 1 1 1
63	1 0 1 1 1 0 0 0 0 1 1 1 1 0 1 1 1 1 0 0

(a) Translate it to a symbolic microprogram as in Table 7.2. (FETCH is in address 64 and INDRCT in address 67.)

(b) List all the things that will be wrong when this microprogram is executed in the computer.

Solution:

(a) 60	:	CLRAC, COM	*U*	JMP	INDRCT
61	:	WRITE, READ	*I*	CALL	FETCH
62	:	ADD, SUB	*S*	RET	63 (VEXT)
63	:	DRTAC, INCDR	*Z*	MAP	60

(b) 60 : Cannot increment and complement *AC* at the same time. With a JMP to INDRCT, control does not return to 61.

61 : Cannot read and write all the same time. The CALL behaves as a JMP since there is no return from FETCH.

62 : Cannot add and subtract at the same time. The RET will be executed independent of *S*.

63 : The MAp is executed irrespective of *Z* or 60.

Q.14. Add the following instructions to the computer of Sec 7.3 (*EA* is the effective address). Write the symbolic microprogram for each routine as in Table 7.2. (Note that *AC* must not change in value unless the instruction specifies as change in *AC*.)

Symbol	Opcode	Symbolic Function	Description
AND	0100	$AC \leftarrow AC \wedge M[EA]$	AND
SUB	0101	$AC \leftarrow AC - M[EA]$	Subtract
ADM	0110	$M[EA] - [EA] + AC$	Add to memory
BTCL	0111	$AC \leftarrow AC \wedge M[EA]$	Bit clear
BZ	1000	If $(AC = 0)$ then $(PC \leftarrow EA)$	Branch if AC zero
SEQ	1001	If $(AC = M[EA])$ then $(PC \leftarrow PC + 1)$	Skip if equal
BPNZ	1010	If $(AC > 0)$ then $(PC \leftarrow EA)$	Branch if positive and nonzero

Solution:

	ORG 16			
AND :	NOP	I	CALL	INDRCT
	READ	U	JMP	NEXT
ANDOP :	AND	U	JMP	FETCH
	ORG 20			
SUB :	NOP	I	CALL	INDRCT
	READ	U	JMP	NEXT
	SUB	U	JMP	FETCH
	ORG 24			
ADM :	NOP	I	CALL	INDRCT
	READ	U	JMP	NEXT

	DRTAC, ACTDR	U	JMP	NEXT
	ADD	U	JMP	Exchange + 2 (Table 7.2)
	ORG 28			
BTCL :	NOP	I	CALL	INDRCT
	READ	U	JMP	NEXT
	DRTAC, ACTDR	U	JMP	NEXt
	COM	U	JMP	ANDOP
	ORG 32			
BZ :	NOP	Z	JMP	ZERO
	NOP	U	JMP	FETCH
ZERO :	NOP	I	CALL	INDRCT
	ARTPC	U	JMP	FETCH
	ORG 36			
SEQ :	NOP	I	CALL	INDRCT
	READ	I	JMP	NEXT
	DRTAC, ADCTDR	U	JMP	NEXT
	XOR (or SUB)	U	JMP	BEQ1
	ORG 69			
BEQ1 :	DRTAC, ACTDR	Z	JMP	EQUAL
	NOP	U	JMP	FETCH
EQUAL :	INCPC	U	JMP	FETCH
	ORG 40			
BPNZ :	NOP	S	JMP	FETCH
	NOP	Z	JMP	FETCH
	NOP	I	CALL	INDRCT
	ARTPC	U	JMP	FETCH

Q.15. Write a symbolic microprogram routine for the ISZ (increment and skip if zero) instruction defined in Chap. 5 (Table 5.4). use the microinstruction format of ????? 7.3. Note that *DR* = 0 status condition is not available in the *CD* field of the computer defined in Sec. 7.3. However, you can exchange *AC* and *DC* and check if *AC* = 0 with the *Z* bit.

Solution:

ISZ :	NOP	I	CALL	INDRCT
	READ	U	JMP	NEXT
	INCDR	U	JMP	NEXT
	DRTAC, ACTDR	U	JMP	NEXT (or Fast INDRCT)
	DRTAC, ACTDR	Z	JMP	ZERO
	WRITE	U	JMP	FETCH
ZERO :	WRITE, INCPC	U	JMP	FETCH

Q.16. Write the symbolic micprogram routines for the DBA (branch and save address) instructions defined in Chap. 5 (Table 5.4). Use the microinstruction format of Sec. 7.3. Minimize the number of microinstructions.

Solution: BSA :

NOP	I	CALL	INDRCT
PCTDR, ARTPC	U	JMP	NEXT
WRITE, INCPC	U	JMP	FETCH

Q.17. Show how output 5 and 6 of decoder F3 in Fig. 7.7 are to be connected to the program ????? *FC*.

Solution: From Table 7.1 :

F3 = 101 (5) $PC \leftarrow PC + 1$

F3 = 110 (6) $PC \leftarrow AR$

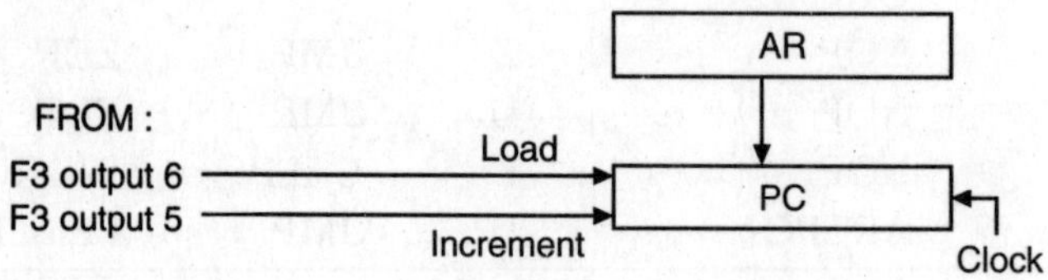

Fig. 6.7.

Q.18. Show how a 9-bit microoperation field in a microinstruction can be divided into subficds to specify 46 microoperations. How many microoperations can be specified in one microinstruction?

Solution: A field of 5 bits can specify $2^5 - 1 = 31$ microoperations

A field of 4 bits can specify $2^4 - 1 = 15$ microoperations

9   46 microoperations

Q.19. A computer has 16 registers, an ALU (arithmetic logic unit) with 32 operations, and a shifter with eigth operations, all connected to a common bus system.

(*a*) Formulate a control word for a microoperation.

(*b*) Specify the number of bits in each field of the control word and give a general encoding scheme.

(*c*) Show the bits of the control word that specify the microoperation $R4 \leftarrow R5 + R6$.

Solution: See Fig. 8.2 (*b*) for control word example.

(*a*) 16 register need 4 bits, ALU need 5 bits, and

(*b*) The shifter need 3 bits. to encode all operations.

(*c*)

4	4	4	5	3	= 20 bits total
SRC1	SRC2	DEST	ALU	SHIFT	
R5	R6	R4	ADD	SHIFT	$R4 \leftarrow R5 + R6$
0101	0110	0100	00100	000	

Q.20. Assume that the input logic of the microprogram sequencer of Fig. 7.8 has four inputs, I_2, I_3, I_0, *T* (test), and three outputs, S_1, S_0 and *L*. Thne operations that are performed in the unit are listed in the following table. Design the input logic circuit using a minimum number of gates.

I_2	I_3	I_0	Operation
0	0	0	Increment CAR if $T = 1$, jump to *AD* if $T = 0$
×	0	1	Jump to *AD* unconditionally
1	0	0	Increment CAR unconditionally

0	**1**	**0**	**Jump to *AD* if *T* = 1, increment CAR if *T* = 0**
1	**1**	**0**	**Call subroutine if *T* = 1, increment CAR if *T* = 0**
0	**1**	**1**	**Return from subroutine unconditionally**
1	**1**	**1**	**map external address unconditionally**

Solution:

I_2	I_1	I_0	T	S_1	S_0	L	
0	0	0	0	0	1	0	AD (1)
0	0	0	1	0	0	0	INC (0)
0	0	1	0	0	1	0	AD (1)
0	0	1	1	0	1	0	AD (1)
0	1	0	0	0	0	0	INC (0)
0	1	0	1	0	1	0	AD (1)
0	1	1	0	1	0	0	PET (0)
0	1	1	1	1	0	0	RET (0)
1	0	0	0	0	0	0	INC (0)
1	0	0	1	0	0	0	INC (0)
1	0	1	0	0	1	0	A (1)
1	0	1	1	0	1	0	AD (1)
1	1	0	0	0	0	0	INC (0)
1	1	0	1	0	1	1	CALL (1)
1	1	1	0	1	1	0	MAP (3)
1	1	1	1	1	1	0	MAF (3)

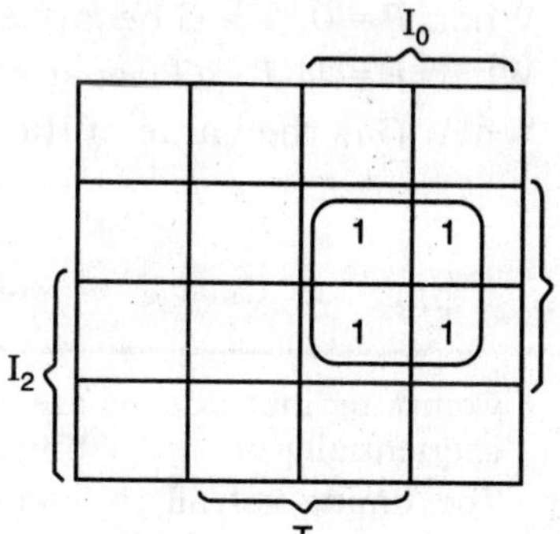

$S_0 = I_1 I_0$

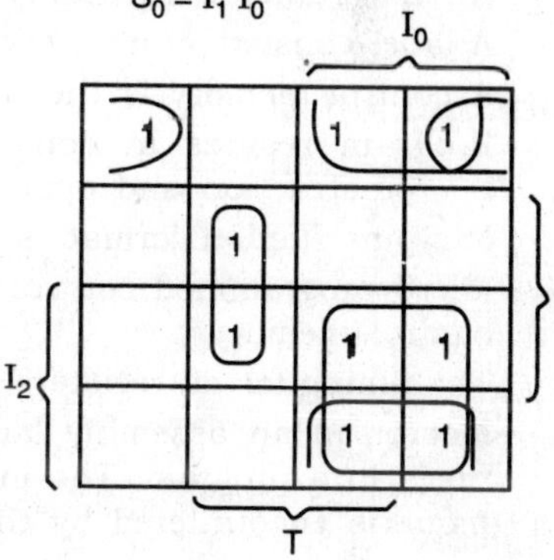

$$S_0 = I_2 I_0 + I_1' I_0 + I_1 I_0' + I_2' I_1' T$$
$$L = I_2 I_1 I_0' T$$

Q.21. Design a 7-bit combinational circuit incrementer for the microprogram sequencer of Fig. 7.89 (see Fig. 4.8). Modify the incrementer by including a control input *D*. When *D* = 0, the circuit increments by one, but when *D* = 1, the circuit increments by two.

Solution: (*a*) See Fig. 4.8 (Chapter 4.5).

(*b*)

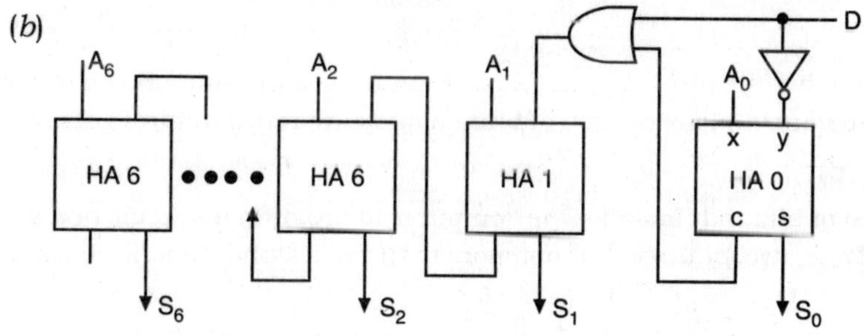

Q.22. Insert an exclusive-OR gate between MUX 2 and the input logic of Fig. 7.8. One input to the gate comes from the test output of the multiplexer. The other input to the gate comes from a bit labeled *P* (for polarity) in the microinstruction from control memory. The output of ther gate to the input *T* of the input logic. What does the polarity control *P* accomplish?

Solution: P is used to determine the polarity of the selected status bit.

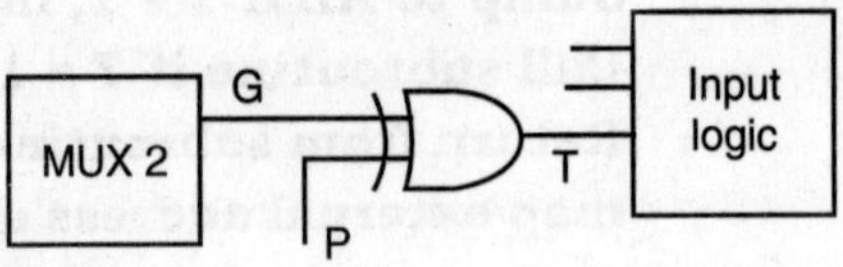

When $P = 0$, $T = G$ because $G \oplus 0 = G$

When $P = 1$, $T = G'$ because $G \oplus 1 = G$

When G is the value of the selected bit in MUX 2.

SUMMARY

1. Computer instructions are normally stored in consecutive memory locations and are executed sequentially one at a time.
2. The timing got all registers in the basic computer controlled by a clock.
3. An instruction is a group of bits that instruct the computer to program a specific operation.
4. A label consists of one, two or three, but not more than three alpha numeric characters.
5. A control memory is the unit which handles micro-instructions.
6. Program resides in control memory is called micro-program, instruction is called microinstruction and operation is called micro-operation.
7. Microinstruction format is of 20-bits.
8. Micro programmed control unit will have two sperate memories– a main memory and a control memory.
9. The timing for all registers in the basic computer is controlled by master clock generator.
10. Macros in an assembly language are used to repeat a sequence of instructions several times in a program. The instructions are stored in a macro body and whenever a macro name is encountered by the assembler it is replaced by the instruction.

EXERCISE

(I) Fill in the Blanks (Solved)

1. Computer .... are normally stored in consecutive memory locations and are executed sequentially one at a time.
2. The timing for all registers in the basic computer is controlled by a ....

(II) True or False (Solved)

1. A two phase consists of one, two, or three, but not more than three alphanumeric characters.
2. An interrupt cycle is a hardware implementation of branch and save return address operation.

(III) Fill in the Blanks (Unsolved)

1. An ....is a group of bits that instruct the computer to program a specific operation.
2. A .... consists of one, two, or three, but not more than three alphanumeric characters.

(IV) True or False (Unsolved)

1. The memory that is a part of a control unit is referred to as an accumulator.
2. The data register is sometimes called a buffer.

(V) Detailed Questions

1. What is the difference between a direct and an indirect address instruction? How many references to memory are needed for such type of instruction to bring an operand into a processor register?
2. Explain why each of the following micro-operations cannot be executed during a single clock pulse.

   (*a*) IR M[PC] (*b*) AC AC + TR (*c*) DR DR + AC (AC does not change)
3. Describe the control gates associated with the program counter PC in the basic computer.
4. Can the letter I be used as a symbolic address in the assembly language program defined for the basic computers? Justify the answer.
5. Write a program loop, using a pointer and a counter that clears to 0 the contents of hexadecimal location 500 through 5FF.
6. Is it possible to have a hardwired command associated with a control memory?
7. Define the following :

   (*a*) Micro operation (*b*) Micro instruction (*c*) Micro program (*d*) Micro code
8. What is the difference between a microprocessor and a microprogram? Is it possible to design a microprocessor without a microprogram? Are all microprogrammed computers also microprocessors?

Answers

(I) Fill in the Blanks (Solved)

1. instructions 2. clock

(II) True or False (Solved)

1. False 2. True

Chapter

7

Memory Organization

7.0 INTRODUCTION

The memory is that part of computer where programs and data are stored. The basical concept is the following :

- **Bits** : The basic unit of memory is the binary digit called a bit. A bit may contain a 0 or 1. It is the simplest possible unit.
- **Memory addresses:**
  - Memories consist of a number of cells or locations each of which can store a piece of information. Each location has a number called its address, by which program can refer to it. The cells is the smallest addressable
  - If an address has m bits, the maximum number of cells addressable is 2 m.
  - **Byte** : 8-bits.
  - Bytes are grouped into words. The significance of word is that most instruction operate on entire word. A computer with a 32-bit/word has 4-bytes/word.
- **Byte ordering :**
  - The bytes in a word can be numbered from left-to-right or right-to-left.
  - The former system, where the numbering begin at the "big" (*i.e.*, high-order) end is called a big endian

Chapter Outline

computer, such as the SPARC or the big IBM mainframes. In contras it is a little endian computer, such as the Intel family using right-to-left numbering for the representation of a 32 bit computer.

7.1. CHARACTERISTICS OF MEMORY SYSTEM

- **Capacity :** The amount of information that can be contained in a memory unit — usually in terms of words or bytes
- **Memory word :** The natural unit of organization in the memory, typically the number of bits used to represent a number.
- **Addressable unit :** The fundamental data element size that can be addressed in the memory — typically either the word size or individual bytes.
- **Unit of transfer :** The number of data elements transferred at a time – usually bits in main memory and blocks in secondary memory.
- **Transfer rate :** Rate at which data is transferred to/from the memory device.
- **Access time :**
  - For RAM, the time to address the unit and perform the transfer
  - For non-random access memory, the time to position the R/W head over the desired location
- **Memory cycle time :** Access time plus any other time required before a second access can be started.
- **Access technique :** How are memory contents accessed
  - **Random access :**
    - Each location has a unique physical address.
    - Locations can be accessed in any order and all access times are the same.
    - What we term "RAM" is more aptly called read/write memory since this access technique also applies to ROMs as well.
    - **Example :** main memory.
  - **Sequential access :**
    - Data does not have a unique address.
    - Must read all data items in sequence until the desired item is found.
    - Access times are highly variable.
    - **Example :** tape drive units.
  - **Direct access :**
    - Data items have unique addresses.
    - Access is done using a combination of moving to a general memory "area" followed by a sequential access to reach the desired data item.
    - **Example :** disk drives.
  - **Associative access :**
    - A variation of random access memory.
    - Data items are accessed based on their contents rather than their actual location.
    - Search all data items in parallel for a match to a given search pattern.
    - All memory locations searched in parallel without regard to the size of the memory.

— Extremely fast for large memory sizes.
— Cost per bit is 5-10 times that of a "normal" RAM cell.
— **Example** : some cache memory units.

7.2. TYPES OF MEMORY

Computer memory system consists a various types of memory. Manufactures produce a number of different types of memory devices having a variety of technologies. The technology affect not only the operating chracteristics but also the manufacturing cost. In the section following we present an overviews of types of memory. You can see the study in detail of memory in the modules 10, 11 and 12.

- Main Memory ("Internal" memory components)
  - RAM (read-write memory): Static RAM, Dynamic RAM
  - ROM (Read Only Memories) : ROMs, PROMs, EPROMs, EEPROMs, Flash Memory.
- Cache memory

The cache memories are high-speed buffers for holding recently accessed data and neighboring data in main memory. The organization and operations of cache provide an apparently fast memory system.

- External Memory
  - Magnetic disks
  - RAID technology disks
  - Optical disks
  - Magnetic tape

7.3. MEMORY SYSTEM ORGANIZATION

No matter how big the main memory, how we can organize effectively the memory system in order to store more information than it can hold. The traditional solution to storing a great deal of data is a memory hierarchy.

- Major design objective of any memory system :
  - To provide adequate storage capacity at.
  - An acceptable level of performance.
  - At a reasonable cost.
- Four interrelated ways to meet this goal
  - Use a hierarchy of storage devices.
  - Develop automatic space allocation methods for efficient use of the memory.
  - Through the use of virtual memory techniques, free the user from memory management tasks.
  - Design the memory and its related interconnection structure so that the proces.

7.4. MULTILEVEL MEMORIES ORGANIZATION

Three key characteristics increase for a memory hierarchy. They are the access time, the storage capacity and the cost. The memory hierarchy is illustrated in Fig. 7.1.

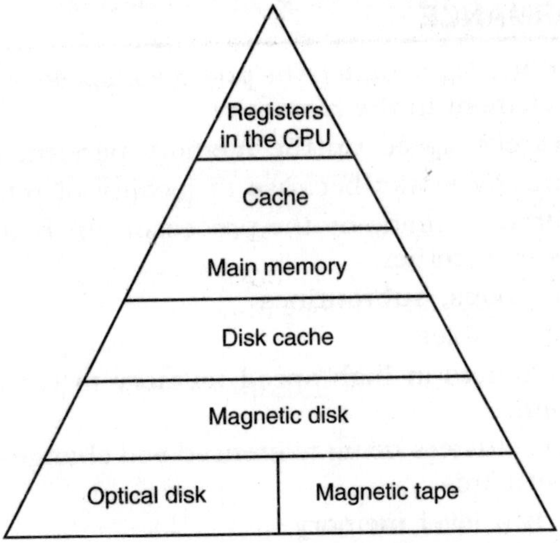

Fig. 7.1. The memory hierarchy

We can see the memory hierarchy with six levels. At the top there are CPU registers, which can be accessed at full CPU speed. Next commes the cache memory, which is currently on order of 32 KByte to a few Mbyte. The main memory is next, with size currently ranging from 16 MB for entry-level systems to tens of Gigabytes. After that come magnetic disks, the current work horse for permanent storage. Finally we have magnetic tape and optical disks for archival storage.

- Basis of the memory hierarchy
  - Registers internal to the CPU for temporary data storage (small in number but very fast)
  - External storage for data and programs (relatively large and fast)
  - External permanent storage (much larger and much slower)

Memory Type	Technology	Size	Access Time
Cache	Semiconductor	128-512 KB	10 ns
Main memory	Semiconductor	4-128 MB	50 ns
Magnetic Disk	Hard Disk	Gigabyte	10ms, 10MB/sec
Optical Disk	CD-ROM	Gigabyte	300 ms, 600 KB/sec
Magnetic Tape	Tape	100sMB	Sec-min, 10 MB/min

Fig. 7.2. Typical Memory Parameters

- Characteristics of the memory hierarchy
  - Consists of distinct "levels" of memory components
  - Each level characterized by its size, access time, and cost per bit
  - Each increasing level in the hierarchy consists of modules of larger capacity, slower access time, and lower cost/bit

7.5 MEMORY PERFORMANCE

Goal of the memory hierarchy. Try to match the processor speed with the rate of information transfer from the lowest element in the hierarchy.

- The memory hierarchy speed up the memory performance.
  The memory hierarchy works because of locality of reference.
  - Memory references made by the processor, for both instructions and data, tend to cluster together
    + Instruction loops, subroutines
    + Data arrays, tables
  - Keep these clusters in high speed memory to reduce the average delay in accessing data.
  - Over time, the clusters being referenced will change — memory management must deal with this.
- Performance of a two level memory
  Example : Suppsose that the processor has access to two level of memory :
  - Two-level memory system
  - Level 1 access time of 1 us
  - Level 2 access time of 10 us
  - Ave access time = H(1) + (1 – H)(10) ns
    where : H is a fraction of all memory access that are found in the faster memory (*e.g.* cache).

7.6 VIRTUAL MEMORY

In computing, **virtual memory** is a memory management technique developed for multitasking kernels. This technique virtualizes a computer architecture's various hardware memory devices (such as RAM modules and disk storage drives), allowing a program to be ***designed as though***:

- there is only one hardware memory device and this "virtual" device acts like a RAM module.
- the program has, by default, sole access to this virtual RAM module as the basis for a contiguous working memory (an address space).

Systems that employ virtual memory :

- use hardware memory more efficiently than systems without virtual memory.
- make the programming of applications easier by:
  - hiding fragmentation.
  - delegating to the kernel the burden of managing the memory hierarchy; there is no need for the program to handle overlays explicitly.
  - obviating the need to relocate program code or to access memory with relative addressing.

Memory virtualization is a generalization of the concept of virtual memory.

Virtual memory is an integral part of a computer architecture; all implementations (excluding[dubious – *discuss*] emulators and virtual machines) require hardware support, typically in the form of a memory management unit built into the CPU. Consequently, older operating systems (such as DOS[1] of the 1980s or those for the mainframes of the 1960s) generally have no virtual memory functionality[dubious – *discuss*], though notable exceptions include the Atlas, B5000, IBM System/360 Model 67, IBM System/370 mainframe systems of the early 1970s, and the Apple Lisa project circa 1980.

Embedded systems and other special-purpose computer systems that require very fast and/or very consistent response times may opt not to use virtual memory due to decreased determinism; virtual memory systems trigger unpredictable interrupts that may produce unwanted "jitter" during input/output operations. This is because embedded hardware costs are often kept low by implementing all such operations with software (a technique called bit-banging) rather than with dedicated hardware. In any case, embedded systems usually have little use for multitasking features or complicated memory hierarchies.

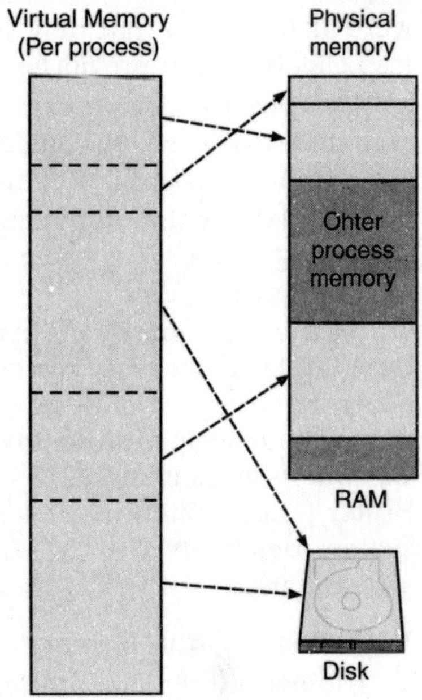

Fig. 7.3.

In computing, **virtual memory** is a memory management technique developed for multitasking kernels. This technique virtualizes a computer architecture's various hardware memory devices (such as RAM modules and disk storage drives), allowing a program to be ***designed as though***:

- there is only one hardware memory device and this "virtual" device acts like a RAM module.
- the program has, by default, sole access to this virtual RAM module as the basis for a contiguous working memory (an address space).

Systems that employ virtual memory :

- use hardware memory more efficiently than systems without virtual memory.
- make the programming of applications easier by:
  - hiding fragmentation.
  - delegating to the kernel the burden of managing the memory hierarchy; there is no need for the program to handle overlays explicitly.
  - obviating the need to relocate program code or to access memory with relative addressing.

To allow for multiprogramming and multitasking, many early systems already had the ability to divide the memory between multiple programs even without providing virtual memory; for instance, early models of the PDP 10 provided memory access through what are called *base and bounds* registers.

Paging was first developed at the University of Manchester as a cheap way to extend the Atlas Computer's working memory by transparently supplementing its 16 thousand words of primary core memory with the additional 96 thousand words of secondary drum memory. Although the first Atlas was commissioned in 1962, working prototypes of paging had reportedly been developed by 1959.[2](p2)[3][4] Also, Burroughs independently released in 1961 the first commercial computer with virtual memory, the B5000,[5][6] which used segmentation rather than paging.

Like many technologies in the history of computing, virtual memory was not accepted without challenge. Before it could be implemented in mainstream operating systems, many models, experiments, and theories had to be developed to overcome numerous problems. Dynamic address translation required expensive specialized hardware that was difficult to build. Moreover, initial implementations slightly slowed down access to memory. There were also worries that new system-wide algorithms utilizing secondary storage would be less effective than previously used application-specific algorithms.

By 1969 the debate over virtual memory for commercial computers was over. An IBM research team led by David Sayre showed that their virtual memory overlay system consistently worked better than the best manually controlled systems.

Possibly the first minicomputer to introduce virtual memory was the Norwegian NORD-1. During the 1970s, other minicomputers implemented virtual memory, notably VAX models running VMS.

Virtual memory was introduced to the x86 architecture with the protected mode of the Intel 80286 processor; however, its segment swapping technique was discovered to scale poorly to larger segment sizes. The Intel 80386 introduced support for paging underneath the existing segmentation layer, and the page fault exception could be chained with other exceptions without causing a double fault. The 80386 did not have a Translation Lookaside Buffer (TLB), which made loading segment descriptors expensive and caused operating system designers to rely strictly on paging rather than a combination of paging and segmentation.

7.6.1. Paged Virtual Memory

Almost all implementations of virtual memory divide the virtual address space of an application program into pages; a page is a block of contiguous virtual memory addresses. Pages are usually at least 4 KiB (4×1024 bytes) in size, and systems with large virtual address ranges or large amounts of real memory generally use larger page sizes.

7.6.1.1. Page Tables

Almost all implementations use page tables to translate the virtual addresses seen by the application program into physical addresses (also referred to as "real addresses") used by the hardware to process instructions. The hardware that handles this specific translation or mapping function between logical addresses and physical addresses is often known as the Dynamic Address Translation (DAT) box or as the Memory Management Unit (MMU). Each entry in the page table contains a mapping for a virtual page to either a subsidiary page table, the real memory address at which the page is stored, or an indicator that the page is currently held in a disk file. (Although most do, some systems may not support use of a disk file for virtual memory.)

Systems can have one page table for the whole system, a separate page table for each application, a separate page table for each segment, a tree of page tables for large segments or some combination of these. If there is only one, different applications which are running at the same time share a single virtual address space, *i.e.* they use different parts of a single range of virtual addresses. Systems that use multiple page or segment tables provide multiple virtual address spaces—concurrent applications think they are using the same range of virtual addresses, but their separate page tables redirect to different real addresses.

7.6.1.2. Dynamic Address Translation

If, while executing an instruction, a CPU fetches an instruction located at a particular virtual address, or fetches data from a specific virtual address or stores data to a particular virtual address, the virtual address must be translated to the corresponding physical address. This is done by a hardware component, sometimes called a memory management unit, which looks up the real address (from the page table) corresponding to a virtual address and passes the real address to the parts of the CPU which execute instructions. If the page tables indicate that the virtual memory page is not currently in real memory, the hardware raises a page fault exception (special internal signal) which invokes the paging supervisor component of the operating system (see below).

7.6.1.3. Paging Supervisor

This part of the operating system creates and manages page tables. If the address translation hardware raises a page fault exception, the paging supervisor accesses secondary storage, returns the page containing the required virtual address, updates the page tables to reflect the physical location of the virtual address and finally tells the dynamic address translation mechanism to restart the request. When all physical memory is already in use as is typical, the paging supervisor must free an area in primary storage to hold the swapped-in page. Freeing memory minimally requires updating the paging table to say that the page is in secondary storage. The supervisor saves time by not re–swapping pages that are already present in secondary storage.

Paging supervisors generally choose the page that has been least recently used, guessing that such pages are less likely to be requested. Every time the dynamic address translation hardware matches a virtual address with a real physical memory address, it time-stamps the page table entry for that virtual address.

7.6.1.4. Permanently Resident Pages

Operating Systems have memory areas that are "pinned" — *i.e.*, never swapped to secondary storage. For example :

- Interrupt mechanisms generally rely on an array of pointers to their handlers (input/output completion, timer event, program error, page fault, etc.). If the pages containing these pointers or the code that they invoke were pageable, interrupt-handling would become far more complex and time-consuming; particularly in the case of page fault interrupts.
- At least some part of the page table structures are almost always not pageable.
- Data buffers that are accessed directly, for example by peripheral devices that use Direct Memory Access (DMA) or by input/output channels. Usually such devices and the buses (connection paths) to which they are attached use physical memory addresses rather than virtual memory addresses. Even on buses with an IOMMU, which is a special memory management unit that can translate virtual addresses used on an input/output bus to physical addresses, the transfer cannot be stopped if a page fault occurs and then restarted when the page fault has been processed. So pages containing locations to which or from which a peripheral device is transferring data are either permanently pinned down or pinned down while the transfer is in progress.
- Timing-dependent kernel/application areas cannot tolerate the varying response time caused by paging. In particular the paging supervisor code or drivers for secondary storage devices must not be swapped out.

Virtual = Real Operation

In OS/VS1, OS/VS2 (SVS), MVS, z/OS, and similar OSes, some parts of systems memory are managed in virtual=real mode, where every virtual address corresponds to a real address. Those are :

- interrupt mechanisms
- paging supervisor and page tables, in older systems
- application programs that use non-standard methods of managing input/output, e.g., reading into an active channel program.

IBM's z/OS has 3 modes, V=V (virtual=virtual; fully pageable), V=R and V=F (virtual = fixed, *i.e.* "pinned down" but with DAT operating).

7.6.2. Segmented Virtual Memory

Some systems, such as Burroughs B5500, do not use paging to implement virtual memory. Instead, they use segmentation, that divide virtual address spaces into variable-length segments. A virtual address consists of a segment number and an offset within the segment.

Notably, the Intel 80286 supports a similar segmentation scheme as an option, but it is little used.

It is possible to combine segmentation and paging, dividing each segment into pages. In systems that combine them, such as Multics, IBM System/38 and IBM System i machines, virtual memory is usually implemented with paging, with segmentation providing memory protection.[9][10][11]

In the Intel 80386 and later IA-32 processors, the segments reside in a 32-bit linear, paged address space. Segments can be moved in and out of that space, and pages in that space can "page" in and out of main memory, providing two levels of virtual memory; however, few if any operating systems do so. Instead, they use only paging. Early x86 virtualization solutions (non-hardware-assisted) combined paging and segmentation because x86 paging offers only two protection domains, whereas a VMM/guest OS/guest applications stack (typically running on rings 0/1/3) needs three memory protection domains.

The difference between paging and segmentation systems is not only about memory division. In Multics, System/38 and Prime machines, segmentation is visible to user processes, as part of memory model semantics. In other words, instead of memory that looks like a single large vector, memory is structured into multiple spaces.

This difference has important consequences; a segment isn't just a "page with a variable length", or a simple way to lengthen the address space (as in Intel 80286). In Multics, segmentation that can provide a single-level memory model, in which there is no differentiation between "process memory" and "file system". *I.e.*, a process's *active* address space for both code and data consists of only a list of segments (files) which are mapped into the process's *potential* address space.

This is not the same as the mechanisms provided in TENEX and TOPS-20 (PMAP), modern Unix-like systems (mmap), and Win32 systems (MapViewOfFile), because inter-file pointers don't work when mapping files into semi-arbitrary places. In Multics, a file (or a segment from a multi-segment file) is mapped into a segment in the address space, so files are always mapped at a segment boundary. A file's linkage section can contain pointers that are "unsnapped links"; an attempt to load the pointer into a register or make an indirect reference through it causes a trap (a field in the pointer can have a value that specifies that an attempt to dereference it should cause a trap). The unresolved pointer contains an indication of the name of the segment to which the pointer refers, and an offset within the segment; the handler for the trap will "snap the link" the pointer by mapping the segment into the address space, putting the segment number into the pointer, and changing the tag field in the pointer so that it no longer causes the trap), and return to the code where the trap occurred, re-executing the instruction that caused the trap. This eliminates the need for a linker completely. This also works when different processes map the same file into different places in their private address spaces.

7.6.3. Avoiding Thrashing

When paging is used a potential problem called "thrashing" can occur, in which the computer spends a disproportionate amount of its capacity swapping pages to and from a backing store and therefore performs useful work more slowly. Adding real memory is the simplest response, although improving application design, scheduling, and memory usage can help.

7.7 ADDRESS TRANSLATION

Most general purpose CPUs implement some form of virtual memory. To summarize, each program running on the machine sees its own simplified address space, which contains code and data for that program only. Each program uses this virtual address space without regard for where it exists in physical memory.

Virtual memory requires the processor to translate virtual addresses generated by the program into physical addresses in main memory. The portion of the processor that does this translation is known as the memory management unit (MMU). The fast path through the MMU can perform those translations stored in the translation lookaside buffer (TLB), which is a cache of mappings from the operating system's page table.

For the purposes of the present discussion, there are three important features of address translation :

- **Latency :** The physical address is available from the MMU some time, perhaps a few cycles, after the virtual address is available from the address generator.
- **Aliasing :** Multiple virtual addresses can map to a single physical address. Most processors guarantee that all updates to that single physical address will happen in program order. To deliver on that guarantee, the processor must ensure that only one copy of a physical address resides in the cache at any given time.
- **Granularity :** The virtual address space is broken up into pages. For instance, a 4 GB virtual address space might be cut up into 1048576 pages of 4 KB size, each of which can be independently mapped. There may be multiple page sizes supported;

A historical note : some early virtual memory systems were very slow, because they required an access to the page table (held in main memory) before every programmed access to main memory. With no caches, this effectively cut the speed of the machine in half. The first hardware cache used in a computer system was not actually a data or instruction cache, but rather a TLB.

Caches can be divided into 4 types, based on whether the index or tag correspond to physical or virtual addresses :

- **Physically indexed, physically tagged** (PIPT) caches use the physical address for both the index and the tag. While this is simple and avoids problems with aliasing, it is also slow, as the physical address must be looked up (which could involve a TLB miss and access to main memory) before that address can be looked up in the cache.
- **Virtually indexed, virtually tagged** (VIVT) caches use the virtual address for both the index and the tag. This caching scheme can result in much faster lookups, since the MMU doesn't need to be consulted first to determine the physical address for a given virtual address. However, VIVT suffers from aliasing problems, where several different virtual addresses may refer to the same physical address. The result is that such addresses would be cached separately despite

referring to the same memory, causing coherency problems. Another problem is homonyms, where the same virtual address maps to several different physical addresses. It is not possible to distinguish these mappings by only looking at the virtual index, though potential solutions include: flushing the cache after a context switch, forcing address spaces to be non-overlapping, tagging the virtual address with an address space ID (ASID), or using physical tags. Additionally, there is a problem that virtual-to-physical mappings can change, which would require flushing cache lines, as the VAs would no longer be valid.

- **Virtually indexed, physically tagged** (VIPT) caches use the virtual address for the index and the physical address in the tag. The advantage over PIPT is lower latency, as the cache line can be looked up in parallel with the TLB translation, however the tag can't be compared until the physical address is available. The advantage over VIVT is that since the tag has the physical address, the cache can detect homonyms. VIPT requires more tag bits, as the index bits no longer represent the same address.

Physically indexed, virtually tagged caches are only theoretical as they would basically be useless.

7.8 MEMORY ALLOCATION

In computer science, **dynamic memory allocation** (also known as **heap-based memory allocation**) is the allocation of memory storage for use in a computer program during the runtime of that program. It can be seen also as a way of distributing ownership of limited memory resources among many pieces of data and code.

Dynamically allocated memory exists until it is released either explicitly by the programmer, or by the garbage collector. This is in contrast to static memory allocation, which has a fixed duration. It is said that an object so allocated has a *dynamic lifetime.*

The task of fulfilling an allocation request consists of finding a block of unused memory of sufficient size.

- Problems during fulfilling allocation request
  - Internal and external fragmentation.
    - Reduction needs special care, thus making implementation more complex (see algorithm efficiency).
  - Allocator's metadata can inflate the size of (individually) small allocations;
    - Chunking attempts to reduce this effect.

Usually, memory is allocated from a large pool of unused memory area called **the heap** (also called the **free store**). Since the precise location of the allocation is not known in advance, the memory is accessed indirectly, usually via a pointer reference. The precise algorithm used to organize the memory area and allocate and deallocate chunks is hidden behind an abstract interface and may use any of the methods described below.

7.8.1. Implementations

Fixed-Size-Blocks Allocation

Fixed-size-blocks allocation, also called memory pool allocation, uses a free list of fixed-size blocks of memory (often all of the same size). This works well for simple embedded systems.

Buddy Blocks

In this system, memory is allocated from a large block in memory that is a power of two in size. If the block is more than twice as large as desired, it is broken in two. One of the halves is selected, and the process repeats (checking the size again and splitting if needed) until the block is just large enough.

All the blocks of a particular size are kept in a sorted linked list or tree. When a block is freed, it is compared to its buddy. If they are both free, they are combined and placed in the next-largest size buddy-block list. (When a block is allocated, the allocator will start with the smallest sufficiently large block avoiding needlessly breaking blocks)

7.9 CACHE MEMORY

Cache memory is the level of computer memory hierarchy situated between the processor and main memory. It is a very fast memory the processor can access much more quickly than main memory or RAM. Cache is relatively small and expensive. Its function is to keep a copy of the data and code (instructions) currently used by the CPU. By using cache memory, waiting states are significantly reduced and the work of the processor becomes more effective.

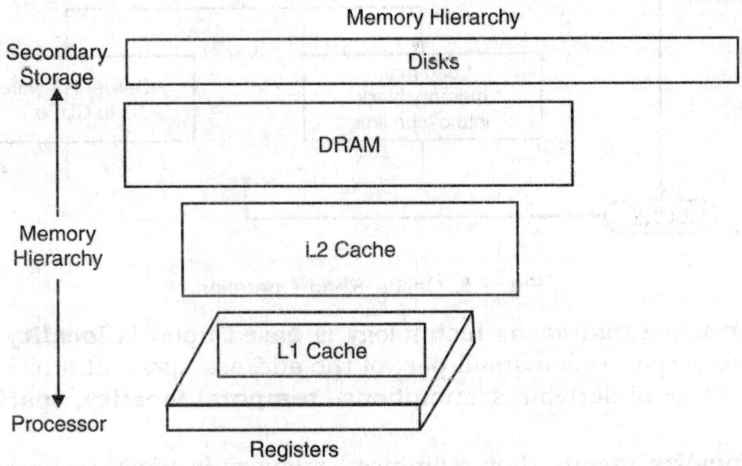

Fig. 7.4. Memory Hierarchy

Cache is much faster than main memory because it is implemented using SRAM (Static Random Access Memory). The problem with DRAM, which comprises main memory, is that it is composed entirely of capacitors, which have to be constantly refreshed in order to preserve the stored information (leakage current). Whenever data is read from the cell, the cell is refreshed. The DRAM cells need to be refreshed very frequently, *i.e.* typically every 4 to 16 ms. This slows down the entire process. SRAM on the other hand consists of flip-flops, which stay in its state as long as the power supply is on. (A flip-flop is an electrical circuit composed of transistors and resistors. See picture) Because of this SRAM need not be refreshed and is over 10 times faster than DRAM.

7.9.1. Principles of Cache Memory

In general cache memory works by attempting to predict which memory the processor is going to need next, and loading that memory before the processor needs it, and saving the results after the processor is done with it. Whenever the byte at a given memory address is needed to be read, the processor attempts to get the data from the cache memory. If the cache doesn't have that data, the processor is halted while it is loaded from main memory into the cache. At that time memory around the required data is also loaded into the cache.

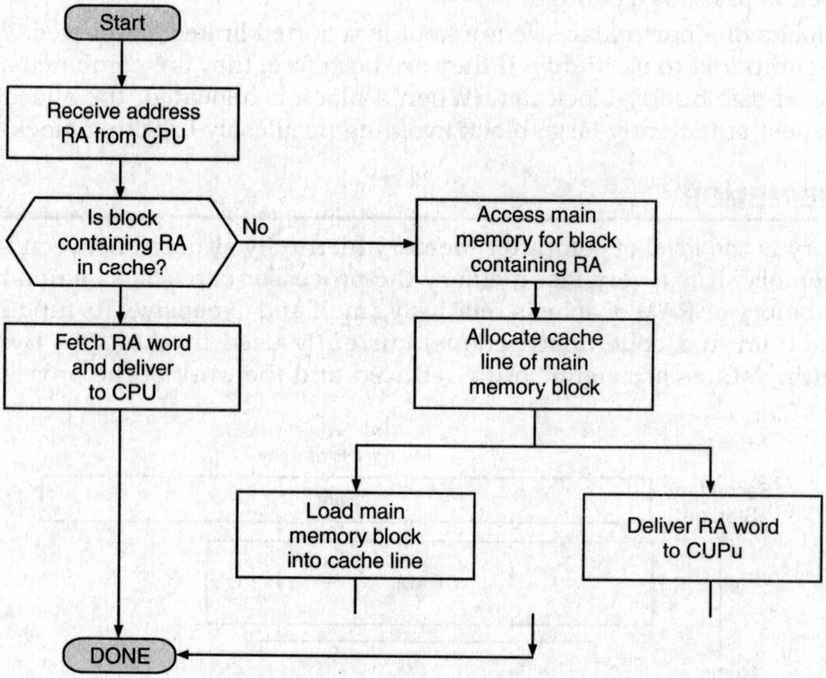

Fig. 7.5. Cache Read Operation

The basic principle that cache technology is based upon is **locality of reference** – programs tend to access only a small part of the address space at a given point in time. This notion has three underlying assumptions– **temporal locality, spatial locality and sequentiality.**

Temporal locality means that referenced memory is likely to be referenced again soon. In other words, if the program has referred to an address it is very likely that it will refer to it again.

Spatial locality means that memory close to the referenced memory is likely to be referenced soon. This means that if a program has referred to an address, it is very likely that an address in close proximity will be referred to soon.

Sequentiality means that the future memory access is very likely to be in sequential order with the current access.

7.9.2. How does Cache Memory Work?

Main memory consists of up to 2^n addressable words, with each word having a unique n-bit address. For mapping purposes, this memory is considered to consist of a number of fixed-length blocks of K words each. Thus, there are $M = 2^n / K$ blocks. The cache is split into C *lines* of K words each, Figure with number of lines considerably smaller than the number of main memory blocks (C << M). At any time, only a subset of the blocks of main memory resides in the lines in the cache. If a word in a block of memory is read, that block is transferred to one of the lines of the cache. Since there are more blocks than lines, an individual line cannot be uniquely and permanently dedicated to a particular block. Therefore, each line includes a tag that identifies which particular block of main memory is currently occupying that line of cache. The tag is usually a portion (number of bits) of the main memory address.

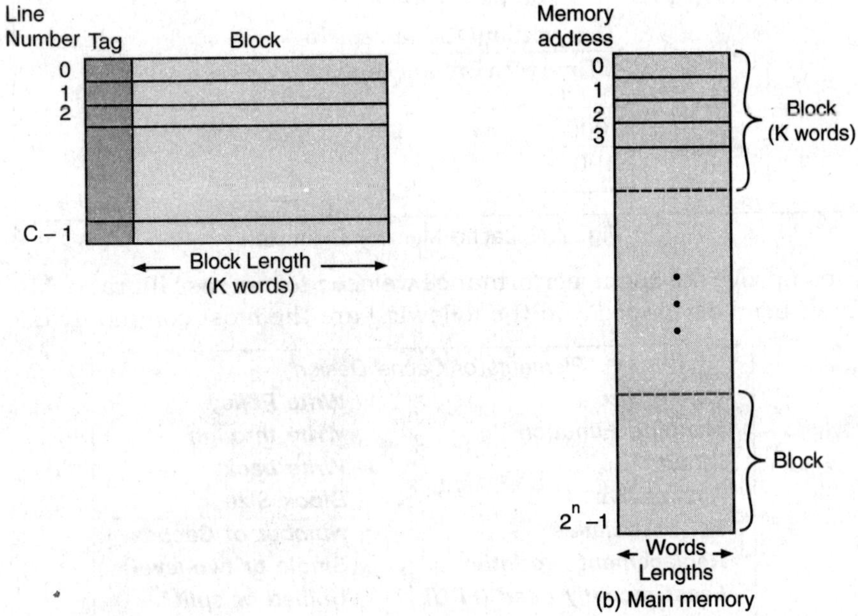

Fig. 7.6. Cache/Main – Memory Structure

7.9.3. Cache's Effectiveness

The effectiveness of the cache is determined by the number of times the cache successfully provides the required data. This is the place to introduce some terminology. A **hit** is the term used when the data, required by the processor is found in the cache. If data is not found, we have a cache **miss**. There are three types of misses.

Compulsory misses – the first access to a block is not in the cache, so the block must be brought into the cache. These are also called cold start misses or first reference misses.

Capacity misses – if the cache cannot contain all the blocks needed during execution of a program, capacity misses will occur due to blocks being discarded and later retrieved. These are misses due to cache size.

Conflict misses – if the block placement strategy is set associative or direct mapped, conflict misses will occur because a block can be discarded and later retrieved if too many blocks map to its set. These are also called collision misses or interference misses.

The time it takes to access cache on a hit is the **hit time**. If however, the CPU does not find the required data in the uppermost level of the memory hierarchy, in other words, when there is a miss, then the data is fetched from the lower levels of memory and replicated into the uppermost level. The time needed to fetch the block from the lower levels, replicate it and then access it, is called **miss penalty**. **Hit ratio** is the percentage of hits from total memory accesses. **Miss ratio** is equal to (1-hit ratio). One way to improve cache effectiveness is to reduce the number of misses and increase the number of hits; in other words we strive for high hit rate and low **miss rate** (*1-hit rate*).

Cache Memory Example :

D1 cache access in 10 ns; main memory access in 100 ns; 90% hit rate

For every 10 memory accesses, 9 will be to cache and 1 to main memory

Time without cach = 10 * 100 ns = 1000 ns

Time with cache = 9 * 10 ns + 1 * 100 ns = 190 ns

$$\text{Speedup} = \frac{\text{Time without Enhancement}}{\text{Time with Enhancement}}$$

$$\text{Speedup} = \frac{1000}{190} \,.\, 5.3$$

Fig. 7.7. Cache Memory Example

In order to improve the cache performance we need to increase hit ratio. Many different techniques have been developed and the following are the most commonly used ones.

Elements of Cache Design

Cache Size

Mapping Function

Direct

Associative

Set associative

Replacement Algorithm

Least recently used (LRU)

First-in-first-out (FIFO)

Least-frequently used (LFU)

Random

Write Policy

Write through

Write back

Block Size

Number of Caches

Single or two-level

Unified or split

Fig. 7.8. Elements of Cache Design

7.9.4. Cache Size

One way to decrease miss rate is to increase the cache size. The larger the cache, the more information it will hold, and the more likely a processor request will produce a cache hit, because fewer memory locations are forced to share the same cache line. However, applications benefit from increasing the cache size up to a point, at which the performance will stop improving as the cache size increases. When this point is reached, one of two things have happened: either the cache is large enough that the application almost never have to retrieve information from disk, or, the application is doing truly random accesses, and therefore increasing size of the cache doesn't significantly increase the odds of finding the next requested information in the cache. The latter is fairly rare - almost all applications show some form of locality of reference.

In general the cache has to be small enough so that the overall average "cost per bit" is close to that of main memory alone and large enough so that the overall average access time is close to that of the cache alone. There are several other reasons to minimize the size of the cache. The larger the cache, the larger the number of gates involved in addressing the cache. The result is that large caches tend to be slightly slower than small ones – even when using the same integrated circuit technology and put into the same place on the chip. Cache size is also limited by the available space on the chip.

Typical cache sizes today range from 1K words to several mega-words. Multitasking systems tend to favor 256 K words as nearly optimal size for main memory cache. The performance of the cache is very sensitive to the nature of the workload, making it impossible to arrive at a single "optimum" cache size.

7.9.5. Mapping Function

The mapping function gives the correspondence between main memory blocks and cache lines. Since each cache line is shared between several blocks of main memory (the number of memory blocks >> the number of cache lines), when one block is read in, another one should be moved out. Mapping functions minimize the probability that a moved-out block will be referenced again in the near future. There are three types of mapping functions: direct, fully associative, and n-way set associative.

Direct Mapping : In direct mapped cache each main memory block is assigned to a specific line in the cache according to the formula $i = j$ modulo C, where i is the cache line number assigned to main memory block j, and C is the number of lines in the cache. Direct mapping cache interprets a main memory address (comprising of $s + w$ bits) as 3 distinct fields.

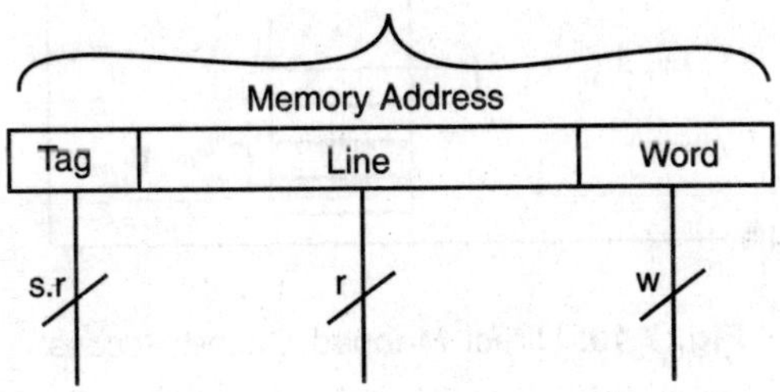

Fig. 7.9. Direct Mapped Cache Address

- Tag (the most significant $s - r$ bits) – a unique identifier for each memory block. It is stored in the cache along with the data words of the line specify one of the 2s blocks.
- Line number (middle r bits) – specifies which line will hold the referenced address.Line number identify one of the C = 2r lines of cache.
- Word (least significant w bits) – specifies the offset of a byte within a line or block. Word specifies the specific byte in a cache line that is to be accessed.[1]

For every memory reference that the CPU makes, the tag of the cache line that is supposed to hold the required block is checked to see if the correct block is in the cache. Since no two blocks that map into the same line have the same tag, the cache controller can determine if we have a cache hit or a cache miss.

Direct mapped cache is the simplest form of cache. It is easy and relatively inexpensive to implement, and determining whether a main memory block can be found in cache is simple and quicker than with other mapping functions. However, it has one significant disadvantage - each main memory block is mapped to a specific cache line. Through locality of reference, it is possible to repeatedly reference blocks that map to the same line number. These blocks will be constantly swapped in and out of cache, causing the hit ratio to be low. Although such swapping is rare in a single-tasking system, in multi-tasking systems it can occur quite often and thus slow down a direct-mapped cache. In general the performance is worst for this type of mapping.

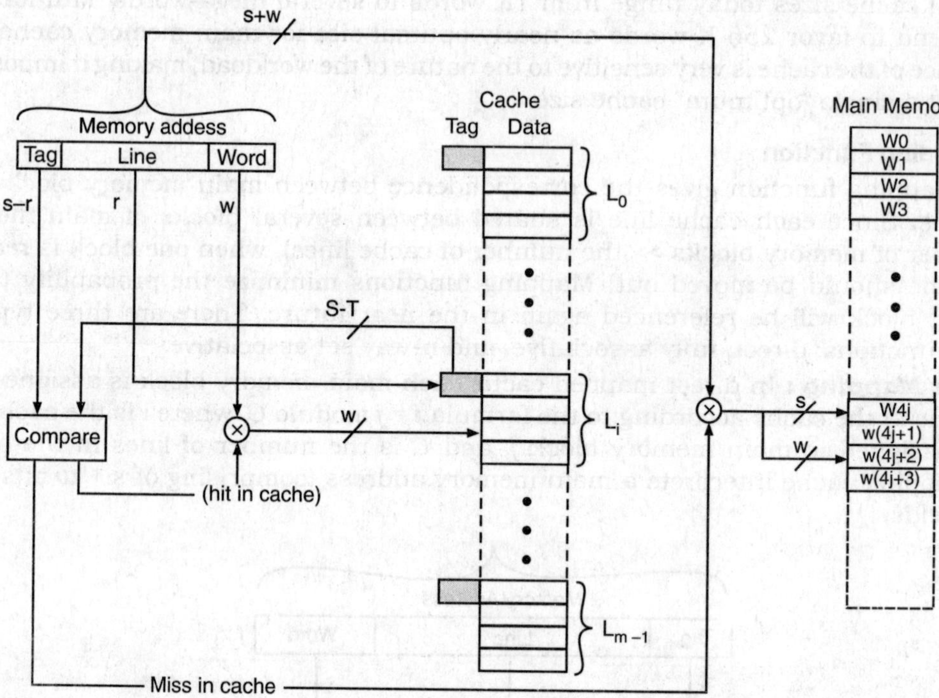

Fig. 7.10. Direct Mapped Cache Access

Fully Associative Mapping: Fully associative cache allows each block of main memory to be stored in any line of the cache. Main memory address ($s + w$ bits) is divided into two distinct fields:tag (s bits) and word offset (w bits).

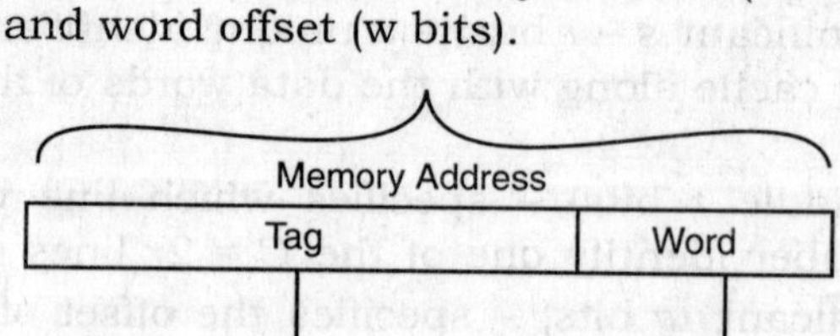

Fig. 7.11. Fully Associative Cache Address

The cache controller must now check the tag of every line in the cache to determine if a given memory address is currently resident in the cache. This is not done using common searching techniques but rather through the use of associative memory (also called Content Addressable Memory (CAM)). CAM basically allows the entire tag memory to be searched in parallel. Unlike typical RAM, CAM associates logic with each memory cell in the memory. This allows for the contents of all memory cells in the CAM to be checked in a single clock cycle. Thus access to the CAM is based upon content not address as with ordinary RAM. CAMs are considerably more expensive in terms of gates than ordinary access by address memories (RAMs) and this limits their use (at least with current technologies) to relatively small memory systems such as cache.

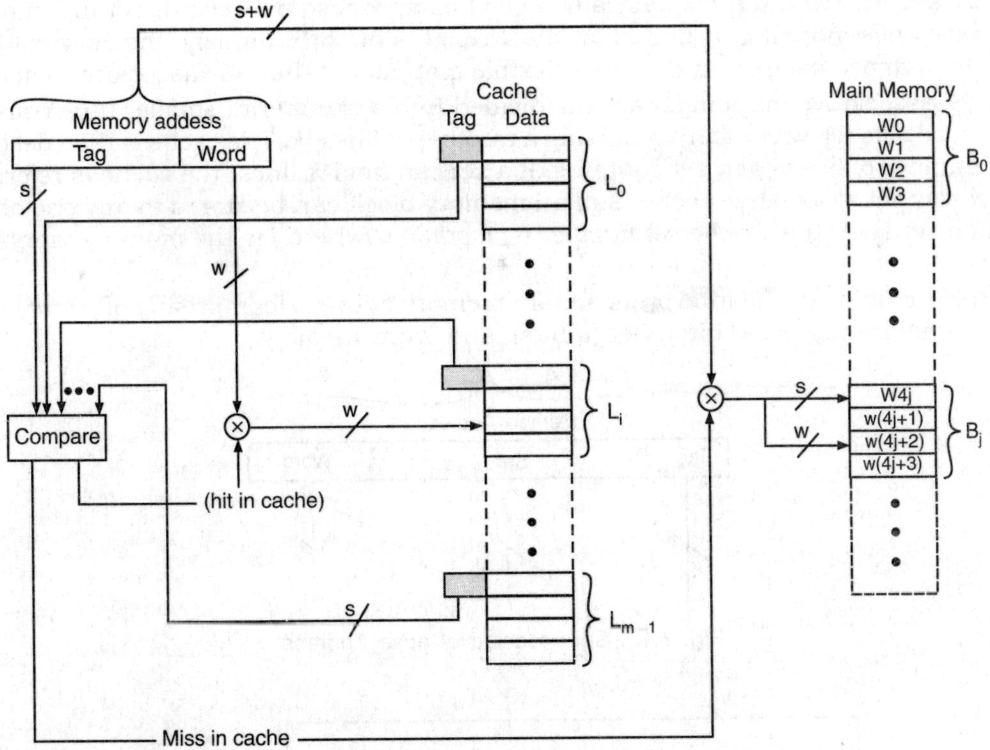

Fig. 7.12. Fully Associative Cache Access

The cache controller will need to uniquely identify every main memory block, which will require s bits since there are 2^s blocks in the main memory. As before, within the block, w bits will be required to uniquely identify a particular byte within a specific block. Since any cache line can hold any main memory block at any time, the cache controller needs to perform a simultaneous search over all tags to find the desired line (to see if it is in the cache). This is where the CAM comes into play. The entire contents of the cache can be compared simultaneously with the CAM. A cache hit is indicated when one of the CAM cells contents matches the search address. If none of the CAM cells contents matches the search address then a cache miss has occurred and the proper block from main memory will need to be loaded into one of the cache lines. At this point the cache controller will invoke the line replacement algorithm, which will select which cache line is to be replaced to make room for the new incoming block. [13]

Fully associative mapping is very flexible and overcomes direct mapping's main weakness. The fully associative cache has the best hit ratio because any line in the cache can hold any address that needs to be cached. However this cache suffers from problems involving searching the cache. Even with specialized hardware to do the searching, a performance penalty is incurred. And this penalty occurs for *all* accesses to memory, whether a cache hit occurs or not. In addition, more logic must be added to determine which of the various lines to use when a new entry must be added (usually some form of a "least recently used" algorithm is employed to decide which cache line to use next). All this overhead adds cost, complexity and execution time. That is why, fully associative cache is very rarely used.

Set Associative Cache is basically a good compromise between direct mapping and fully associative mapping. It builds on the strengths of both: namely, the easy control of the direct mapped cache and the more flexible mapping of the fully associative cache.

In set associative mapping, cache is divided into a number of smaller direct-mapped areas called sets (v), with each set holding a number of lines (k).The cache is then described in the number of lines each set contains. If a set can hold X lines, the cache is referred to as an X-way set associative cache. A main memory block can be stored in any one of the k lines in a set such that *cache set number* = *j modulo v* (where j is the main memory block number).

Direct mapping cache interprets a main memory address (comprising of $s + w$ bits) as 3 distinct fields: Tag ($s - d$ bits), Set (d bits), and Word (w bits).

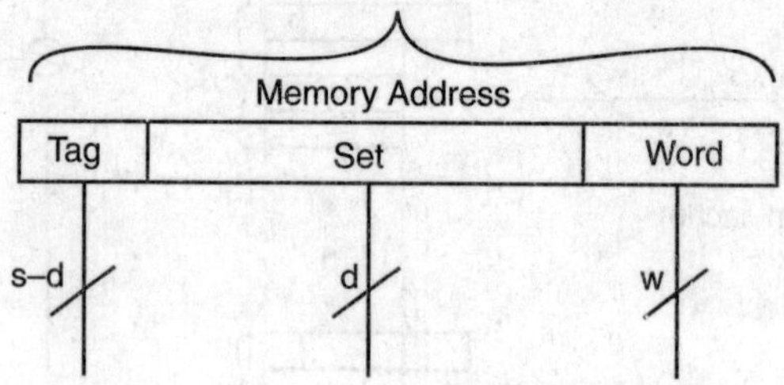

Fig. 7.13. Set-AssociativeCache Address

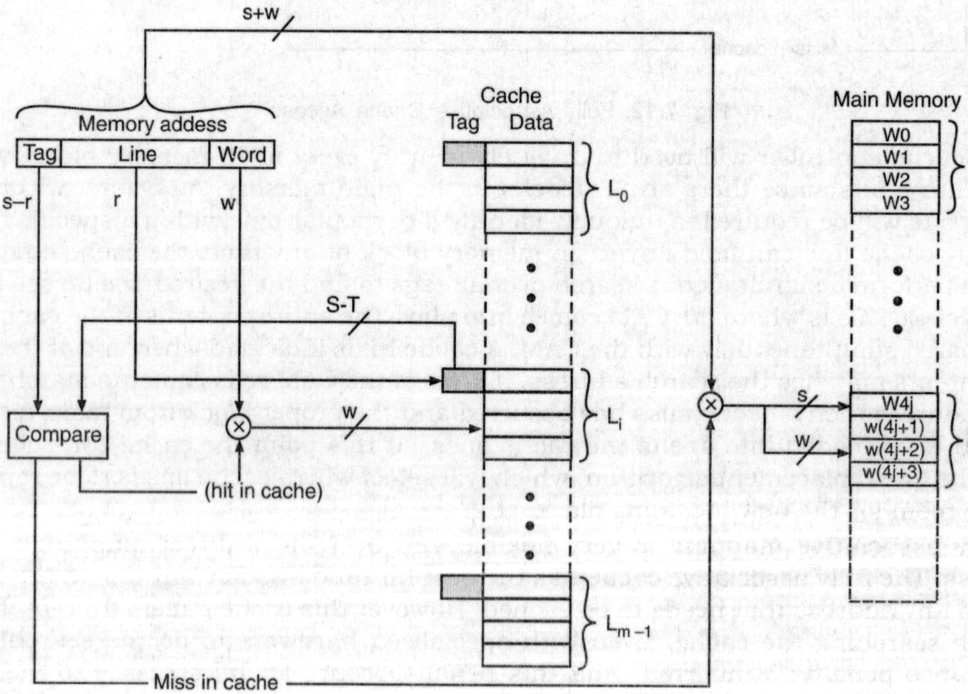

Fig. 7.14. Set-Associative Cache Access

For each main memory group, the cache is capable of holding v different main memory blocks simultaneously. A four-way set associative cache could hold up to four main memory blocks from the same group simultaneously, while an eight-way set associative cache would be capable of holding up to eight main memory blocks from the same group simultaneously, and so on whenever the CPU issues a block request, the cache controller for the k-way set associative cache will need to check the contents of k different cache lines for the particular group to which the required block belongs. Once again, the use of CAM will allow the contents of all cache lines to be checked in parallel. If one of the "*k*" cache lines contains the requested block, a cache hit occurs, otherwise a cache miss occurs and one of the two cache lines belonging to that group will need to be selected for replacement. In general, set associative mapping has better hit ratio than direct mapped cache and is faster than fully associative, therefore providing better performance than the other two mapping functions. 2 and 4 way associative caches give the best ratio between hit ratio and search speed.

Summary : In the "real world", the direct mapped and set associative caches are by far the most common. Direct mapping is used more for level 2 caches on motherboards, while the higher-performance set-associative cache is found more commonly on the smaller primary caches contained within processors.

Cache Type	Hit Ratio	Search Speed
Direct Mapped	Good	**Best**
Fully Associative	**Best**	Moderate
N-Way Set Associative, *N* > 1	Very Good, Better as *N* Increases	Good, Worse as *N* Increases

Fig. 7.15. Mapping Function Comparison Table

7.9.6. Cache Write Policies

Before a cache line can be replaced, it is necessary to determine if the line has been modified. The contents of the main memory block and the cache line that corresponds to that block are essentially copies of each other and should therefore hold the same data. If cache line X has not been modified since its arrival in the cache, updating the main memory block it corresponds to is not required prior to its replacement. The incoming cache line can simply overwrite the existing cache memory. On the other hand, if the cache line has been modified, at least one write operation has been performed on the cache line and thus the corresponding main memory block must be updated. Basically there are two different policies that can be employed to ensure that the cache and main memory contents are the same: write-through and write-back.

***Write-through* :** Assuming a cache hit (a write hit), the information is written immediately to both the line in the cache and to the block in the lower-level memory (with its normal wait state delays). The advantages of this technique are that the contents of the main memory and the cache are always consistent, it is easy to implement and read miss never results in writes to main memory. On the other hand, the write through policy has a significant drawback. It needs a main memory access, which is slower and results in a more memory bandwidth usage. In spite of this, most Intel microprocessors use a write-through cache design.

***Write-back* (sometimes called a *posted write or copy back* cache) :** On a cache hit, the information is written only to the line in the cache. This allows the processor to immediately resume processing. The modified cache line is written to main memory only when it is replaced. To reduce the frequency of writing back blocks on replacement, a *dirty bit* is commonly used. This status bit indicates whether the block is dirty (modified while in the cache) or clean (not modified). If it is clean the block is not written on a miss. The advantages of the write-back policy are that writes occur at the speed of the cache memory, multiple writes within a block require only one write to main memory, which results in less memory bandwidth usage.

Write-back is a faster alternative to the write-through policy but it has one major disadvantage - the contents of the cache and the main memory are not consistent. This is the cache coherency problem and it is an active research topic. The cache coherency becomes an issue, for example, when a hard disk is read and information is transferred into the main memory through the DMA (Direct Memory Access) system, which does not involve the processor. The cache controller must constantly monitor the changes made in the main memory and ensure that the contents of the cache properly track these changes to the main memory. There are many techniques, which have been employed to allow the cache controller to "snoop" the memory system - but once again, these add complexity and expense. In the PC environment there are special cache controller chips that can be added which basically handle all the responsibilities for supervising the cache system. The cache coherency problem becomes acute in a bus architecture in which more than one device (typically processors) has a cache and main memory is shared.

Writes introduce yet another complication not present in reads. If a write operation has been requested and a cache miss results, again one of two options for handling the write miss can be employed. The line may be brought into the cache, and then updated, which is termed a *write-allocate* policy, or the block may be updated directly in main memory and not brought into the cache which is termed a *write-no allocate* policy. Typically, write-through caches will employ a write-no allocate policy while write-back caches will utilize a write-allocate policy.

Write hit policy	**Write miss policy**
Write Through	Write Allocate
Write Through	No Write Allocate
Write Back	Write Allocate
Write Back	No Write Allocate

Fig. 7.16. Possible Combinations on write

Write Through with Write Allocate : On hits it writes to cache and main memory, on misses it updates the block in main memory and brings the block to the cache. Bringing the block to cache on a miss does not make a lot of sense in this combination because the next hit to this block will generate a write to main memory anyway (according to Write Through policy).

Write Through with No Write Allocate : On hits it writes to cache and main memory, on misses it updates the block in main memory not bringing that block to the cache; Subsequent writes to the block will update main memory because Write Through policy is employed. So, some time is saved not bringing the block in the cache on a miss because it appears useless anyway.

Write Back with Write Allocate : On hits it writes to cache setting "dirty" bit for the block, main memory is not updated, on misses it updates the block in main memory and brings the block to the cache. Subsequent writes to the same block, if the block originally caused a miss, will hit in the cache next time, setting dirty bit for the block. That will eliminate extra memory accesses and result in very efficient execution compared with Write Through with Write Allocate combination.

Write Back with No Write Allocate : On hits it writes to cache setting "dirty" bit for the block, main memory is not updated, on misses it updates the block in main memory not bringing that block to the cache. Subsequent writes to the same block, if the block originally caused a miss, will generate misses all the way and result in very inefficient execution.

7.10. WHAT IS DATA COMMUNICATIONS?

The distance over which data moves within a computer may vary from a few thousandths of an inch, as is the case within a single IC chip, to as much as several feet along the backplane of the main circuit board. Over such small distances, digital data may be transmitted as direct, two-level electrical signals over simple copper conductors. Except for the fastest computers, circuit designers are not very concerned about the shape of the conductor or the analog characteristics of signal transmission.

Frequently, however, data must be sent beyond the local circuitry that constitutes a computer. In many cases, the distances involved may be enormous. Unfortunately, as the distance between the source of a message and its destination increases, accurate transmission becomes increasingly difficult. This results from the electrical distortion of signals traveling through long conductors, and from noise added to the signal as it propagates through a transmission medium. Although some precautions must be taken for data exchange within a computer, the biggest problems occur when data is transferred to devices outside the computer's circuitry. In this case, distortion and noise can become so severe that information is lost.

Data communications concerns the transmission of digital messages to devices external to the message source. "External" devices are generally thought of as being independently powered circuitry that exists beyond the chassis of a computer or other digital message source. As a rule, the maximum permissible transmission rate of a message is directly proportional to signal power, and inversely proportional to channel noise. It is the aim of any communications system to provide the highest possible transmission rate at the lowest possible power and with the least possible noise.

7.10.1. Communications Channels

A communications channel is a pathway over which information can be conveyed. It may be defined by a physical wire that connects communicating devices, or by a radio, laser, or other radiated energy source that has no obvious physical presence. Information sent through a communications channel has a source from which the information originates, and a destination to which the information is delivered. Although information originates from a single source, there may be more than one destination, depending upon how many receive stations are linked to the channel and how much energy the transmitted signal possesses.

In a digital communications channel, the information is represented by individual data bits, which may be encapsulated into multibit message units. A byte, which consists of eight bits, is an example of a message unit that may be conveyed through a digital communications channel. A collection of bytes may itself be grouped into a frame or other higher-level message unit. Such multiple levels of encapsulation facilitate the handling of messages in a complex data communications network.

The message source is the transmitter, and the destination is the receiver. A channel whose direction of transmission is unchanging is referred to as a simplex channel. For example, a radio station is a simplex channel because it always transmits the signal to its listeners and never allows them to transmit back.

Any communications channel has a direction associated with it :

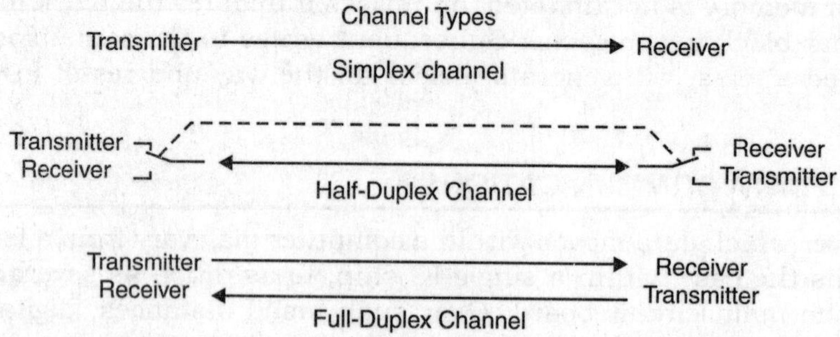

Fig. 7.17.

A half-duplex channel is a single physical channel in which the direction may be reversed. Messages may flow in two directions, but never at the same time, in a half-duplex system. In a telephone call, one party speaks while the other listens. After a pause, the other party speaks and the first party listens. Speaking simultaneously results in garbled sound that cannot be understood.

A full-duplex channel allows simultaneous message exchange in both directions. It really consists of two simplex channels, a forward channel and a reverse channel, linking the same points. The transmission rate of the reverse channel may be slower if it is used only for flow control of the forward channel.

7.10.2. Serial Communications

Most digital messages are vastly longer than just a few bits. Because it is neither practical nor economic to transfer all bits of a long message simultaneously, the message is broken into smaller parts and transmitted sequentially. Bit-serial transmission conveys a message one bit at a time through a channel. Each bit represents a part of the message. The individual bits are then reassembled at the destination to compose the message. In general, one channel will pass only one bit at a time. Thus, bit-serial transmission is necessary in data communications if only a single channel is available. Bit-serial transmission is normally just called serial transmission and is the chosen communications method in many computer peripherals.

Byte-serial transmission conveys eight bits at a time through eight parallel channels. Although the raw transfer rate is eight times faster than in bit-serial transmission, eight channels are needed, and the cost may be as much as eight times higher to transmit the message. When distances are short, it may nonetheless be both feasible and economic to use parallel channels in return for high data rates. The popular centronics printer interface is a case where byte-serial transmission is used. As another example, it is common practice to use a 16-bit-wide data bus to transfer data between a microprocessor and memory chips; this provides the equivalent of 16 parallel channels. On the other hand, when communicating with a timesharing system over a modem, only a single channel is available, and bit-serial transmission is required. This figure illustrates these ideas :

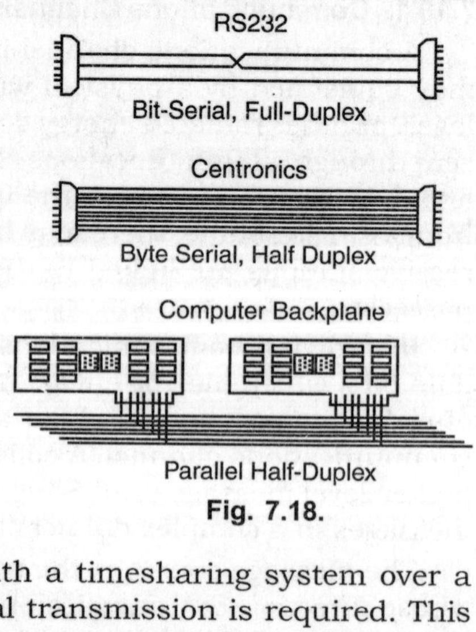

Fig. 7.18.

The baud rate refers to the signalling rate at which data is sent through a channel and is measured in electrical transitions per second. In the EIA232 serial interface standard, one signal transition, at most, occurs per bit, and the baud rate and bit rate are identical. In this case, a rate of 9600 baud corresponds to a transfer of 9,600 data bits per second with a bit period of 104 microseconds (1/9600 sec.). If two electrical transitions were required for each bit, as is the case in non-return-to-zero coding, then at a rate of 9600 baud, only 4800 bits per second could be conveyed. The channel efficiency is the number of bits of useful information passed through the channel per second. It does not include framing, formatting, and error detecting bits that may be added to the information bits before a message is transmitted, and will always be less than one.

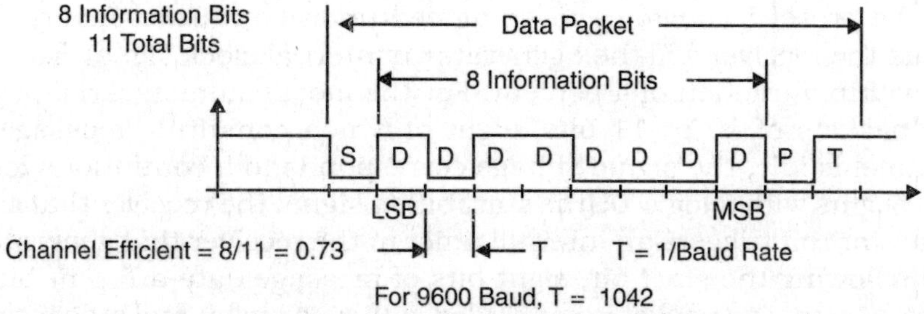

Fig. 7.19.

The data rate of a channel is often specified by its bit rate (often thought erroneously to be the same as baud rate). However, an equivalent measure channel capacity is bandwidth. In general, the maximum data rate a channel can support is directly proportional to the channel's bandwidth and inversely proportional to the channel's noise level.

A communications protocol is an agreed-upon convention that defines the order and meaning of bits in a serial transmission. It may also specify a procedure for exchanging messages. A protocol will define how many data bits compose a message unit, the framing and formatting bits, any error-detecting bits that may be added, and other information that governs control of the communications hardware. Channel efficiency is determined by the protocol design rather than by digital hardware considerations. Note that there is a tradeoff between channel efficiency and reliability - protocols that provide greater immunity to noise by adding error-detecting and -correcting codes must necessarily become less efficient.

7.10.3. Asynchronous vs. Synchronous Transmission

Serialized data is not generally sent at a uniform rate through a channel. Instead, there is usually a burst of regularly spaced binary data bits followed by a pause, after which the data flow resumes. Packets of binary data are sent in this manner, possibly with variable-length pauses between packets, until the message has been fully transmitted. In order for the receiving end to know the proper moment to read individual binary bits from the channel, it must know exactly when a packet begins and how much time elapses between bits. When this timing information is known, the receiver is said to be synchronized with the transmitter, and accurate data transfer becomes possible. Failure to remain synchronized throughout a transmission will cause data to be corrupted or lost.

Two basic techniques are employed to ensure correct synchronization. In synchronous systems, separate channels are used to transmit data and timing information. The timing channel transmits clock pulses to the receiver. Upon receipt of a clock pulse, the receiver reads the data channel and latches the bit value found on the channel at that moment.

The data channel is not read again until the next clock pulse arrives. Because the transmitter originates both the data and the timing pulses, the receiver will read the data channel only when told to do so by the transmitter (via the clock pulse), and synchronization is guaranteed.

Techniques exist to merge the timing signal with the data so that only a single channel is required. This is especially useful when synchronous transmissions are to be sent through a modem. Two methods in which a data signal is self-timed are nonreturn-to-zero and biphase Manchester coding. These both refer to methods for encoding a data stream into an electrical waveform for transmission.

In asynchronous systems, a separate timing channel is not used. The transmitter and receiver must be preset in advance to an agreed-upon baud rate. A very accurate local oscillator within the receiver will then generate an internal clock signal that is equal to the transmitter's within a fraction of a percent. For the most common serial protocol, data is sent in small packets of 10 or 11 bits, eight of which constitute message information. When the channel is idle, the signal voltage corresponds to a continuous logic '1'. A data packet always begins with a logic '0' (the start bit) to signal the receiver that a transmission is starting. The start bit triggers an internal timer in the receiver that generates the needed clock pulses. Following the start bit, eight bits of message data are sent bit by bit at the agreed upon baud rate. The packet is concluded with a parity bit and stop bit. One complete packet is illustrated below :

Typical Format
1 St art Bit
8 data Bits
1 parity Bit
1 Stop Bit

Baud Rate	T
110	9.09 ms
300	3.33 ms
1200	833 μs
2400	417 μs
4800	208 μs
9600	104 μs
19200	52 μs

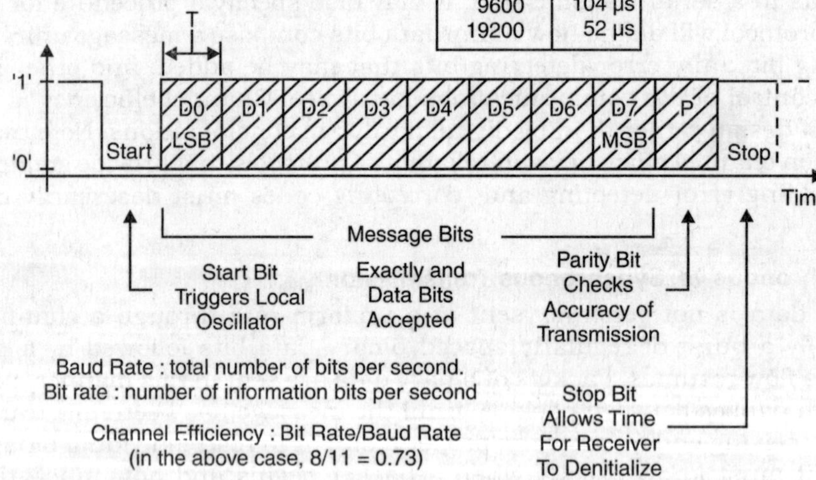

Baud Rate : total number of bits per second.
Bit rate : number of information bits per second

Channel Efficiency : Bit Rate/Baud Rate
(in the above case, 8/11 = 0.73)

Fig. 7.20.

The packet length is short in asynchronous systems to minimize the risk that the local oscillators in the receiver and transmitter will drift apart. When high-quality crystal oscillators are used, synchronization can be guaranteed over an 11-bit period. Every time a new packet is sent, the start bit resets the synchronization, so the pause between packets can be arbitrarily long. Note that the EIA232 standard defines electrical, timing, and mechanical characteristics of a serial interface. However, it does not include the asynchronous serial protocol shown in the previous figure, or the ASCII alphabet described next.

7.11 BUS CONTROL

The basic operational design of a computer system is called its *architecture*. John Von Neumann a pioneer in computer design is given credit for the architecture of most computers in use today. For example the 80 × 86 family uses the *Von Neumann architecture* (VNA). A typical Von Neumann system has three major components: the *central processing unit* (or *CPU*) *memory* and *input/output* (or *input/output*). The way a system designer combines these components impacts system performance:

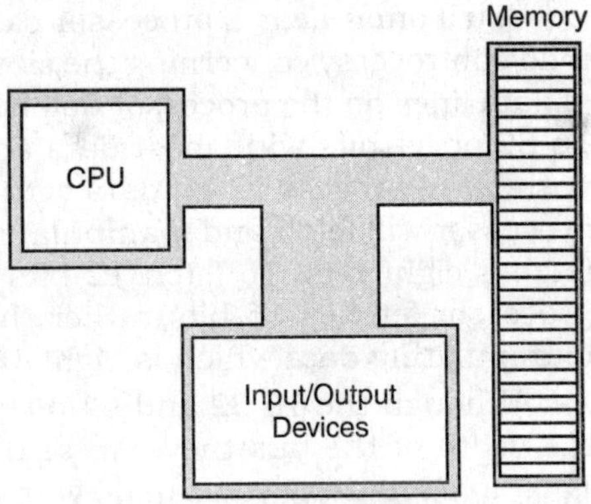

Fig. 7.21.

In VNA machines like the 80 × 86 family the CPU is where all the action takes place. All computations occur inside the CPU. Data and CPU instructions reside in memory until required by the CPU. To the CPU most I/O devices look like memory because the CPU can store data to an output device and read data from an input device. The major difference between memory and I/O locations is the fact that I/O locations are generally associated with external devices in the outside world.

7.11.1. The System Bus

The *system bus* connects the various components of a VNA machine. The 80 × 86 family has three major busses: the *address* bus the *data* bus and the *control* bus. A bus is a collection of wires on which electrical signals pass between components in the system. These busses vary from processor to processor. However each bus carries comparable information on all processors; *e.g.* the data bus may have a different implementation on the 80386 than on the 8088 but both carry data between the processor input/output and memory.

A typical 80x86 system component uses *standard TTL logic levels.* This means each wire on a bus uses a standard voltage level to represent zero and one. We will always specify zero and one rather than the electrical levels because these levels vary on different processors (especially laptops).

7.11.2. The Data Bus

The 80x86 processors use the *data bus* to shuffle data between the various components in a computer system. The size of this bus varies widely in the 80x86 family. Indeed this bus defines the "size" of the processor.

On typical 80 × 86 systems the data bus contains eight 16 32 or 64 lines. The 8088 and 80188 microprocessors have an eight bit data bus (eight data lines). The 8086 80186 80286 and 80386SX processors have a 16 bit data bus. The 80386DX 80486 and Pentium Overdrive' processors have a 32 bit data bus. The Pentium' and Pentium Pro processors have a 64 bit data bus. Future versions of the chip (the 80686/80786?) may have a larger bus.

Having an eight bit data bus does not limit the processor to eight bit data types. It simply means that the processor can only access one byte of data per memory cycle (see "The Memory Subsystem" for a description of memory cycles). Therefore the eight bit bus on an 8088 can only transmit half the information per unit time (memory cycle) as the 16 bit bus on the 8086. Therefore processors with a 16-bit bus are naturally faster than processors with an eight bit bus. Likewise processors with a 32-bit bus are faster than those with a 16 or eight bit data bus. The size of the data bus affects the performance of the system more than the size of any other bus.

You'll often hear a processor called an *eight 16 32 or 64 bit processor.* While there is a mild controversy concerning the size of a processor most people now agree that the number of data lines on the processor determines its size. Since the 80 × 86 family busses are eight 16 32 or 64-bits wide most data accesses are also eight 16 32 or 64-bits. Although it is possible to process 12-bit data with an 8088 most programmers process 16-bits since the processor will fetch and manipulate 16-bits anyway. This is because the processor always fetches eight bits. To fetch 12-bits requires two eight bit memory operations. Since the processor fetches 16-bits rather than 12 most programmers use all 16-bits. In general manipulating data which is eight 16 32 or 64 bits in length is the most efficient.

Although the 16 32 and 64 bit members of the 80 × 86 family *can* process data up to the width of the bus they can also access smaller memory units of eight 16 or 32-bits. Therefore anything you can do with a small data bus can be done with a larger data bus as well; the larger data bus however may access memory faster and can access larger chunks of data in one memory operation. You'll read about the exact nature of these memory accesses a little later (see "The Memory Subsystem").

80 × 86 Processor Data Bus Sizes

Processor	Data Bus Size
8088	8
80188	8
8086	16
80186	16
80286	16
80386sx	16
80386dx	32
80486	32
80586 class/ Pentium (Pro)	64

7.11.3. The Address Bus

The data bus on an 80 × 86 family processor transfers information between a particular memory location or input/output device and the CPU. The only question is "*Which memory location or input/output device?*" The address bus answers that question. To differentiate memory locations and input/output devices the system designer assigns a unique memory address to each memory element and input/output device. When the software wants to access some particular memory location or input/output device it places the corresponding address on the address bus. Circuitry associated with the memory or input/output device recognizes this address and instructs the memory or input/output device to read the data from or place data on the data bus. In either case all other memory locations ignore the request. Only the device whose address matches the value on the address bus responds.

With a single address line a processor could create exactly two unique addresses: zero and one. With *n* address lines the processor can provide 2***n* unique addresses (since there are 2***n* unique values in an *n*-bit binary number). Therefore the number of bits on the address bus will determine the *maximum* number of addressable memory and I/O locations. The 8088 and 8086 for example have 20-bit address busses. Therefore they can access up to 1 048 576 (or 2**20) memory locations. Larger address busses can access more memory. The 8088 and 8086 for example suffer from an anemic address space - their address bus is too small. Later processors have larger address busses:

80 × 86 Family Address Bus Sizes

Processor	Address Bus Size	Max Addressable	Memory In English!
8088	20	1 048 576	One Megabyte
8086	20	1 048 576	One Megabyte
80188	20	1 048 576	One Megabyte
80186	20	1 048 576	One Megabyte
80286	24	16 777 216	Sixteen Megabytes
80386sx	24	16 777 216	Sixteen Megabytes
80386dx	32	4 294 976 296	Four Gigabytes
80486	32	4 294 976 296	Four Gigabytes
80586 / Pentium (Pro)	32	4 294 976 296	Four Gigabytes

Future 80 × 86 processors will probably support 48-bit address busses. The time is coming when most programmers will consider four gigabytes of storage to be too small much like they consider one megabyte insufficient today. (There was a time when one megabyte was considered far more than anyone would ever need!) Fortunately the architecture of the 80386 80486 and later chips allow for an easy expansion to a 48-bit address bus through *segmentation.*

7.11.4. The Control Bus

The control bus is an eclectic collection of signals that control how the processor communicates with the rest of the system. Consider for a moment the data bus. The CPU sends data to memory and receives data from memory on the data bus. This prompts the question "Is it sending or receiving?" There are two lines on the control bus *read* and *write* which specify the direction of data flow. Other signals include system clocks interrupt lines status lines and so on. The exact make up of the control bus varies among processors in the 80 × 86 family. However some control lines are common to all processors and are worth a brief mention.

The *read* and *write* control lines control the direction of data on the data bus. When both contain a logic one the CPU and memory-input/output are not communicating with one another. If the read line is low (logic zero) the CPU is reading data from memory (that is the system is transferring data from memory to the CPU). If the write line is low the system transfers data from the CPU to memory.

The byte enable lines are another set of important control lines. These control lines allow 16 32 and 64-bit processors to deal with smaller chunks of data. Additional details appear in the next section.

The 80 × 86 family unlike many other processors provides two distinct address spaces: one for memory and one for input/output. While the memory address busses on various 80x86 processors vary in size the input/output address bus on all 80 × 86 CPUs is 16-bits wide. This allows the processor to address up to 65 536 different input/output *locations*. As it turns out most devices (like the keyboard printer disk drives etc.) require more than one input/output location. Nonetheless 65 536 input/output locations are more than sufficient for most applications. The original IBM PC design only allowed the use of 1 024 of these.

SOLVED PROBLEMS

Q.1. Define Multiprogramming.

Solution: Multiprogramming.

Many operating systems are designed to enable the CPU to process a number of independent programs concurrently, this concept, called multiprogramming refers to the existance of two or more programs in different parts of the memory hierarchy at the same time.

Q.2. What is associate memory? Make the diagram of associative memory.

Solution: The time required to find an item stored in memory can be reduced considerably if stored data can be identified fo access by the content of the data itself rather than by an address. A memory unit accessed by content is caled an associate memory or Content Addressable Memory (CAM). This type of memory is accessed simultaneously in parallel on the basis of data content rather than by specific address or location. When a world is written in an associative memory, no address is given. The memory is capable of finding an empty unused location to store the word. When a word is to be read from an associative memory, the content of the word, or part of the word, is specified.

An associative memory is more expensive than RAM, because each cell must have storage capability as well as logic circuits for matching its content wtih an external argument.

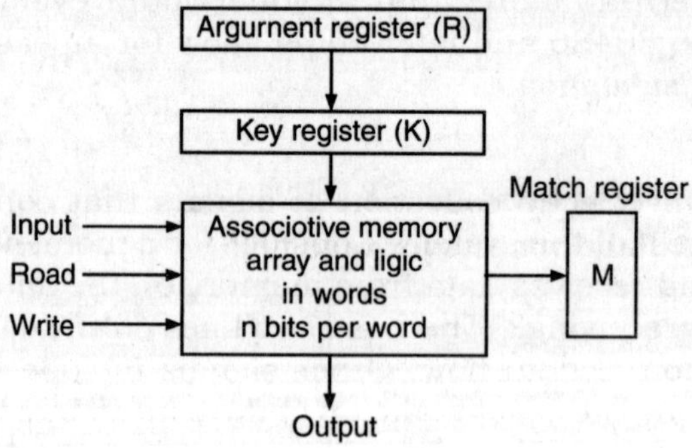

Fig. 7.22. Block diagram of associative memory

Q.3. Define:

Solution: (*a*) Locality of reference

(*b*) Hit ratio

(*c*) Mapping.

Solution: (*a*) Location of reference : Analysis of a large number of typical programs has shown that the references to memory at any given interval of time tend to be confined within a few localized areas in memory. This phenomenon is known as the property of locality of reference.

(b) Hit ratio : The performance of cache memory is frequently measured in terms of a quantity called hit ratio. When the CPU refers to memory and finds the words in cache, it is said to produce hit. If word is not found in cache, it is in main memory and it counts as a miss. The ratio of the number of hits divided by the total CPU references to memory (hits + misses) is the hit ratio. hit ratio of 0.9 and higher have been reported. This high ratio verifies the validity of the locality of reference property.

(c) Mapping : The transformation of data from main memory to cache memory is referred to as mapping process.

There are three types of mapping procedure are of partical interest when considering the organization of cache memory:

(*i*) Associative mapping

(*ii*) Direct mapping

(*iii*) Set associative mapping

Q.4. Describe the direct mapping.

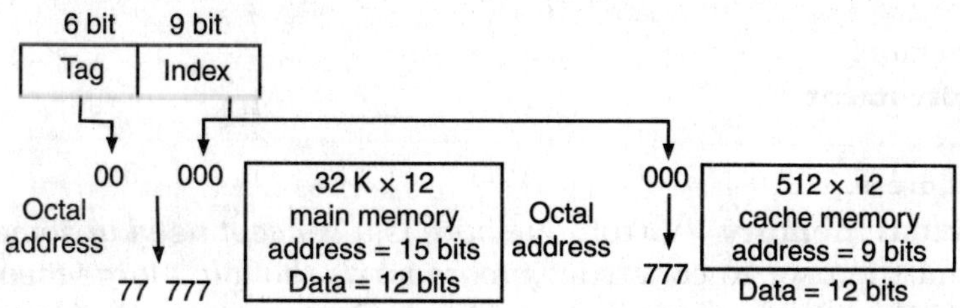

Fig. 7.23.

Solution: The CPU address of 15-bits is divided into two fields. The nine least significant bits constitutes the index field and remaining six bits from the tag field.

Figure shows that main memory needs an address that includes both the tag and the index bits. The number of 1-bit in the index field is equal to the number of address bits required to access the cache memory.

In the general case, there are 2^h words in cache memory and 2^n words i main memory. The n bit memory address is divided into two fields:

k bits for the index field and nk bits for tage field,

The direct mapping cache organization uses the n-bit address to access the main memory and the k-bit index to access the cache, when a new words is first brought into cache, the tag bits are stored along side the data bits. When the CPU generates the memory requests, the index field is used for the address to access the cache the tag field opf the CPU address as compared with the tag in the word read frm the cache. If two tages are matched, there is a hit and desired data world is in cache. If there is no match, there is a miss and then the required word is read from main memory. It is then stored in the cache together with the new tag replacing the previous value.

The word at address zero is presently stored in the cache (index = 000, tag, = 00, data = 1220). Suppose that the CPU now wants to access the word at address 02000. The index address is 000, so it is used to access the cache. If two tags are then compared, the cache tag is 00. The address tag is 02, which does not produce match. Therefore, the main memory is accessed and the data word 5670 is transferred to the CPU. The cache word at index address 000 is then replaced with a tag of 02 and data of 5670.

Direct mapping organization operates are shown below :

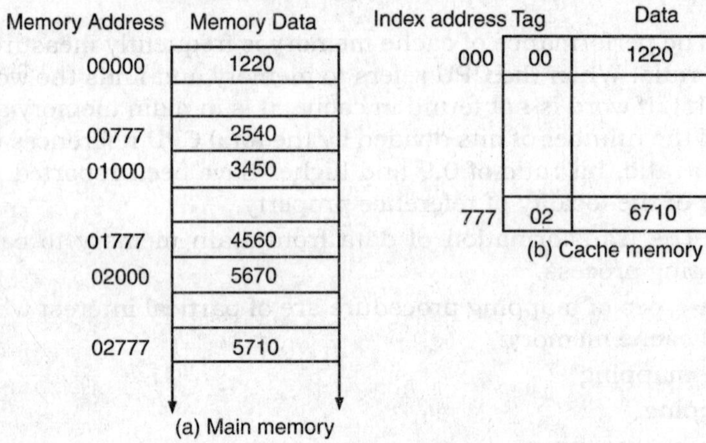

Fig. 7.24. Direct Mapping cache organization

Q.5. Define:

(a) Virtual memory

(b) Page Replacement

(c) Page Fault

(d) Logical address.

Solution: (a) Virtual memory : Virtual *memory is a concept used in some large computer system* that permit the user to construct *programs as though a large memory space were available,* equal to the totality of auxiliary memory. Each address that is referenced by the CPU goes through an *address mapping from the so-called virtual address to a physical address in main memory.*

A virtual memory system provides a mechanism for translating program generated *addresses into correct main memory locations. This is done dynamically, while programs are being executed in the CPU.*

(b) Page Replacement : A virtual memory system is a combination of shardware and software techniques. The memory management software system handles all the software operations for the efficient utilization of memory space. It must decide :

1. which page in main memory ought to be removed to make room for a new page.
2. which a new page is to be transferred from auxiliarly memory to main memory.
3. where the page is to be placed in main memory.

(c) Page Fault : When a program starts execution, one or more pages are transferred into main memory and the page table is set to indicate their position. The program is executed from main memory until it attempts to reference a page that is still in auxilary memory. This condition is called page fault when page fault occurs, the execution of the present program is suspended until the required page is brought into main memory.

(d) Logical address : The address generated by a segmented program is called a logical address. The logical address may be larger than the physical memory address as in virtual memory, but it may also be equal and sometimes even smaller than the length of the physical memory address.

Q.6. (a) How many 128 × 8 RAM chips are needed to provide a memory capacity of 2048 bytes?

(b) How many lines of the address bus must be used to access 2048 bytes of memory? How many of these lines will be common to all chips?

(c) How many lines must be decoded for chip and select? Specify the size of the decoders.

Solution:

(a) $\frac{2048}{128} = 16$ chips

(b) $2048 = 2^{11}$ 11 lines to address 2048 bytes

$128 = 2^7$ 7 lines to address each chip and 4 lines to decoder for selecting 16 chips

(c) 4×16 decoder

Q.7. A computer uses RAM chips of 1024 × 1 capacity.

(a) How many chips are needed, and how should their address lines be connected to provide a memory capacity of 1024 bytes?

(b) How many chips are needed to provide a memory capacity of 16K bytes? Explain in words how the chips are to be connected to the address bus.

Solution: (a) 8 chips are needed with address lines connected in parallel.

(b) $16 \times 8 = 128$ chips. Use 14 address lines ($16\text{ K} = 2^{14}$)

7 lines specify the chip address.

7 lines are decoded into chip-select inputs.

Q.8. A ROM chip of 1024 × 8 bits has four select inputs and operates from a 5-volt power supply. How many pins are needed for the IC package. Draw a block diagram and label all input and output terminate in the ROM.

Solution: 10 pins for inputs, 4 for chip-select, 8 for outputs, and 2 for power. Total 24 pins are needed.

Q.9. Extend the memory sytem of figure given below to 4906 byts of RAM and 4096 bytes of ROM. List the memory-address map and indicate what size decoders are needed?

Solution: $\frac{4096}{128} = 32$ RAM chips,

$\frac{4096}{512} = 8$ ROM chips.

$4096 = 2^{12}$,

There are 12 common address lincs and 1 line to select between RAM and ROM.

Component	**Address**	**16 15 14 13**	**12 11 10 9**	**8 7 6 5**	**4 3 2 1**
RAM	0000 – 0FFF	0 0 0 0	5 × 32 decoder	x x x x	x x x x
ROM	1000 – 1FFF	0 0 0 1	3 × 8 decoder x	x x x x	x x x x
		to $\overline{CS2}$ ↑			

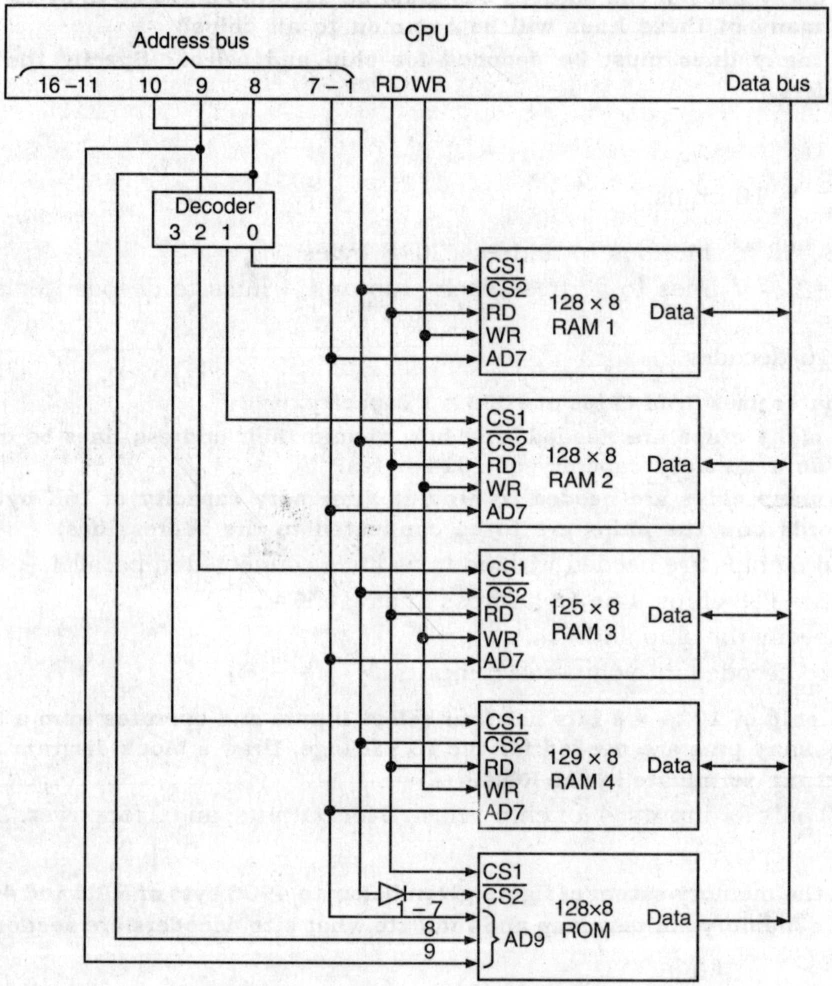

Fig. 7.25. Memory connection to the CPU

Q.10. A computer employs RAM chips of 256 × 8 and ROM chips of 1024 × 8. The computer system needs 2 K bytes of RAM, 4 K bytes of ROM and four interface units, each with four registers. A memory mapped I/O configuration is used. The two higehst order bits of the address bus are assigned 00 for RAM, 01 for ROM, and 10 for interface registers.

(*a*) How many RAM and ROM chips are needed?

(*b*) How a memory-address map for the system?

(*c*) Give the address range in hexadecimal for RAM, ROM and interface.

Solution: RAM 2048/256 = 8 chips, $2048 = 2^{11}$, $256 = 2^8$

ROM 4096/1024 = 4 chips, $4096 = 2^{12}$, $1024 = 2^{10}$

Interface 4 × 4 = 16 registers; $16 = 2^4$

Component	**Address**	**16 15 14 13**	**12 11 10 9**	**8 7 6 5**	**4 3 2 1**
RAM	0000 – 07FF	0 0 0 0	0.3 × 8 decoder	x x x x	x x x x
ROM	4000 – 4FFF	0 0 0 1	2 × 4 dec. x x	x x x x	x x x x
Interface	8000 – 800F	1 0 0 0	0 0 0 0	0 0 0 0	x x x x

Q.11. An 8-bit computer has a 16-bit address bus. The first 15 lines of the address are used to select a bank of 32 K bytes of memory. The high order bit of the address is used to select a register which receives the contents of the data bus. Explain how this configuration can be used to extend the memory capacity of the system to eight banks of 32 K bytes each, for a total of 256 bytes of memory.

Solution: The processor selects the external register with an address 8000 hexadecimal. Each bank of 32 K bytes aer selected by adderss 0000.7FFF. The processor loads an 8-bit number into th register with a single 1 and 70'S. Each output of the register selects one of the 8 banks of 32 bytes through 2 chip-select input. A memory bank cn be changed by changing in number in the register.

Q.12. A magnetic dish system has the following parameters.

T_s = average time to position the magnetic head over a track
R = rotation speed of disk in revolutions per second
N_t = number of bits per track
N_s = number of bits per sector
Calcualte the average time T_α that it will take to read one sector.

Solution: Average time, T_α = Ts + time for half revolution + time to read a sector.

$$T_\alpha = T_s + \frac{1}{2R} + \frac{N_S}{N_t} \times \frac{1}{R}$$

Q.13. What is the transfer rate of an eight truck magnetic tape whose speed is 120 inches per second and whose density is 1600 bits per inch?

Solution: An eight-track tape reads 8-bits (one character) at the same time.

Transfer rate = 1600 × 120 = 192,000 character/sec.

Q.14. Obtain the complement function for the match logic of one wold in an associative memory. in other words, show that M_is the sum of exclusive-OR functions. Draw the logic diagram for M_i and terminate it with an inverter to obtain M.

Solution:

$$M_i = \prod_{j=1}^{n} [(A_j \oplus F_{ij})' + k_j']$$

$$M_i = \sum_{j=1}^{n} (A_j \oplus F_{ij}) + k_j$$

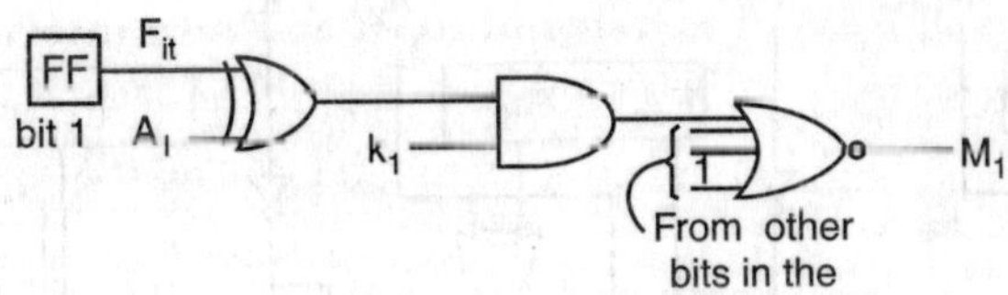

Fig. 7.26.

Q.15. Obtain the Boolean functions for the match logic of one word in an associative memory taking into consideration a tag bit that indicats whether the word is active or inactive.

Solution: A match occurs if T_i = 1

$$\text{match} = M_i T_i$$

Q.16. What additional logic is required to give a no-match result for a word in an associative memory when all key bits zeros?

Solution: $$M_i = M\left(\prod_{j=1}^{n} A_jF_{ij} + A_j'\,F_{ij}' + k_j'\right)(K_1 + K_2 + K_3 ... + K_n)$$

At least one key bit k_i must be equal to 1.

Q.17. Describe in words and by means of a block diagram how multiple matched words can be read out from an associative memory.

Solution:

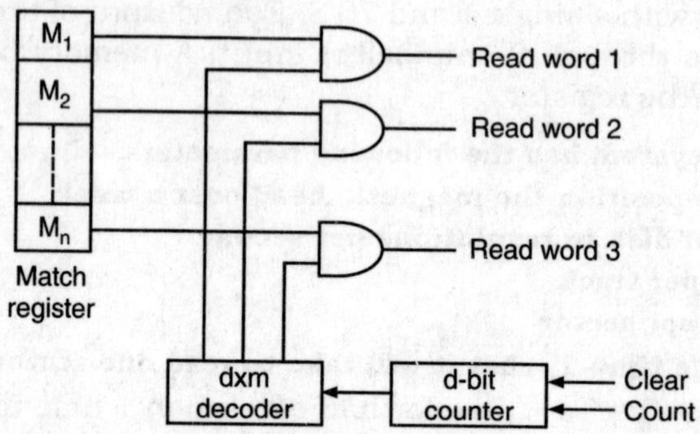

Fig. 7.27.

A d-bit counter drives a d-to-m line decoder where $2^d = m$ where, m = number of words in memory.

For each count, the M_i bit is checked and if 1, the corresponding read signal for work i is activated.

Q.18. Determine the logic of one cell and of an enture word for an associative memory that has an output indicator when the unmasked argument is greater than (but not equal to) the word in the associative memory.

Solution: Let $x_j = A_j F_{ij} + A_i F_{ij}$ (argument bit = memory word bit)

Output indicator $G_i = 1$ if :

$A_1 F_{i1} = 1$ and K = 1 (First bit is $A = 1$ while $F_{i1} = 0$)

or if $x_1 A_2 F_{i2} = 1$ and $K_2 = 1$ (First pair of bits are equal and second bit is $A = 1$ while $F_{j2} = 0$)

$$G_i = (A_1 F_{i1} + K_1)(x_1 A_2 F_{i2} + K_2)(x_1 x_2 F_{i3} + K_3) ... (x_1 x_2 ... x_{n-1} A_n F_{in} + k_n)$$

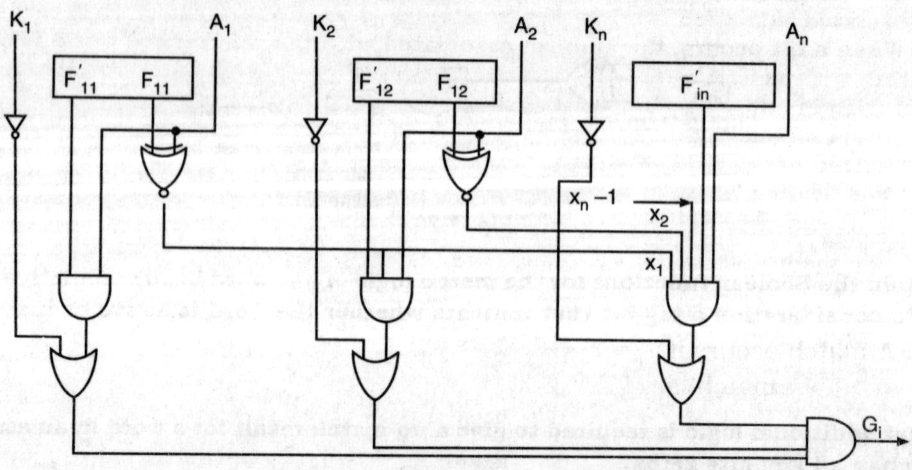

Fig. 7.28.

Q.19. A two way set associative cache memory uses blocks of four words. The cache can accomodate a total of 2048 words from main memory. The main memory size is 128 K × 32.

(*a*) Formulate all partinent information required to construct the cache memory.

(*b*) What is the size of the cache memory?

Solution: $128\ K = 2^{17}$

For a set size of 2, index address has 10 bits to accomodate $\frac{2048}{2} = 1024$ words of cache.

(a)

7-bits	10-bits
TAG	INDEX
← Block →	Word ←
8-bits	2-bits

(b)

	Tag1	Data 1	Tag 2	Data 2
	0 ← 7 →	← 32 bits →	← 7 →	← 32 bit 3 →
1021				

Size of cache memory $= 1024 \times 2\ (7 + 32) = 1024 \times 78$

Q.20. The access of a cache memory is 100 ns and that of main memory 1000 ns. It is estimated that 80 percent of the memory requests are for read and the remaining 20 percent for write. The hit ratio for read accesses only is 0.9. A wirte through procedure is used.

(*a*) What is the average access time of the system considering only memory read cycles?

(*b*) What is the average access time of the system for both read and write requests?

(*c*) What is the ratio taking into consideration the write cycles?

Solution: (*a*) $0.9 \times 100 + 0.1 \times 11000 = 90 + 110 = 200$ nsec.

cache access memory access

(*b*) $0.2 \times 1000 + 0.8 \times 200 = 200 + 160 = 360$ nsec.

read access from (α)

(*c*) Hit ratio $= 0.8 \times 0.9 = 0.72$

Q.21. A four-way set associative cache memory has four words in each set. A replacement procedure base on the least recently used (LRU) algorithm is implemented by means of 2-bit counters associated with each word in the set. A value in the range 0 to 3 in thus recorded for each work. When a hit occurs, the counter associated with the referrenced world is set to 0, those counters with values originally lower than teh referenced one are incremented by 1, and all other remain unchanged. If a miss occurs, the word with counter is set to 0. The other three counters are incremented by 1. Show that this procedure works for the following sequence of word reference:

***A, B, C, D, B, E, D, A, C, E, C, E.* (Start with *A, B, C, D* as the initial four words, with word *A* being the least recently used).**

Solution: Sequence: *A B C D E D A C E C E*

LRU

Count value = 32 10

Initial words = *A B C D*

B is a hit *A C D B*

E is a miss *C D E F*

D is a hit *C B E D*

A is a miss	$B\,E\,D\,A$
C is a miss	$E\,D\,A\,C$
E is a hit	$D\,A\,C\,E$
C is a hit	$D\,A\,E\,C$
E is a hit	$D\,A\,C\,E$

Q.22. A digital computer has a memory unit of 64 K × 16 and a cache memory of 1K words. The cache uses direct mapping with a block size of four words.

(a) How many bits are thee in the tag, index, block, and word fields of the address format?

(b) How many bits are there in each word of cache, and how are they divided into functions? Include a valid bit.

(c) How many blocks can the cache accommodate?

Solution: 64 K × 16 : 16-bit address; 16-bit data

(a) 6 8 2 = 16-bit address

TAG	BLOCK	WRD

Index = 10-bit cache address

(a) 1 6 16 = 23-bits in each word of cache

V	TAG	DATA

(c) 2^8 = 256 blocks of 4 words each

Q.23. An address space is specified by 24-bits and the corresponding memory space by 16-bits.

(a) How many words are there in the address space?

(b) How many words are there in the memory space?

(c) If a page consists of 2K words, how many pages and blocks are there in the system?

Solution: (a) Address phase = 24-bits 2^{24} = 164 K words

(b) Memory space = 16-bits 2^{16} = 24 K words

(c) $\frac{16\,K}{2\,K} = 8\,K$ pages $\frac{64\,K}{2\,K}$ = 32 blocks

Q.24. A virtual memory has a page size of 1 K words. There are eight pages and four blocks. The associative memory page table cocntains the following entries:

Page	Block
0	3
1	1
4	2
6	0

Make a list of all virtual addresses (in decimal) that will cause a-page fault if used by the CPU.

Solution: The pages which are not in main memory are :

Page	Address	Address that will cause fault
2	2 *K*	2048 – 3071
3	3 *K*	3072 – 4095
5	5 *K*	5120 – 6143
7	7 *K*	7168 – 8191

Q.25. A vitual memory system has an address space of 8 K words, a memory space of 4 K, words and page block sizes of 1K words (see figure, given below). The following page reference changes occur during a given time iterval. (Only page changes are listed. F the same page is referenced again, it is not listed twice.)

4 2 0 1 2 6 1 4 0 1 0 2 3 5 7

Determine the four pages that are resident in main memory after each page reference change if the replacement algorithm used is

(*a*) FIFO

(*b*) LRU

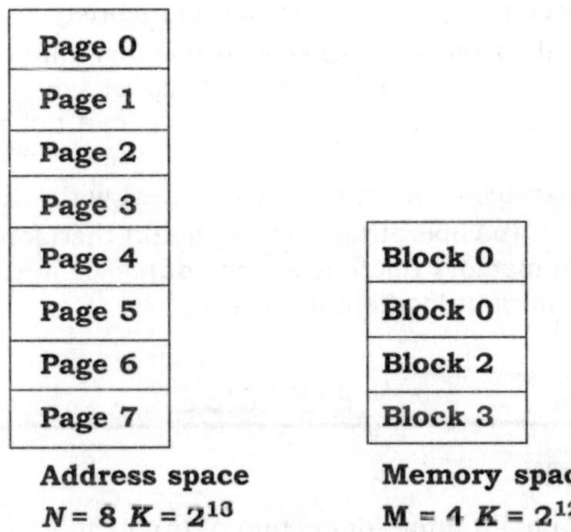

Fig. 7.28. Address space and memory space split into groups of 1K words

Solution: 4 2 0 1 2 6 1 4 0 10 2 3 5 7

Page reference	*Pages in main memory*	*Contents First in of FIFC*	*Pages in memory*	*LRU Most recently used*
initial	0124	4201	0124	4201
2	0124	4201	0124	4012
6	0126	2016	0126	0126
1	0126	2016	0126	0261
4	0146	0164	0246	2614
0	0146	0164	0146	6140
1	0146	0164	0146	6401
0	0146	0164	0146	6410
2	1246	1642	0124	4102
3	2346	6423	0123	1023
5	2345	4235	0235	0235
7	2357	2357	2357	2357

SUMMARY

1. Memory management is the act of managing computer memory. In its simpler forms, this involves providing ways to allocate portions of memory to programs at their request, and freeing it for reuse when no longer needed.
2. The management of main memory is critical to the computer system.
3. Virtual memory systems separate the memory addresses used by a process from actual physical addresses, allowing separation of processes and increasing the effectively available amount of RAM using disk swapping.
4. The quality of the virtual memory manager can have a big impact on overall system performance.
5. Garbage collection is the automated allocation and deallocation of computer memory resources for a program.
6. This is generally implemented at the programming language level and is in opposition to manual memory management, the explicit allocation and deallocation of computer memory resources.
7. Region-based memory management is an efficient variant of explicit memory management that can deallocate large groups of objects simultaneously.
8. In systems with virtual memory, programs in memory must be able to reside in different parts of the memory at different times. This is because when the program is swapped back into memory after being swapped out for a while it can not always be placed in the same location.
9. The virtual memory management unit must also deal with concurrency.
10. Memory management in the operating system should therefore be able to relocate programs in memory and handle memory references and addresses in the code of the program so that they always point to the right location in memory.

EXERCISE

(I) Fill in the Blanks (Solved)

1. A .... system is an inter connection of two or more CPUs with memory and input-output equipment.
2. The advantage of biased exponents is that they contain only .... numbers.

(II) True or False (Solved)

1. There are many different topologies for multistorage switching networks to control processor communication. One of them is omega switching network.
2. A priority interrupt system is a communication of branch and loop techniques.

(III) Fill in the Blanks (Unsolved)

1. There are some similarities between multiprocessor and .... system since both support concurrent operations.
2. The .... needs the usual circuits of an interface to communicate with the CPU and input/output device.

(IV) Detailed Questions

1. Show the step-by-step multiplication process using booth algorithm, when the following binary numbers are multiplied. Assume 5-bit registers to hold signed numbers.

   (*a*) + 10 *x* + 14 (*b*) + 10 x – 14
2. Derive an algorithm for evaluating the square root of a binary fixed-point number.
3. Why should the sign of the remainder after a division be the same as the sign of the dividend?

4. Show that there can be no mantissa overflow after a multiplication operation.
5. Show the hardware to be used for the addition and subtraction.
6. List four peripheral devices that produce an acceptable output for a person to understand.
7. What is the difference between isolated I/O and memory mapped I/O? What are the advantages and disadvantages of each?
8. Give at least six status conditions for the setting of individual bits in the status register of an asynchronous communication interface.
9. What programming steps are required to check when a source interrupts the computer while it is still being serviced by a previous interrupt request from the same source?
10. Why does DMA have priority over the CPU when both request a memory transfer?
11. What is the advantage of using interrupt - initiated data transfer over transfer under program control without an interrupt?

Answers

(I) Fill in the Blanks (Solved)

1. computer
2. positive

(II) True or False (Solved)

1. True
2. False

Chapter

8

Input/Output Organization

8.0 INTRODUCTION

Peripheral Devices: In computer's hardware language the peripheral devices can be defined as the devices that are connected to the computer in order to get most of the advantage out of it,if you in to detail description of peripheral devices it can be explained as the devices that are optional, and which are not required in principle. In earlier ages when computers were too much expensive and personal computer were very hard to afford, at that time the Motherboard, CPU (Central Processing Unit) and Memory (RAM (Random Access Memory) & ROM (Read Only Memory)) were considered to be the main components of the computer and any other device additionally attached to the computer were considered as peripheral device so this means that the keyboard, mice, etc. at that time were considered as peripheral devices but now a days generally they are not considered peripheral devices.

The common peripheral devices are microphones, cameras, disk drives, scanners, etc.

There is a common understanding that among people that the internal devices such as Sound Cards, etc. are not peripheral because they are added in the computer case which is wrong following is the peripheral devices list for a general idea.

Peripheral devise are the computer devices that are connected to the computer externally such as

Chapter Outline

printer, scanner, keyboard, mouse, tape device, microphone and external modem. They can be internal such as CD-ROM or internal modem and external as the scanner, printer and Zip drive. Peripheral devices can be classified according to their functions.

Input

Input devices are the type of the computer devices that are used to provide the control signals to the computer. Keyboard and the Mouse are the examples of the input devices.

Output

Output devices are the devices that are used to display the results. Printer, scanner, speaker and the monitor are the examples of the output devices.

Storage

A storage device is a device that is used to store the information such as hard disk drive, flash drive, floppy disk and the tape drive.

1. CD-ROM
2. CD-R
3. DVD-ROM
4. USB flash drive
5. Graphics Cards
6. Sound Cards
7. Modem
8. Network Card
9. Doking Station

Are some of the examples of peripheral devices.

Examples of Computer Peripheral Devices

A peripheral device is an external object that gives input and output for the computer. A few common input devices include :

1. kcyboard
2. mouse
3. joystick
4. pen tablet
5. MIDI keyboard
6. scanner
7. digital camera
8. video camera
9. microphone

A few common output devices include :

1. monitor
2. projector
3. TV screen
4. printer
5. plotter
6. speakers

There are also devices that function as both input and output devices, for instance :

1. external hard drives

2. media card readers
3. digital camcorders
4. digital mixers
5. MIDI equipment

8.1 INPUT/OUTPUT PORTS

The picture below shows an internal view of some of the input/output ports on the right and a covered view on the left. Input/Output stands for Input and Output. The most common device for input is the keyboard. When you type, you are putting information into the computer, which is known as input. The most common device for output is the monitor. After the information has made its way through the computer, it is sent out to the monitor for us to see. This is known as output. On the back of computers are several input/output, (or input/output), ports. Above, on the very top are two PS/2 ports, normally used for mouse and keyboard connections. Below that are the USB, (or Universal Serial Bus), ports. Below those are two serial ports beside a long parallel port that is often used to connect to a printer. On the bottom right is a game port for joysticks or other game controllers. On the bottom left is a microphone hook up, a speaker hook up, and an additional hook up for another sound input device like a musical keyboard. Below is a picture of the input/output ports on a more recent computer.

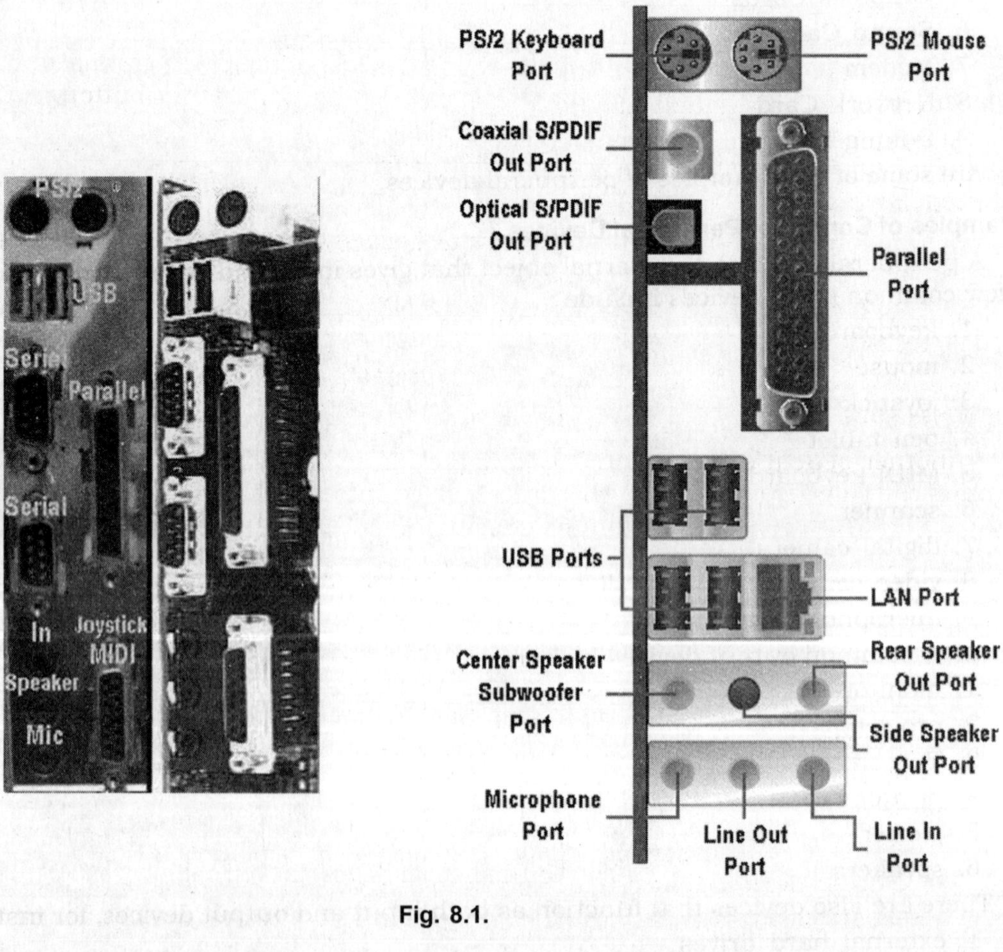

Fig. 8.1.

8.2. INTERRUPTS

A interrupt is a signal that stops the current program forcing it to execute another program immediately. The interrupt does this without waiting for the current program to finish. It is unconditional and immediate which is why it is called an interrupt.

The whole point of a interrupt is that the main program can perform a task without worrying about an external event.

Example

For example if you want to read a push button connected to one pin of an input port you could read it in one of two ways either by polling it or by using interrupts.

Polling

Polling is simply reading the button input regularly. Usually you need to do some other tasks as well, *e.g.* read an RS232 input so there will be a delay between reads of the button. As long as the delays are small compared to the speed of the input change then no button presses will be missed.

If however you had to do some long calculation then a keyboard press could be missed while the processor is busy. With polling you have to be more aware of the processor activity so that you allow enough time for each essential activity.

Hardware Interrupt

By using a hardware interrupt driven button reader the calculation could proceed with all button presses captured. With the interrupt running in the background you would not have to alter the calculation function to give up processing time for button reading.

The interrupt routine obviously takes processor time – but you do not have to worry about it while constructing the calculation function.

You do have to keep the interrupt routine small compared to the processing time of the other functions – its no good putting tons of operations into the ISR. Interrupt routines should be kept short and sweet so that the main part of the program executes correctly, e.g. for lots of interrupts you need the routine to finish quickly ready for the next one.

Hardware Interrupt or polling ?

The benefit of the hardware interrupt is that processor time is used efficiently and not wasted polling input ports. The benefit of polling is that it is easy to do.

Another benefit of using interrupts is that in some processors you can use a wake-from-sleep interrupt. This lets the processor go into a low power mode, where only the interrupt hardware is active, which is useful if the system is running on batteries.

Hardware Interrupt Common Terms

Terms you might hear associated with hardware interrupts are ISR, interrupt mask, non maskable interrupt, an asynchronous event, interrupt vector and context switching.

ISR	Interrupt Service Routine.
Interrupt vector	The address that holds the location of the ISR.
Interrupt mask	Controls which interrupts are active.
NMI	Non Maskable Interrupt - an interrupt that is always active.
Asynchronous event	An event that could happen at any time.
Context switching	Saving/restoring data before & after the ISR.

ISR

An ISR is simply another program function. It is no different to any other function except that it does context switching (saving/restoring processor registers) and at the end of the function it re-enables interrupts.

After the ISR finishes program execution returns to the original program and it continues exactly from where it was interrupted. The orriginal program will have no idea that this has happened.

Hardware Interrupt Vector

This is a fixed address that contains the location of your ISR - for a PIC micro it is usually address 4. In other micros there may be an interrupt vector for each vector - you have to make an interrupt routine for each one that you want to use. For PIC micros you have just one interrupt and you have to detect which interrupt triggered by examining the interrupt flag register(s).

You program the interrupt address with the address of your interrupt routine. Whenever the interrupt is triggered (and if the interrupt is unmasked) program operation jumps to the location of your interrupt routine.

Note : high level language compilers take care of all of this for you - in 'C' you just declare the function using the keyword **interrupt** (as the type returned from the function). It then puts the address of this routine in the interrupt vector.

A Basic Interrupt System

This is a general description of an interrupt system (biased slightly to PIC micros) which is true for any interrupt module and it is useful for understanding how to control and use interrupts.

Hardware Interrupt Block Diagram

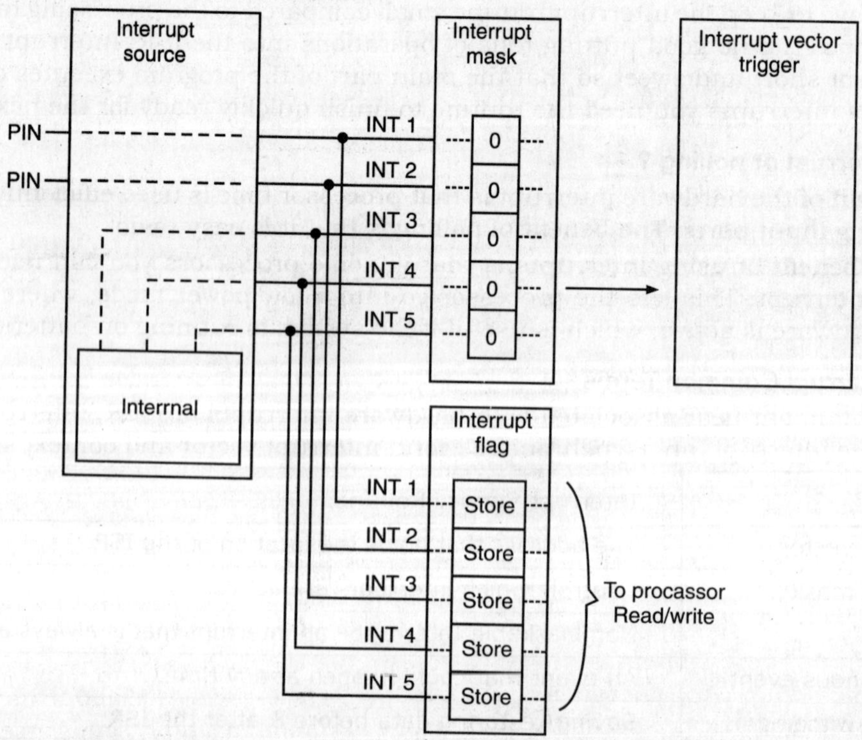

Input Sources

These are signals that start an interrupt. They can be external pins (where a rising or falling edge of a input triggers the interrupt) or internal peripheral interrupts, e.g. ADC completed, Serial Rx data received, timer overflow etc. (The 16F877 has 15 interrupt sources).

> **Note :** You can ususlly control which edge (rising or falling) is used by setting bits in yet another register!

Interrupt Flags

Each hardware interrupt source has an associated interrupt flag – whenever the interrupt is triggered the corresponding interrupt flag is set. These flags are usually stored as bits within an interrupt register.

The processor can read from and write to the interrupt register, reading from it to find out which interrupts occurred and writing to it to clear the interrupt flags.

Interrupt Mask

The interrupt mask has a set of bits identical to those in the interrupt register. Setting any bit (or unmasking) lets the corresponding signal source generate an interrupt - causing the processor to execute the ISR.

> **Note :** When a bit in the mask is clear (masked) the ISR is not activated for that signal source but the interrupt register is still set by the signal source. So you can still detect what the hardware is doing by polling the interrupt register.

Hardware Interrupt Vector

The interrupt vector is a location in memory that you program with the address of your interrupt service routine (ISR). Whenever an unmasked interrupt occurs program execution starts from the address contained in the interrupt vector.

8.3 INTERRUPTS AND EXCEPTIONS

Much of the functionality embedded inside a personal computer is implemented by hardware devices other than the processor. For example, serial communication is handled by an Intel 8251A chip (or a 16550AF in newer computers), which is located on the computer's motherboard (usually inside an ASIC) while the keyboard is controlled by a 8042A microprocessor.

Since each device operates at its own pace, a method is needed for synchronizing the operation of the processor with these devices. One solution is for the processor to sit in a tight loop, asking each device about its current state (a method known as polling). When data is available in one of the devices, the processor can then read and process the incoming bytes. While this method works, it has many disadvantages. First, it is very wasteful in terms of processing power, since the processor is constantly busy reading the status of the attached devices instead of executing some useful code. And second, when the rate of data transfer is extremely high, the processor might lose data bytes arriving from the hardware devices.

So, a different approach was devised. Instead of polling hardware devices to wait for their response, each device is responsible for notifying the processor about its current state. When a hardware device needs the processor's attention, it simply sends an electrical signal (hardware interrupt) through a dedicated pin in the interrupt controller chip (located on the computer's motherboard).

The interrupt controller serves as an intermediate between the hardware devices and the processor. Its responsibility is to alert the processor when one of the hardware devices needs its immediate attention. In this case, the processor stops its current activity and jumps to execute a function (interrupt handler) which was previously associated with the calling device (or more accurately, associated with the interrupt vector of the device).

While hardware interrupts have a significant importance, software interrupts also play a major role in the normal operation of the PC. These interrupts are synchronous rather than asynchronous, since they are generated by the processor itself when it encounters an INT instruction inside the code stream. Software interrupts are usually used by the operating system to publish its internal functions, thereby allowing applications to take advantage of these services.

Exceptions belong to a special type of software interrupts. They are generated by the processor itself whenever some unexpected critical event occurs. For instance, a page fault exception (interrupt 14) is triggered when the processor attempts to access a page, which is marked as not-present. The exception handler can then reload the page from disk (virtual memory) and restart the instruction which generated the exception.

Three types of exceptions can be generated by the processor: faults, traps and aborts. When a fault exception occurs, the CS and (E)IP registers which are pushed on the stack, point to the address of the instruction, which generated the exception. This gives the exception handler a chance to fix the condition which caused the exception to occur, before restarting the faulting instruction. Traps are similar to interrupts in the sense that they make the processor push the address of the next instruction to the stack, while aborts neglect to specify the location of the faulting instruction, since they are usually used to indicate severe errors (such as hardware errors or illegal system tables) which are not recoverable.

8.3.1. Software Interrupts

If you've ever programmed for DOS, you probably know that the INT 21h interface has a salient importance in linking applications with operating system services. For instance, calling an operating system function for opening a file, simply requires issuing an INT 21h interrupt with the correct parameters set in the processor registers.

The BIOS (Basic Input Output System) uses a similar software communication mechanism. The INT 10h, 13h and 16h are all interfaces to internal BIOS functions, which control the screen, the disk controller and the keyboard. Instead of accessing hardware devices directly, DOS uses the BIOS services to control the operation of the system. Modern operating systems such as Windows NT drop their reliance upon the BIOS in favor of faster mechanisms (device drivers) for accessing the PC's hardware.

When the processor bumps into an INT instruction, it pushes the address of the next instruction (CS and IP registers) and the contents of its flags register to the stack and then jumps to execute the interrupt handler. This ensures that when the handler code ends, the processor returns automatically to the original code stream (the code which was executing before the interrupt occured).

Software interrupts are still widely used by protected-mode operating systems in order to publish their internal services to the external world. However, most of these interfaces are now wrapped by functions and cannot be accessed directly (although it's possible, it's not recommended). For instance, Windows NT uses the undocumented interrupt 2Eh to allow control transfers between its user mode and kernel mode components. Whenever a non-privileged Windows application issues a call to an API function, the system routes the call until it reaches an INT 2Eh instruction. The IDT gate corresponding to this interrupt points to a handler inside ntoskrnl, which performs the required privileged operations on behalf of the application. Windows 95 and 98 use a similar technique to hop between user and kernel mode, but rely upon INT 30h instead of using INT 2Eh.

Here are the precise steps taken by the processor when it encounters an INT instruction :

1. Push Flags, CS and IP to the stack (in this order).
2. Multiply interrupt number by 4 and use the resulting number as an offset into the interrupt table (located at the beginning of the physical address space).
3. Get far address (CS and IP) of the interrupt handler from the table entry.
4. Disable interrupts (note that if you wish hardware interrupts to be serviced even when the processor executes inside an interrupt handler, you must issue a STI instruction at the beginning of the handler code).
5. Jump to execute the handler.
6. When the handler code ends (an IRET instruction is executed), pop CS, IP and the flags from the stack so that control returns to the currently active application.

Many applications (especially TSRs - Terminate and Stay Resident programs) hook software interrupts to supply additional services on top of those offered by DOS and the BIOS. One good example is Netware 3.1, which hooks INT 21h in order to add support for network file operations (such as reading a file located on a remote computer). Other pieces of code (mostly device drivers) hook hardware interrupts to monitor various devices and ensure their proper operation.

To hook an interrupt, a program need only to replace a specific interrupt table entry with the address of its own interrupt handler. Whenever this interrupt occurs, the handler is automatically invoked by the processor. A good programming practice is to place a call to the previous handler inside the interrupt handling function. This ensures that all previously installed hooks will get a chance to handle the incoming interrupt.

Note that when loading an application which uses software interrupts to call operating system services, there is no need to patch the application code with the actual memory addresses of the OS services. Therefore, the operating system can freely relocate its services in memory by simply updating the corresponding interrupt table entries with the new addresses.

8.3.2. Hardware Interrupts

Hardware interrupts are not very different in behavior. But still, when considering the path traveled by a hardware interrupt from the instant it leaves the hardware device until it reaches the processor, you must take into account the actions taken by the interrupt controller (or the 8259A if you prefer).

When the processor executes an INT instruction, it retrieves the interrupt code from the opcode itself. This code is later used by the processor to index the **IVT** (interrupt vector table) and find the address of the interrupt handler. On the other hand, when a hardware interrupt is detected, the interrupt controller sends the interrupt code to the processor via the data bus. When the code is finally acquired by the processor (either from the INT opcode or from the interrupt controller) the steps taken are equivalent to those discussed earlier.

Hardware interrupts in protected-mode behave in the same manner as described in the real-mode section. The only major difference is that when executing in protected-mode, the processor consults its IDT rather than searching the real-mode IVT for the address of the interrupt handler.

Table 1 presents the interrupt vectors occupied by the interrupt controller chips under different operating systems :

Table 8.1. Interrupt vectors occupied by the 8259A chips

OS	Vectors occupied by the master 8259A	Vectors occupied by the slave 8259A
DOS	8h - Fh	70h - 77h
Windows 95 / 98	50h - 57h	58h - 5Fh
Windows NT	30h - 37h	38h - 3Fh

8.3.3. Exceptions

Most of the information which was presented in the previous sections, applies to protected-mode as well. However, exception 13 (General Protection Fault) has a wider meaning in protected-mode and can indicate additional conditions than those defined for real-mode.

That's it. More information about the implementation of interrupts and exceptions in protected-mode can be obtained from the "Pentium Processor User's Manual" available from Intel.

When executing in real-mode, the following exceptions are supported :

Table 8.2. Real-Mode Exceptions

Exception	Vector	Condition
Division by Zero	0	Attempting to execute a DIV or an IDIV instruction with a divisor which equals zero.
Debug / Single Step	1	Used in conjunction with the debug registers to indicate a breakpoint hit. The processor also issues this interrupt after executing every instruction when the TRAP flag is set.
Breakpoint	3	The INT 3 instruction generates this exception. Since the opcode of this instruction is only one byte long, it is often used by debuggers to set a breakpoint in application code. The debugger needs only to replace the first byte of the instruction with the INT 3 opcode (11001100b) and wait for the breakpoint exception to occur. After the exception handler resumes execution and before returning to the debugged application, the debugger sets the INT 3 opcode to the previously overwritten byte and issues an IRET (interrupt return) instruction.
Overflow	4	When performing arithmetic instructions with signed operands, the processor set the OF flag to indicate an overflow. The INTO instruction tests this flag and if it is set - generates an exception.

Bounds Check	5	The BOUND instruction is used to verify that an array index does not exceed a certain limit. If the limit is exceeded, an exception is generated.
Invalid Opcode	6	Occurs when the processor executes one of the reserved opcodes or uses the LOCK prefix improperly. Can also indicate an invalid operand following an opcode.
Device Not Available	7	Attempting to execute a floating-point instruction when there is no coprocessor installed. The EM bit of the CR0 register can be cleared to disable this exception.
Double Fault	8	An interrupt occurs which has no corresponding entry in the IVT or a second exception is generated while the processor is executing a previously activated exception handler.
Stack Exception	12	Stack operation exceeds offset FFFFh or a selector pointing to a non-present segment is loaded into SS.
CS, DS, ES, FS, GS Segment Overrun	13	Word memory access at offset FFFFh or an attempt to execute past the end of the code segment.
Floating-Point Error	16	An error with the numeric coprocessor (Divide-by-Zero, Underflow, Overflow...).

Other exceptions exist only in **Virtual 8086 Mode** :

Table 8.3. Virtual-Mode exceptions

Exception	**Vector**	**Condition**
Invalid Task State Segment	10	Indicates that one of the checks made during a task switch failed.
Segment Not Present	11	Loading one of the segment registers with a selector to a segment marked as not-present.
Page Fault	14	Accessing a supervisor page from user privileged code or attempting to access a page which is marked as not-present.
Alignment Check	17	Memory access to an unaligned memory location (can only occur while executing ring 3 code).

8.3.4. Protected-Mode

Handling interrupts in protected-mode requires setting up a table in memory known as the IDT (Interrupt Descriptor Table). The **IDT** is somewhat different from the interrupt vector table that exists in real-mode. Instead of being "stuck" at physical address 0, the protected-mode IDT can float around in the linear address space with absolute freedom (although it is possible to change the address of the IVT while in real-mode, it is incompatible with the implementation of the 8086 processor). The linear address of the IDT is determined by a value set into the IDTR register using the LIDT instruction. Each entry in the IDT is 8-bytes long (as opposed to the 4-bytes entries of the IVT) and can contain one of three gate descriptors :

- A task gate - Causes a task switch to occur.
- An interrupt gate - Control is transferred to the interrupt handler with interrupts disabled.

- A trap gate - Control is transferred to the interrupt handler (the interrupt flag remains unchanged).

The following Fig. 8.3. presents the internal structure of a trap gate entry :

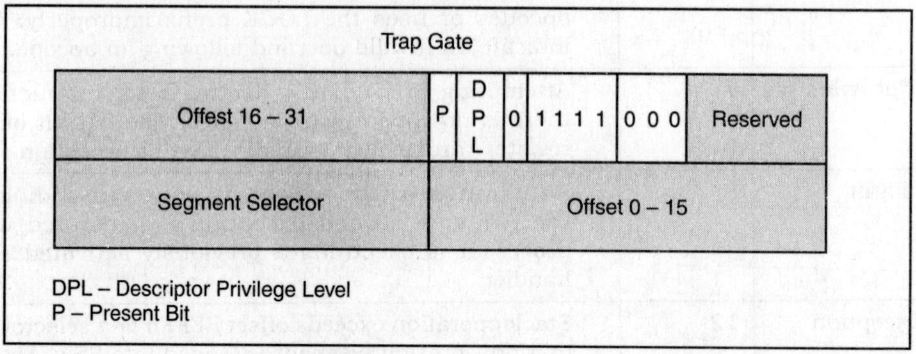

Fig. 8.3. Structure of a trap gate entry

Trap and interrupt gates have much in common since they are both used for control transfers between different privilege levels. The selector and offset fields specified in these gate descriptors are used to locate and execute the interrupt handler code. A control transfer is only permitted to a code residing in the same or a higher privilege level but never to a lower privileged one.

8.4 INPUT–OUTPUT INTERFACE

All Input–output devices, including hard disk drives, are much slow in access time and speed of transferring information as compared to CPU. Thus, an input output interface is required that provides a way of transferring information between internal storage and external I/O devices. This show in Fig. 8.4. All Input/Output devices have some form of controller/interface unit and the CPU communication with input/output divides via this controller. The purpose of inter face unit is to supervise and sysnchronize the input-output transfer there by resolving the differences between the computer system and the peripheral devices.

For example :

(*a*) A conversion of signal values may be required as most or all of the peripheral devices are electromagnetic or electromechanical in nature where as CPU and memory are electronic devices.

(*b*) The Peripheral devices are much slower as compared to CPU. Thus, there is a need for synchronization mechanism.

(*c*) The Peripheral devices and CPU/memory may differ in their word formats. In addition to interface unit, each peripheral may hare its own controller to super wise its operations.

Data buffering is an important aspect of input/output controllers. Each controller has a memory area associated with it called as Buffer. Data buffering is a solution to the problem of mismatch in transfer rates between the memory and the CPU and peripheral devices. The CPU sends all output information to buffer for retransmission by the peripheral controller to the peripheral devices.

For example while sending text to a printer, the characters are transferred at a fast rate to the buffer associated with the printer controller which will transfer them on to the printer at the Printer's transfer rate. Similarly input from the devices is sent to the buffer from where it is transferred by the CPU to main memory at high speed.

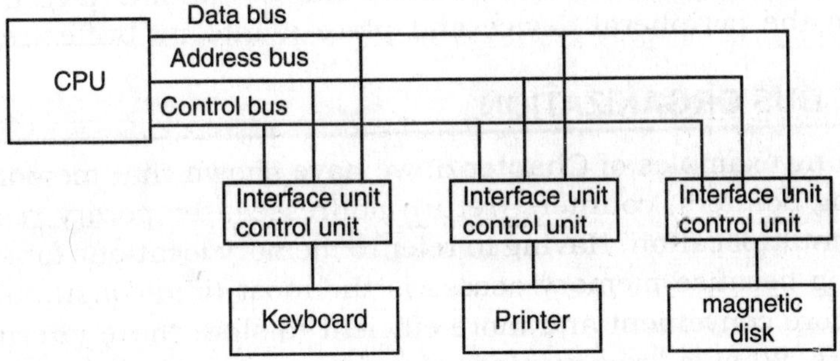

Fig. 8.4.

8.4.1 Input/Output Bus and Interface Module

All communication between the individual major components of the computer system takes place through the system bus. The bus is a cable, which is capable of carrying signals that represent data from one place to another. The bus used in a particular computer may be specific to that computer or may be an industry-standard bus. If an industry standard bus is used then it becomes easy to upgrade the computer by buying a component from an independent manufacturer. Most modern PCS use the PCI bus where as the older machine used the ISE bus.

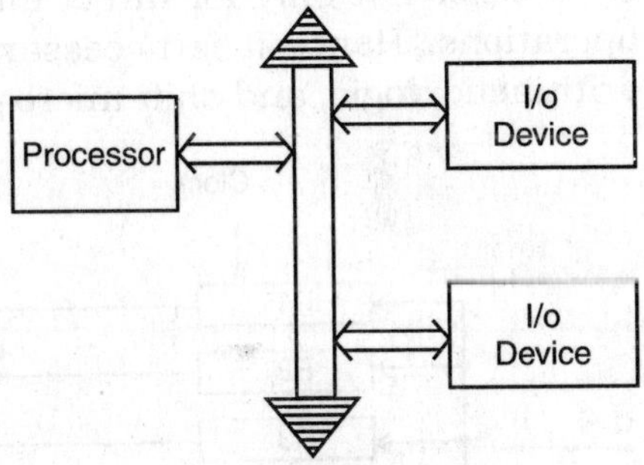

Fig. 8.5.

When the data is to be transferred from memory to a printer when it will be sent through the system bus. The control signals required to access memory and to activate the printer are also sent by the CPU via the system control bus.

In Fig. 8.5, it is shown that I/O bus consists of data bus, address bus and control bus. When CPU has to communicate with a particular device, it places the device address on the address bus together with the necessary control signals on the control bus. Each interface unit decodes the address and the control signals received from the I/O bus, interprets them and provides corresponding signals to the peripheral controller. All other peripheral devices whose device address does not match with the address placed by the Processor are disabled by their interfaces. The control information contained by the control bus is referred as an I/O command, which is actually an instruction and is executed in the interface and its attached peripheral unit. The I/O commands are classified into Four major types –

(*a*) **Control command :** It is used to activate the particular device and inform it about the operation to be performed. Each Peripheral has its own sequence of control commands depending on its mode of operation. For example magnetic tape may be instructed to start the tape moving inform ward motion or to rewind the tape etc. as desired.

(*b*) **Status command :** It is used by the Processor to check various status conditions in the interface unit and the Peripheral device. Basically, it is used to detect the errors that might occur during a transfer. These error conditions are indicated by setting corresponding bits in the status register which can be read by the Processor at certain intervals.

(c) **Data output command :** This command is issued to transfer the data contained in the data bus into one of the registers in the interface unit.

(d) **Data input command :** This command causes the interface unit to receive the data from the peripheral device and place it into its buffer register.

8.5 MULTIPLE BUS ORGANIZATION

In the programming examples of Chapter 6, we have shown that memory location as are needed for storing pointers, counters, return addresses, temporary results, and partial products during multiplication. Having to refer to memory locations for such applications is time consuming because memory access is the most time-consuming operation in a computer. It is more convenient and more efficient to store these intermediate values in processor registers. When a large number of registers are included in the CPU, it is most efficient to connect them through a common bus system. The registers communicate with each other not only for direct data transfers, but also while performing various micro-operations. Hence it is necessary to provide a common unit that can perform all the arithmetic, logic, and shift micro-operations in the processor.

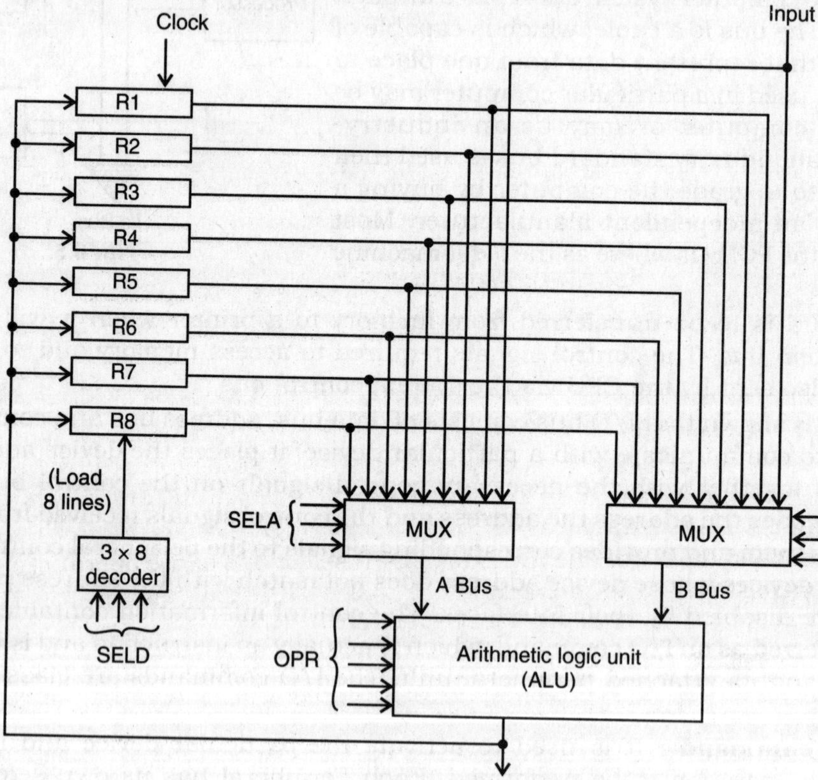

(a) Block diagram

3	3	3	5
SELA	SELB	SELD	OPR

(b) Control word

Fig. 8.6. Register set with common ALU

A bus organization for seven CPU registers is shown in Fig. 8.6. The output of each register is connected to two multiplexers (MUX) to form the two buses a and B. The selection lines in each multiplexer select one register or the input data for the particular bus. The A and B buses from the inputs to a common arithmetic logic unit (ALU). The operation selected in the ALU determines the arithmetic or logic micro-operation that is to be performed. The result of the micro-operation is available for output data and also goes into the inputs of all the registers. The register that receives the information from the output bus is selected by a decoder. The decoder activates one of the register load inputs, thus providing a transfer path between the data in the output bus and the inputs of the selected destination register.

The control unit that operates the CPU bus system directs the information flow through the registers and ALU by selecting the various components in the system. For example, to perform the operation the control must provide binary selection variables to the following selector inputs :

$$R1 \leftarrow R2 + R3$$

1. MUX A selector (SELA) : to place the content of R2 into bus *A*.
2. MUX *B* selector (SELB) : to place the content of R3 into bus *B*.
3. ALU operation selector (OPR) : to provide the arithmetic addition $A + B$.
4. Decoder destination selector (SELD) : to transfer the content of the output bus into *R*1.

The four control selection variables are generated in the control unit and must be available at the beginning of a clock cycle. The data from the two source registers propagate through the gates in the multiplexers and the ALU, to the output bus, and into the inputs of the destination register, all during the clock cycle interval. Then, when the next clock transition occurs, the binary information from the output bus is transferred into *R*1. To achieve a fast response time, the ALU is constructed with high-speed circuits. The buses are implemented with multiplexers or three-state gates, as shown in Sec. 4–3.

Control Word

There are 14 binary selection inputs in the unit, and their combined value specified a *control word*. The 14-bit control word is defined in Fig. 8.6 (*b*). It consists of four fields. Three fields contain three bits each, and one field has five bits. The three bits of SELA select a source register for the *A* input of the ALU. The three bits of SELB select a register for the *B* input of the ALU. The three bits of SELD select a destination register using the decoder and its seven load outputs. The five bits of OPR select one of the operations in the ALU. The 14-bit control word when applied to the selection inputs specify a particular micro-operation.

Table 8.4. Encoding of Register Selection Fields

Binary code	SELA	SELB	SELD
000	Input	Input	None
001	*R*1	*R*1	*R*1
010	*R*2	*R*2	*R*2
011	*R*3	*R*3	*R*3
100	*R*4	*R*4	*R*4
101	*R*5	*R*5	*R*5
110	*R*6	*R*6	*R*6
111	*R*7	*R*7	*R*7

binary code listed in the first column of the table specifies the binary code for each of the three fields. **The register selected by fields SELA, SELB, and SELD is the one whose decimal number is equivalent to the binary number in the code. When SELA or SELB is 000, the corresponding multiplexer selects the external input data. When SELD = 000, no destination register is selected but the contents of the output bus are available in the external output.**

The ALU provides arithmetic and logic operations. In addition, the CPU mist provide shift operations. The shifter may be placed in the input of the ALU to provide a per shift capability, or at the output of the ALU to provide post shifting capability. In some cases, the shift operations are included with the ALU. An arithmetic logic and shift unit was designed in Sec. 4.7. The function table for this ALU is listed in Table 4.8. The encoding of the ALU operations for the CPU is taken from Sec. 4.7 and is specified in Table 8.5. **The OPR field has five bits and each operation is designated with a symbolic name.**

Table 8.5. Encoding of ALU Operations

OPR code	*Operation*	*Symbol*
00000	Transfer *A*	TSFA
00001	Increment *A*	INCA
00010	Add *A* + *B*	ADD
00101	Subtract *A* – *B*	SUB
00110	Decrement *A*	DECA
01000	AND *A* and *B*	AND
01010	OR *A* and *B*	OR
01100	XOR *A* and *B*	XOR
01110	Complement *A*	COMA
10000	Shift right *A*	SHRA
11000	Shift left *A*	SHLA

Example of Micro-operations

A control word of 14 bits is needed to specify a micro-operation in the CPU. The control word for a given micro-operation can be derived from the selection variables. For example, the subtract micro-operation can be derived from the selection variables. For example, the subtract micro-operation given by the statement

$$R1 \leftarrow R2 - R3$$

specified *R*2 for the *A* input of the ALU, *R*3 for the *B* input of the ALU, *R*1 for the destination register, and an ALU operation to subtract *A* – *B*. Thus the control word is specified by the four fields and the corresponding binary value for each field is obtained from the encoding listed in Tables 8.4 and 8.5. The binary control word for the subtract micro-operation is 010 011 001 00101 and is obtained as follows :

Field :	SELA	SELB	SELD	OPR
Symbol :	R2	R3	R1	SUB
Control word :	010	011	001	00101

The control word for this micro-operation and a few others are listed in Table 8.3.

The increment and transfer Micro-operations do not use the *B* input of the ALU. For these cases, the B field is marked with a dash. We assign 000 to any unused field when formulating the binary control word, although any other binary number may be used. To place the content of a register into the output terminals we place the content of the register into the *a* input of the ALU, but none of the registers are selected to accept the data. The ALU operation TSFA places the data from the register, through the ALU, into the output terminals. The direct transfer from input to output is accomplished with a control word of all 0's (making the *B* field 000). A register can be cleared to 0 with an exclusive operation. This is because $x \oplus x = 0$.

	Symbolic Designation				
Micro-operation	***SELA***	***SELB***	***SELD***	***OPR***	***Control Word***
$R1 \leftarrow R2 - R3$	R2	R3	R1	SUB	010 011 001 00101
$R4 \leftarrow R4$ V $R5$	R4	R5	R4	OR	100 101 100 01010
$R6 \leftarrow$ R6 + 1	r6	—	R6	INCA	110 000 110 00001
$R7 \leftarrow R1$	R1	—	R7	TSFA	001 000 111 00000
Input $\leftarrow R2$	R2	—	None	TSFA	010 000 000 00000
Output $\leftarrow$ Input	Input	—	None	TSFA	000 000 000 00000
$R4 \leftarrow$ sh1 $R4$	R4	—	R4	SHLA	100 000 100 11000
$R5 \leftarrow 0$	R5	R5	R5	XOR	101 101 101 01100

It is apparent from these examples that many other Micro-operations can be generated in the CPU. The most efficient way to generate control words with a large number of bits is to store them in a memory unit. A memory unit that stores control words is referred to as a control memory. By reading consecutive control words from memory, it is possible to initiate the desired sequence of microoperations for the CPU. This type of control is referred to as micro-programmed control. A micro-programmed control unit is shown in Fig. 8.7. The binary control word for the CPU will come from the outputs of the control memory marked "micro-ops."

8.6 ASYNCHRONOUS DATA TRANSFER

In digital system, the internal operations are synchronized by using clock pulses generated by a common pulse generator. The clock pulses are applied to all the registers within a unit and thus all the data transfers between internal registers take place simultaneously on the occurrence of the clock pulse. CPU and I/O interface are two separate units designed independent of each other. Now, the data transfer among these two units is said to be synchronous if the interface registers and the CPU registers share a common clock. On the other hand, if the internal timing in each unit is independent of the other, *i.e.*, each unit uses its own clock. For internal registers, the transfer is asynchronous. Most communications b/w computers and I/O devices are asynchronous.

Asynchronous data transfer between two units requires to transmit control signals between he units in order to indicate the time at which data transmission is to start. This can be done in two ways.

8.6.1 Strobe Control

In this method, a strobe pulse is supplied by one of the units to the other to indicate when the transfer has to start.

Handshaking : Here, each data being transferred is accompanied by a control signal to indicate the presence of data on the bus. The unit that receives the data acknowledges the receipt of data by sending another control signal to the source of data.

Strobe Control

In strobe control method of data transfer s single control signal called as strobe is used to time each transfer. This strobe pulse can be activated either by the source or destination.

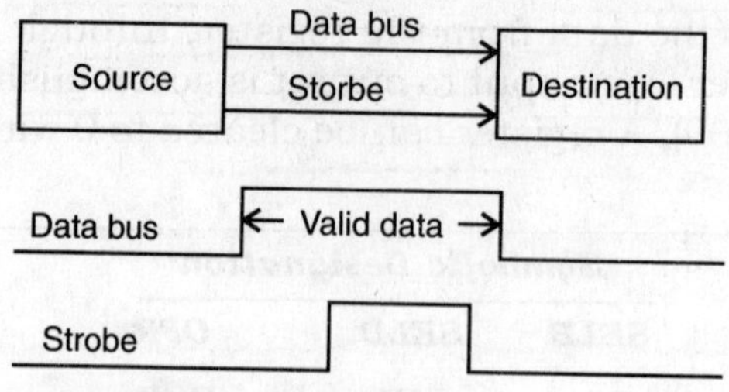

Fig. 8.7.

Source initiated strobe for data transfer

When activated by the source, strobe informs the destination unit that a valid data is present in the bus. The data bus transfers the information from the source unit to the destination unit this is shown in Fig. 8.7. The source unit first places the data on the data bus and then asserts the strobe pulse after a short delay. This delay is there because the data takes sometime to settle down to a steady value. The data in the data bus and the strobe pulse are held in active state for a sufficient amount of time to allow the destination unit to receive the data. The source unit removes the data from the data bus after disabling the strobe signal. Actual, there is no need for the source to remove the data on the bus is no longer valid. New valid data is available only when the strobe pulse goes high (enabled) again. This entire sequence of operations will be clear from the timing diagram. The strobe here may be a memory–write signal issued by CPU to memory. The CPU acting as a source places the data on the bus and informs the memory (destination unit) about a memory write operation through the strobe.

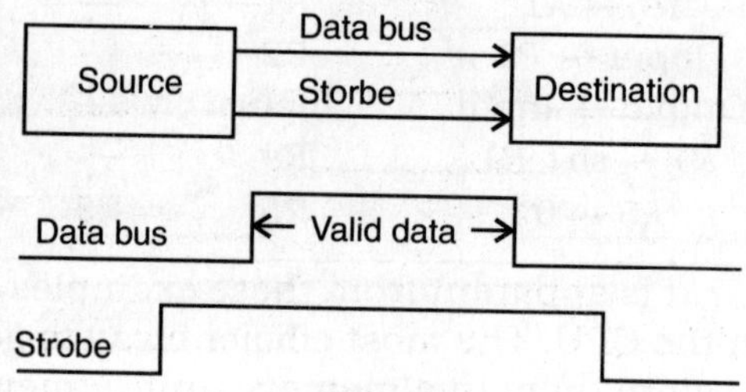

Fig. 8.8.

Destination–initiated strobe for data transfer

In this case, the destination unit first asserts the strobe pulse to inform the source to place the data on the data bus. In response to the strobe sent by the destination, the source places the requested data on the bus. This is known in Fig. 8.8. The data in the data bus and the strobe pulse are held in active state for a sufficient amount of time to allow the destination unit to receive the data. The destinations unit, after receiving the data removes the strobe signal and after a brief delay, the source removes the data from the bus. This entire sequence of operations will be clear from the timing diagram given in Fig. 8.9.

The strobe here can be a memory-read signal issued by the CPU to memory. The CPU acting as a destination generates the strobe to inform the memory (source unit) to place the specified data on bus.

There is no way to inform the source that initiates the transfer that the destination has actually received the data and similarly there is no way to inform the destination that initiates the transfer that the source has actually placed the data on bus.

A set of handshaking lines is required to handle the communication between an interface unit and an input/output device.

8.6.2 Hand Shaking

The problems associated with the strobe method can be solved by using a second control signal to serve as a reply to the unit initiating the transfer. This method of data transfer is known as handshaking. In this method one control line is used by the source that informs the destination about the valid data in the bus. The second control line is used by the destination that informs the source about whether it is ready to accept the data or not. The sequence of control signals generated is dependent on the unit initiating the transfer.

Source initiating Data transfer

This mode of data transfer uses two handshaking signals, data valid, issued by the source and data accepted produced by the destination. This is shown in Fig. 8.9. Following is the sequence of steps followed in the transfer :–

(*a*) The source initiates the transfer by :
   (*i*) Placing the data on the data bus
   (*ii*) Activating the data valid line

(*b*) the destination accepts the data from the bus and then enables the data accepted line.

(*c*) In response, the source disables the data valid line to indicate that the bus does not contain valid data.

(*d*) Finally, the destination disables the data accepted line and the system returns to the initial state.

The entire sequence is listed in Fig. 8.8.

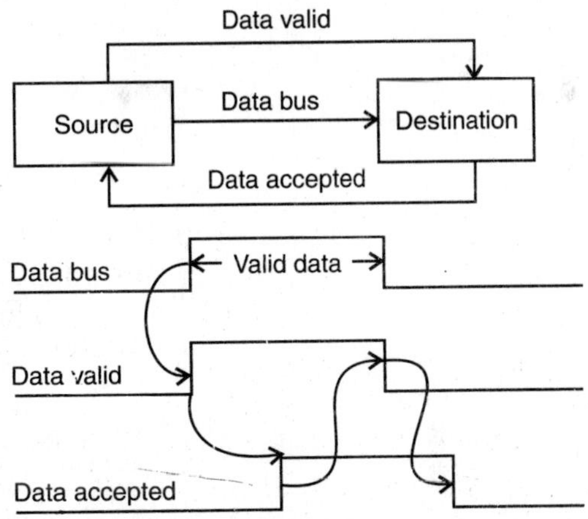

Fig. 8.9.

Destination initiated data transfer

The mode of data transfer, uses two handshaking signals, ready for data, issued by the destination and data valid, produced by the source. This is shown in Fig. 8.9. Following is the sequence of steps :

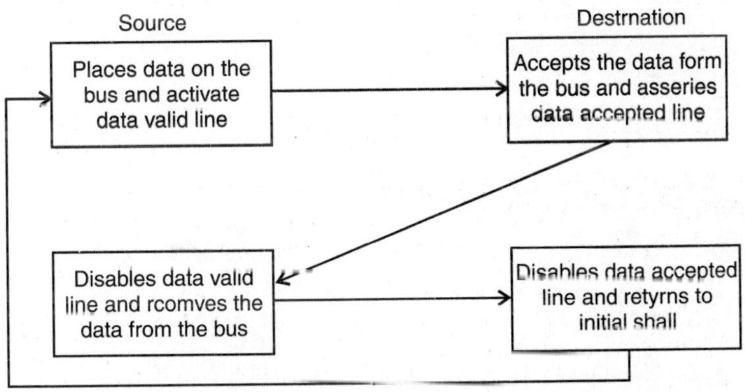

Fig. 8.10.

(*a*) The destination initiates the transfer by generating ready for the data, issued by the destination and data valid. Produced by the source.

(*b*) The source after receiving the ready for data signal places the data on the bus.

(*c*) The destination accepts the data from the bus and then disables the ready for the data line.

(*d*) In response to this, the source disables the data valid line to indicate that the bus does not contain valid data and the system returns to the initial state. The entire sequence is listed in Fig. 8.11.

into a single packet then multiple packets may be used and are numbered accordingly, Generally it has a fixed size.

The tail filed contains check sum information, which is used by the receiver to check whether, not the packet was corrupted during the transmission.

In asynchronous signal each data element just like a character is prefixed with a start and stop bit. In synchronous signals, each groups or block of characters is prefixed with start and stop codes. This mean using synchronous signals can provide higher speeds (more character per second) than synchronous signals.

8.8 MODES OF TRANSFER

The ultimate source or destination for the information transferred to and from an I/O device is the memory unit, CPU acts as an intermediate path. The problem with I/O is synchronizing a very fast CPU with very slow input/output device. The transfer of data between the central computer and input output devices can be handled in number of ways. In some methods, CPU is intermediate unit while in others data is transferred directly between the memory unit and input/output device. Any of the following methods may be used to handle data transfer to and from the peripherals.

8.8.1 Programmed input/output

All the input/output operations are explicitly handled by CPU. In these case these are actual input/output instructions written in the computer program that initiate data transfer. Usually the transfer is between a CPU register and Peripheral device other set of instructions is required for the transfer of data between CPU and memory. In this method, input/output is directly under program control and requires the CPU to monitor the interface unit until it indicates that it is ready for data transfer, which is time consuming process.

8.8.2 Interrupt Initiated input/output

The Programmed Input–output method keeps the CPU busy needlessly. This can be avoided by using an facility. In this method the interface unit monitors the input/output device to determine if it is ready for data transfer. In the mean time CPU continues with another program. The interface unit issues interrupt request signal to the CPU when the device is ready. In response to the interrupt signal, CPU momentarily suspends the current task and invokes a sub routine to handle input/output and later returns to the task it was originally processing following are the two variants of this type :

(*a*) **Device Polling :** the interrupt handler checks each device to determine the source of interrupt.

(*b*) **Vector Interrupt :** A "numbered" interrupt contains information about which device generated the interrupt request for example. Intel 80 × 86 is like this.

8.8.3 Direct Memory Access (DMA)

Here the interface unit transfers the data to and from the memory unit via memory bus. This method allows blocks of memory to be read/written. The CPU programs the DMA controller by giving the memory address, the buffer and the number of bytes to be read or written and then proceeds with some other task. The DMA controller then reads or writes data directly to the main memory without involving CPU as the intermediate unit. The CPU simply delays its memory access operation to allow for the direct memory I/O transfer. When the transfer is done, the DMA controller interrupts the CPU to indicate that the input/output operation has been completed.

Many computer systems include an input/outputProcessor (IOP) that combines interface logic together with the requirements for DMA. The IOP is capable of handling several peripheral devices using DMA and interrupt facility. Such a system devices the computer into three modules, the main memory, the Central Processing Unit (CPU) and input/output Processor (IOP). CPU may transfer data to or from a number of peripheral devices. The operation treated the input/ output subsystem for reading and writing in the same manner as memory, using address bus, data bus and Wr, RD control lines, this requires CPU intervention and is thus costly in time.

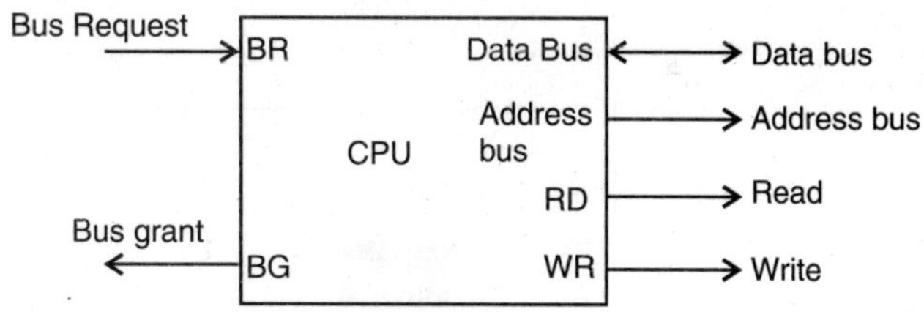

Fig. 8.16.

DMA is the capability provided by some computer bus architectures that allows data to bc transferred directly from a peripheral device to the memory. The processor is greed from involvement with the data transfer process and thus it speeds up the overall computer operation.

During DMA transfer, the CPU leaves the control of memory buses and a DMA controller takes over the buses to handle the data transfer directly between the peripheral devices and the memory unit. The two control signals in the CPU, bus request (BR) and bus grant (BG) shown in Fig. 8.16, facilitate the DMA transfer. The DMA controller uses BR input signal to request CPU for the control of buses. When CPU receives this signal, it responds by suspending the execution of the current instruction and placing the system bus (address bus, data bus, and read and write lines) into a high impedance state. In high impedance state, the output is disconnected and thus has no logical significance. The CPU then activates the BG line, which informs the DMA controller that the buses are freed by the Processor. The DMA controller then takes the control of buses to mange the data transfer without the Processor controlling every aspect of the transfer. The DMA controller disables the BR line to terminate the transfer. In response to this, the CPU disables the BG line takes back the control of buses and resumes its operation.

Direct memory Access Channels

Direct memory Access (DMA) channels are system pathways that are used by many devices for transforming information directly to and from memory. DMAs are used most commonly today by Floppy disk drives, tape drives and sound cards. As you already know, the processor is considered as the brain of the machine. In older machines the processor really did almost everything. In addition to executing Programs it was also responsible for handling transfer of data to and from peripherals.

Unfortunately, having the Processor involved in these transfers is very inefficient, because it is a time consuming process and the Processor is tied up, unable to do anything else. The Problem lies in synchronizing a very fast CPU with a very slow input/output device, which we have already discussed previous topics.

The invention of DMA enabled the devices to remove the middle man, the CPU from the path allowing the transfer under DMA can be made in jumps of ways :

(*i*) **Burst transfer :** In this technique, a bock sequence of number of memory words is transferred in a continuous burst. This mode in suitable for high speed input/ output like magnetic disks, where it is not possible to stop or slow down the data transmission until an entire bock is transferred.

SOLVED PROBLEMS

between memory and input output devices, without CPU intervention. With the advancement in technology, the input/output module is enhanced to function as a processor with a get of specialized instructions designed for input/output. Thus instead of having each interface unit communicate directly with CPU, a computer may have one or more input/output processors. The CPU initiates an input/output operation by directing this external processor to execute an input/output program in memory and input/output processor **** and executes tense instruction without involving CPU. CPU only specifies a sequence of input/output activities for the processor and is interrupted only on the completion of the entire sequence. CPU is thus relied of takes involved in input/output transfers, improving the speed of operation and hence the performance.

An input/output processor is a special processor designed to handle peripheral devices like terminals, tape and disk units and printers unlike DMA controller that is initialized entirely by CPU, an input/output processor is capable of executing input/output instruction thus executing complete control over input/output operations. Here CPU does not execute input/output instructions but these instructions are stored in memory and are executed by IOP in the input/output channel. The CPU directs the input/output processor to execute an I/O program in memory. This program specialties the device and actions to be performed Fig. 8.19 gives the block diagram of system with one input/output process. The memory unit as shown can communicate with both CPU and IOP. The CPU handles the data processing task while IOP provides a data transfer path between input/output devices and the memory.

The communication between memory and IOP is via direct memory access method. IOP and the peripheral devices connected to it communicate in a way similar to the program control method. In most of the computer systems, CPU acts as the master processor while IOP is a slave processor. The CPU initiates the input/output program and IOP fetches and executes all input/output instructions. POP operates independent of the CPU and controls the transfer of data between peripheral devices and memory. The CPU instructs the IOP to start an input/output transfer and checks the status conditions reflected in the status register needed to make input/output related decisions. In response to this, the IOP places the status word in the specified location in memory.

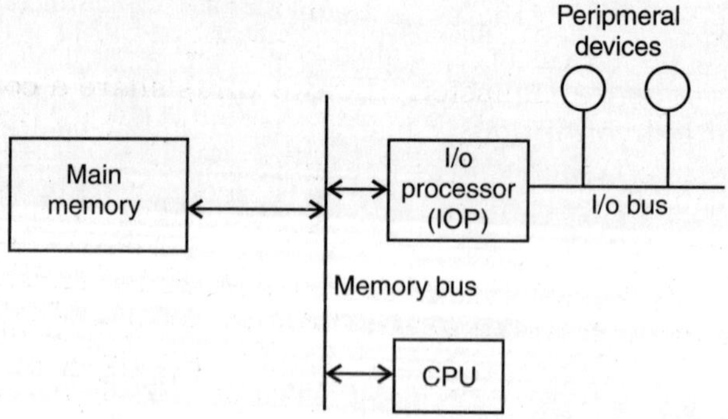

Fig. 8.19.

SOLVED PROBLEMS

Q.1. How do you mean by isolated ? Define memory mapped I/O.

Solution: Many computers use one common bus to transfer information between memory and input/output and the CPU. The distinction between a memory transfer and input/

output transfer "is made through separate read and write lines. The CPU specifies whether-the address on the address line is for a memory word or for an interface register by enabling one of two possible read or write lines. The input/output read and input/output write control lines are enabled during an input/output transfer.

The memory read and memory write control lines are enabled during a memory transfer. This configuration isolates all input/output interface addresses from the addresses assigned to memory and is referred to as the isolated input/output method for assigning addresses in a common bus.

In the isolated input/output configuration, the CPU has distinct input and output instructions, and each of these instructions is associated with the address of an interface register.

The isolated input/output method isolates memory and input/output addresses so that memory address values are not affected by interface address assignment since each has its own address space.

In memory mapped input/output, it use the same address space for both memory and input/output. This is the case in computers that employ only one set of read and write signals and do not distinguish between memory and input/output address.

Computers with memory mapped input/output can use memory type instructions to access input/output data with memory mapped input/output. All instructions that refer to memory are also available for input/output.

Q.2. What is the difference between serial and parallel data transmission ?

Solution: The transfer of data between two units may be done in parallel or serial. In parallel data transmission, each bit of the message has its own path and the total messages is transmitted through *n* separate conductor paths.

In serial data transmission, each bit in the message is sent in sequence one at a time. This method requires the use of one pair of conductors or one conductor and a common ground. Parallel transmission is faster but requires many wires. It is used for short distances and where speed is important. Serial transmission is slower but is less expensive, since it requires only one pair of conductors. Serial transmission can be synchronous or asynchronous.

Q.3. What is the difference between synchronous and asynchronous data transfer in case of serial/transfer?

Solution: In synchronous transmission, the two units share a common clock frequency and bits are transmitted continuously at the rate dictated by the clock pulses. In long distant serial transmission, each unit is driven by a sperate clock of the same frequency. Synchronization signals are transmitted periodically between two units to keep their clocks in step with each other.

In asynchronous transmission, binary information is sent only when it is available and the line remains idle when there is no information to be transmitted. This is contrast to synchronous transmission, where bits must be transmitted continuously to keep clock frequency in both units synchronized with each other.

A serial asynchronous data transmission technique used in many interactive terminals employs special bits that are inserted at both ends of the character code with this technique.

Each character consists of three parts :

a start bit,

a character bits, and

stop bits,

Solution:

$A_7 - A_2$						A_1	A_0
12 = 0	0	0	0	1	1	0	0
13 = 0	0	0	0	1	1	0	1
14 = 0	0	0	0	1	1	1	0
15 = 0	0	0	0	1	1	1	1
To CS						RS1	RS0

$CS = A_2 A_3 A'_4 A'_5 A'_6 A'_7$

$RS1 = A_1$

$RS0 = A_0$

Q. 7. Six interface units of the type shown in figure given in Q.6 are connected to CPU that uses an I/O address of eight bits. Each one of the six chip select (CS) inputs is connected to a different address lines. Thus the high-order address line is connected to the CS input of the first interface unit and the sixth address line is connected to the CS input of the sixth interface unit. The two low-order address lines are connected to the RSI and RSO of all six interface units. Determine the 8-bit address of each register in each interface.

Solution:

Interface	*Port A*	*Port B*	*Control Reg*	*Status Reg*
1	10000000	10000001	10000010	10000011
2	01000000	01000001	01000010	01000011
3	00100000	00100001	00100010	00100011
4	00010000	00010001	00010010	00010011
5	00001000	00001001	00001010	00001011
6	00000100	00000101	00000110	00000111

Q.8. Give peripheral devices that produce an acceptable output for a person to understand.

Solution:

(*i*) Character printer
(*ii*) Line printer
(*iii*) Laser printer
(*iv*) Digital plotter
(*v*) Graphics display
(*vi*) Vcice output
(*vii*) Digital to analog converter
(*viii*) Instrument indicator.

Q.9. Indicate whether the following constitute a control, status or data transfer commands.

(*a*) Skip next instruction if flag is set.
(*b*) Seek a given record on a magnetic disk.
(*c*) Check if input/output device is ready.
(*d*) Move printer paper to beginning of next page.
(*e*) Read interface status register.

Solution:

(*a*) Status command – Checks status of flag bit.
(*b*) Control command – Moves magnetic head in disk.
(*c*) Status command – Checks if device power is on.
(*d*) Control command – Moves paper position.
(*e*) Data input command – Reads value of a register.

Q.10. A CPU with a 20-MHz clock is connected to a memory unit whose access time is 40 ns. Formulate a read and write timing diagrams using a READ strobe and a WRITE strobe. Include the address in the timing diagram.

Solution: 20 MHz = 20×10^6 Hz

$$T = \frac{10^{-6}}{20} = 50 \text{ ns}$$

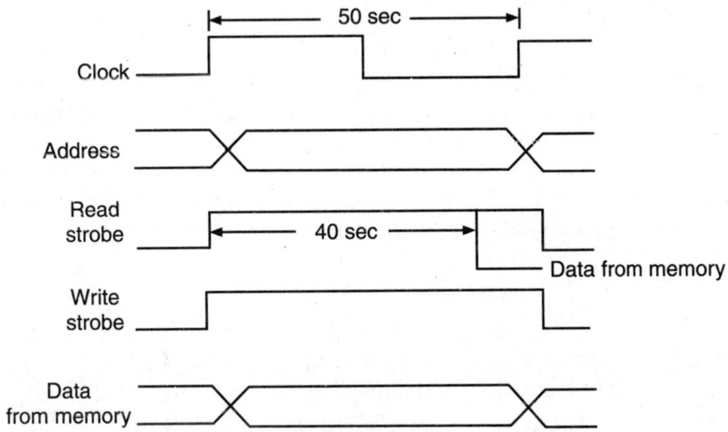

Fig. 8.22.

Q.11. The asynchronous communication interface shown in the figure given below is connected between a CPU and a printer. Draw a flowchart that describes the sequence of operations in the transmitter portion of the interface when the CPU sends characters to be printed.

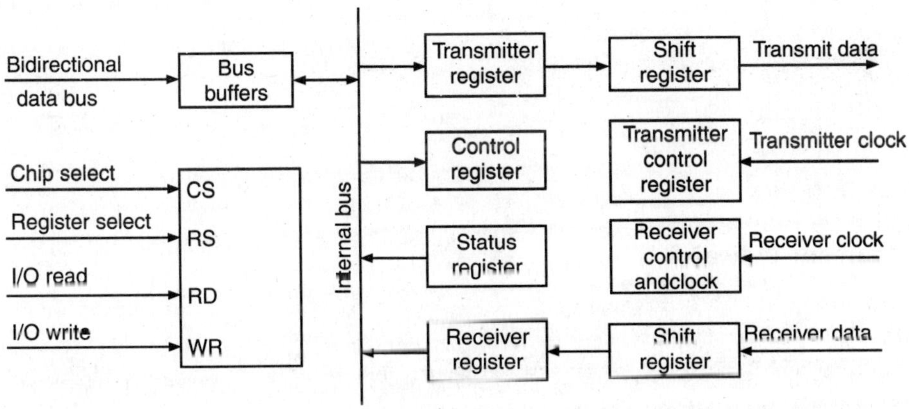

CS	RS	Operation	Register selected
0	x	x	None : data bus in high-impedance
1	0	WR	Transmitter register
1	1	WR	Control register
1	0	RD	Receiver register
1	1	RE	Status register

Fig. 8.23. Block diagram of a typical asynchronous communication interface

Solution:

Initial	$F = 0\,0\,1\,1$	Output $\leftarrow R_4$
After delete = 1	$F = 0\,0\,1\,0$	
After delete = 0	$F = 0\,0\,0\,1$	$R_4 \leftarrow R_3$
After insert = 1	$F = 1\,0\,0\,1$	$R_1 \leftarrow$ Input
(Insert goes to 0)	F = 0 1 0 1	$R_2 \leftarrow R_1$
	F = 0 0 1 1	$R_3 \leftarrow R_2$

Q.17. What are the values of input ready and output ready and control bits F_1 through F_4 in figure given in Q.16 when

(*a*) the buffer is empty ?

(*b*) the buffer is full ?

(*c*) the buffer contains two data items ?

Solution:

	Input ready	Output ready	$F_1 - F_4$
(*a*) Empty buffer	1	0	0 0 0 0
(*b*) Full buffer	0	1	1 1 1 1
(*c*) Two items	1	1	0 0 1 1

Q.18. Show a block diagram similar to figure given below for the data transfer from a CPU to an interface and then to an input/output device. Determine a procedure for setting and clearing the flag bit.

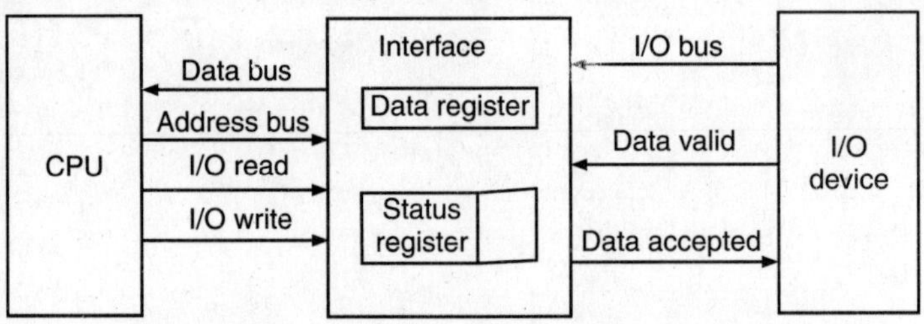

Fig. 8.26. Data transfer from input/output device to CPU

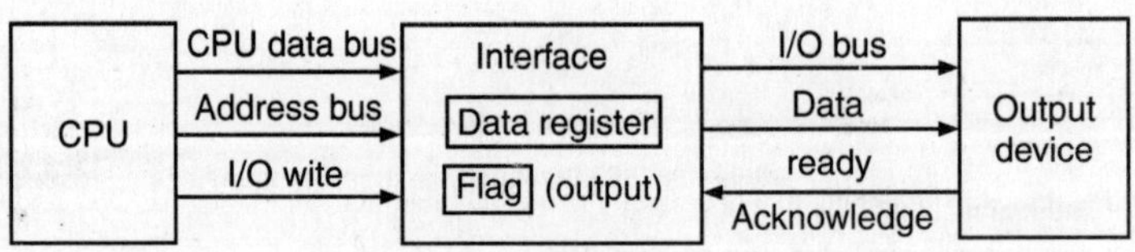

Fig. 8.27.

Solution: Flag = 0 if data register full (After CPU writes data)

Flag = 1 if data register empty (After the transfer to device)

when flag goes to 0, enable "Data ready" and place data on input/output bus. When "Acknowledge"

Q.19. Using the configuration established in Prob. 18, obtain a flowchart for the CPU program to output data.

Solution: CPU Program flow chart.

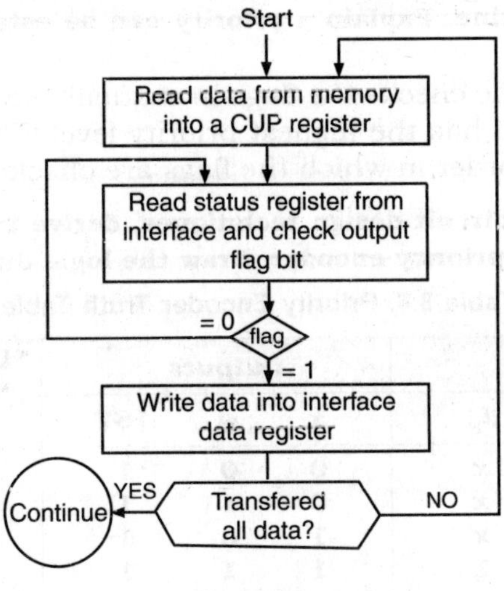

Fig. 8.28.

Q.20. In most computers an interrupt is recognized only after the execution of the instruction. Consider the possibility of acknowledging the interrupt of any time during the execution of the instruction. Discuss the difficulty that may angels.

Solution: If an interrupt is recognized in the middle of an instruction execution, it is necessary to save all the information from control register in addition to processor register. The state of the CPU to be saved is more complex.

Q.21. What happen in the daisy-chain priority interrupt shown in figure given below when device 1 requests an interrupt after device 2 has sent an interrupt request to the CPU but before the CPU responds with the interrupt acknowledge?

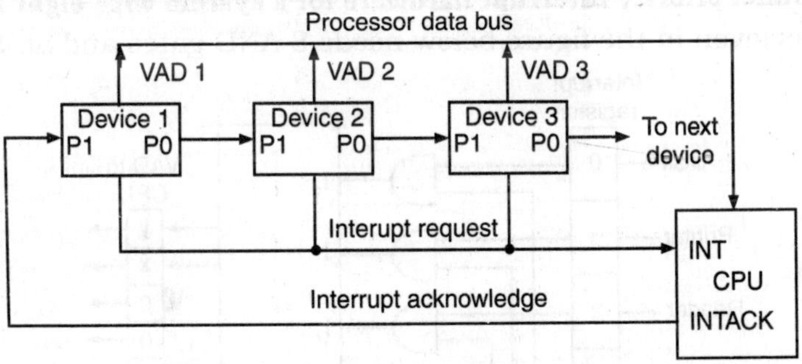

Fig. 8.29. Daisy-chain priority interrupt

Solution:

	Device 1	**Device 2**
1. Initially, device 2 sends an interrupt request	PI = 0; PO = 0; RF = 0	PI = 0; PO = 0; RF = 1
2. Before CPU responds with a acknowledge device 1 sends interrupt request:	PI = 0; PO = 0; RF = 1	PI = 0; PO = 0; RF = 1
3. After CPU sends an acknowledges device 1 has priority:	PI = 1; PO = 0; RF = 1 VAD enable = 1	PI = 0; PO = 0; RF = 1 VAD enable = 0

Q.22. Consider a computer without priority interrupt hardware. Any one of many source can interrupt the computer, and any interrupt results in staring the return address and branching

to a common interrupt routine. Explain a priority can be established in interrupt service program.

Solution: The service routine checks the flags in sequence to determine which one is set. The firs flag that is checked has the highest priority level. The priority level of the other source-corresponds to the order in which the flags are checked.

Q.23. Using combinational circuit design techniques, derive the Boolean expression listed in Table given below for the priority encoder. Draw the logic diagram of the circuit.

Table 8.6. Priority Encoder Truth Table.

Inputs				*Outputs*			
I_0	I_1	I_2	I_3	x	y	$1ST$	*Boolean functions*
1	x	x	x	0	0	1	
0	1	x	x	0	1	1	$x = I'_0 I'_1$
0	0	1	x	1	0	1	$y = I'_0 I'_1 + I'_0 I'_2$
0	0	0	1	1	1	1	$(IST) = I_0 + I_1 + I_2 + I_2$
0	0	0	0	x	x	0	

Solution:

I_0	I_1	I_2	I_3	x	y	$1ST$
1	x	x	x	0	0	1
0	1	x	x	0	1	1
0	0	1	x	1	0	1
0	0	0	1	1	1	1
0	0	0	0	x	x	0

Map Simplification, $x = I'_0 I'_1$ $y = I'_0 I'_1 + I'_0 I'_2$

Q.24. Design a parallel priority interrupt hardware for a system with eight interrupt sources.

Solution: Same as given in the figure below needs 8 AND gates and an 8 × 3 decoder.

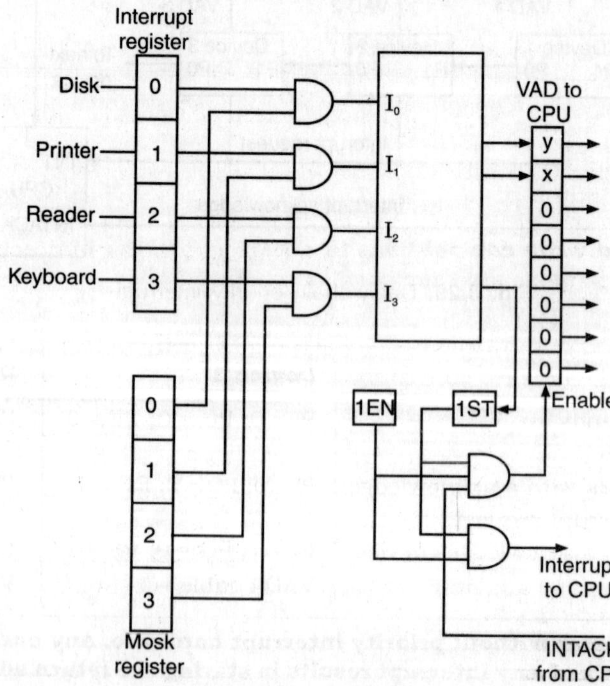

Q.25. Obtain the truth table of an 8 × 3 priority encoder. Assume that the three outputs *xyz* from the priority encoder are used to providing a vector address of the form 101*xyz*00. List the eight vector addresses starting from the one with the highest priority.

Solution:

I_0	I_1	I_2	I_3	I_4	I_5	I_6	I_7	*xyz*	*IST*
1	x	x	x	x	x	x	x	000	1
0	1	x	x	x	x	x	x	001	1
0	0	1	x	x	x	x	x	010	1
0	0	0	1	x	x	x	x	011	1
0	0	0	0	1	x	x	x	100	1
0	0	0	0	0	1	x	x	101	1
0	0	0	0	0	0	1	x	110	1
0	0	0	0	0	0	0	1	111	1
0	0	0	0	0	0	0	0	xxx	0

(b)

Binary	**hexadecimal**
10100000	A0
10100100	A4
10101000	A8
10101100	AC
10110000	B0
10110100	B4
10111000	B8
10111100	BC

Q.26. What should be done in figure given in Q.24 to make the four VAD values equal to the binary equivalent of 76, 77, 78 and 79 ?

Solution: $76 = (01001100)_2$

Replace the six 0's by 0 1 0 0 1 1 *xy*

Q.27. What programming steps are required to check when a source interrupts the computer while it is still being serviced by a previous interrupt request from the same source?

Solution: Set the mask bit belonging to the interrupt source so it can in terrupt again. At the beginning of the service routine check the value of the return address within the source service program, then the same source has interrupted again while being serviced.

Q.28. Why are read and write control lines in a DMA controller bidirectional ? Under what condition and for what purpose are they used as inputs ? Under what condition and for what purpose are they used as outputs ?

Solution: *When* the CPU communicates with the DMA controller the read and write lines are used as inputs from the CPU to the DMA controller.

When the DMA controller communicate with memory, the read and write lines are used as outputs from the DMA to memory.

Q.29. It is necessary to transfer 256 words from a magnetic disk to a memory section starting from address 1230. The transfer is by means of DMA as shown in figure given below.

(a) Give the initial values that the CPU must transfer to the DMA controller.

(b) Give the step-by-step account of the actions taken during the input of the first two words.

Solution: (a) CPU inititties DMA by transferring :

256 to the words count register.
1230 to the DMA address register.
Bits to the control register to specify a write operations.

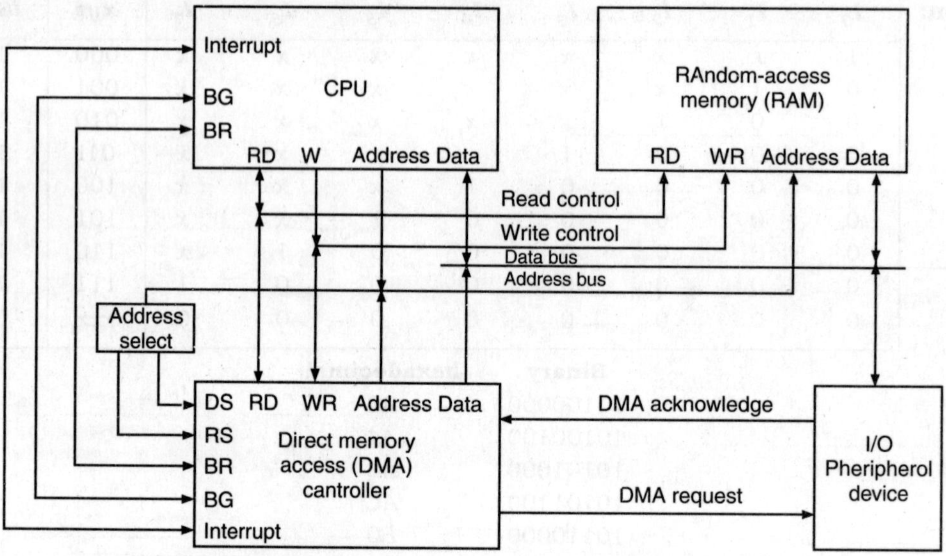

Fig. 8.31. DMA transfer in a computer system

(b) (*i*) input/output device sends a "DMA" request."
(*ii*) DMA sends BR (bus request) to CPU.
(*iii*) CPU responds with *a* BG (bus grant).
(*iv*) Contents of DMA address register are placed in address bus.
(*v*) DMA sends "DMA acknowledge" to input/output device and enables the write control line to memory.
(*vi*) Data word is placed on data bus by input/output device.
(*vii*) Increment DMA address register by 1 and decrement DMA word count register by 1.
(*viii*) Repeat steps 4-7 for each data word transferred.

Q.30. A DMA controller transfers 16-bit words to memory using cycle stealing. The words are assembled from a device that transmits characters at a rate of 2400 characters per second. The CPU is fetching and executing instructions at an average rate of 1 million instructions per second. By how much will the CPU be slowed down because of the DMA transfer?

Solution: CPU refers to memory on the average once (or more) every 2 × 416.6 = 833.3 μ sec.

Characters arrive one every 1/2400 = 416.6 μ sec.
Two characters of 8 bits each are packed into a 16-bit word every 2 × 416.6
= 833.3 μ sec

The CPU is slowed down by no more than (1/833.3) × 100 = O12%.

Q.31. Why does DMA have priority over the CPU when both request a memory transfer?

Solution: The CPU can wait to fetch instructions and data from memory without any damage occurring except loss of time. DMA usually transfer data from a device that cannot be stopped since information continues to flow so loss of data may occur.

Q.32. Draw a flowchart similar to the one is figure given below that describes the CPU-input/output channel communication in the IBM 370.

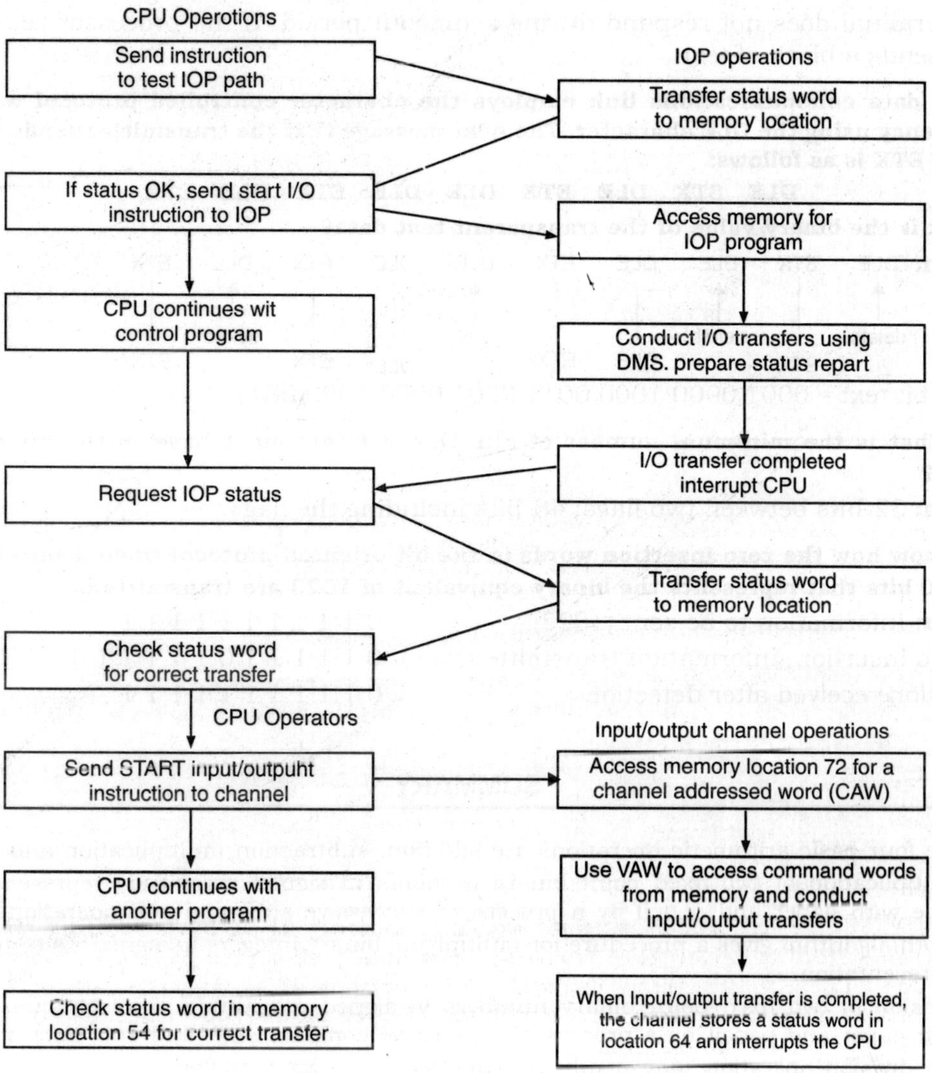

Fig. 8.32.

Q.33. The address of a terminal connected to a data communication processor consists of two letters of the alphabet or a letter followed by one of the 10 numerals. How many different addresses can be formulated.

Solution: There are 26 letters and 10 numerals.

$$26 \times 26 + 26 \times 10 = 936 \text{ possible addresses}$$

Q.34. List a possible line procedure and the character sequence for the communication between a data communication processor and a remote terminal. The processor inquires if the terminal is operative. The terminal responds witrh yes or no. If the response is yes, the processor sends a block of text.

Solution: The processor transmits the address of the terminal followed by ENQ (enquiry) code 00000101.

The terminal responds with either ACK (acknowledge) or NAK (negative Acknowledge) or the terminal does not respond during a timeout period. If the processor receives an ACk, it sends a block of text.

Q.35. A data communications link employs the character controlled protocol with data transparency using the DLE character. The next message that the transmitter sends between STX and ETX is as follows:

DLE STX DLE ETX DLE DLE ETX DLE ETX

What is the binary value of the transparent text data?

Solution:

DLE	STX	DLE	DLE	ETX	DLE	DLE	ETX	DLE	ETX
↑ delete	STX	↑ delete	DLE	ETX	↑ delete	DLE	ETX	↑ delete	ETX

32 – bit text = 0001 0000 1000 0011 0001 0000 1000 0011

Q.36. What is the minimum number of bits that a frame must have in the bit oriented protocol?

Solution: 32-bits between two flags; 48-bits including the flags.

Q.37. Show how the zero insertion words in the bit-oriented protocol when a zero followed by the 10 bits that represents the binary equivalent of 1023 are transmitted.

Solution: Information to be sent (1023) : 0 1 1 1 1 1 1 1 1 1 1

After zero insertion, information transmitted : 0 1 1 1 1 1 0 1 1 1 1 1 0

Information received after detection : 0 1 1 1 1 1 1 1 1 1 1

SUMMARY

1. The four basic arithmetic operations are addition, subtraction multiplication and division.
2. Multiplication of two fixed point binary numbers in signed magnitude representation is done with paper and pencil by a process of successive shift and add operations.
3. Booth algorithm gives a procedure for multiplying binary integers in signed 2's complement representation.
4. Division of two fixed point binary numbers in signed magnitude representation is done with paper and pencil by a process of successive compare shift and sub tract operation.
5. The division operation may result in a quotient with an over flow.
6. Representation of floating point numbers is known as the IEEE 754 floating point standard.
7. The IEEE 754 standard has two main representation
   → Single precision
   → Double precision
8. Devices that are under control of the computer are said to be connected on line.
9. Input-output interface provides a method for transferring information between internal storage and external input/output devices.
10. The input/output bus consists of data lines, address lines and control lines.

11. The are three that computer buses can be used to communicate with memory and input/output
    → Use two sperate buses, one for memory and other for input/output
    → Use one common bus for both memory and input/output but have sperate control lines for each
    → Use one common bus for memory and input/output with common control lines.
12. Every input/output activity is initiated by input/output instruction written in the code.
13. An interrupt generated internally by CPU is sometimes termed as trap.
14. IOP is the input output processor
15. Computer components can be broadly divided into two categories - Hardware and Software.
16. Hardware refers to any physical component of a computer. For example, CPU, Monitor, Keyboard, Hard Disk, Floppy Disk, etc.
17. Software refers to the programs which are required to operate the computer
18. All hardware devices except Mother Board are called peripheral devices
19. Motherboard, also called as system board, is the most important hardware component of a microcomputer.
20. Input devices are used to input data, information and instructions into the RAM.
21. Coming to the display devices which is one of the most important peripherals in computer is the graphic display device for display of graphs, diagrams and pictures to make the visual presentation of information more effective for user interaction and decision making.
22. Graphic display is made up of a series of dots called 'pixels' whose pattern produces the image. Each dot on the screen is defined as a separate unit which can be directly addressed. Since each dot can be controlled individually there is much greater flexibility in drawing pictures. There are three categories of display screen technology.
23. CRT Displays: The main components of a CRT terminal are the electron gun, the electron beam controlled by an electromagnetic field and a phosphor coated display screen.
24. The electron gun emits an electron beam which is directed towards the phosphor coated display by the electromagnetic field in order to create an image. The quality of display is indicated by the resolution of the display device. Resolution is determined by the number of pixels horizontally and vertically.

EXERCISE

(I) Fill in the Blanks (Solved)

1. The .... runs against a server, a multiport computer that contains large amounts of memory, allowing multiple terminals to share the server's resources.
2. .... refers to any enhancement to existing IBM VGA standard.
3. The function of .... chip is to support the CPU chip in processing of mathematical calculations.
4. The .... were used as an input medium in electro-mechanical calculating equipment.
5. The layout of the keyboard is like that of the traditional .... typewriter.

(II) True or False (Solved)

1. Input devices are used to input data, information and instructions into the Hard Disk
2. Digitiser is used to create drawings and pictures using digitiser tablet by a process called digitising.
3. Joystick is a pointing device, which is used to select a displayed menu item or draw pictures on the monitor screen
4. OBR (Optical Bar code Reader) devices can sense marks on computer readable paper.
5. A Drum Printer is a line printer in which a revolving cylinder acts as the printing element.

(III) Fill in the Blanks (Unsolved)

1. .... is also called as system board.
2. The function of math co-processor chip is to support the .... in processing of mathematical calculations.
3. Most keyboards have .... sections of keys.
4. .... is a handy device which can be moved on a smooth surface to a simulate the movement of cursor that is desired on the display screen.
5. .... is a pointing device, which is used to select a displayed menu item or draw pictures on the monitor screen.

(IV) True or false (Unsolved)

1. Both hardware and software are dependent on each other.
2. Motherboard is also called as peripheral.
3. The punched cards were used as an output medium.
4. The keyboard is one of the most common output device for computers.
5. Resolution is determined by the number of pixels horizontally and vertically.

(V) Detailed Questions

1. Explain the term Super VGA.
2. What is the difference between Impact Printing and Non-Impact Printing?

Answers

(I) Fill in the Blanks (Solved)

1. Terminal.
2. Super VGA.
3. Math co-processor.
4. Punched card.
5. QWERTY

(II) True or False (Solved)

1. False.
2. True.
3. False.
4. False.
5. True

☐☐☐

Chapter

9

Introduction to 8085 Microprocessor

Chapter Outline

9.0 INTRODUCTION

The term microcomputer is used to describe a system that includes at minimum a microprocessor, program memory, data memory, and an input-output (input/output) device. Some microcomputer systems include additional components such as timers, counters, and analog-to-digital converters. Thus, a microcomputer system can be anything from a large computer having hard disks, floppy disks, and printers to a single-chip embedded controller.

In this book we are going to consider only the type of microcomputers that consist of a single silicon chip. Such microcomputer systems are also called microcontrollers, and they are used in many household goods such as microwave ovens, TV remote control units, cookers, hi-fi equipment, CD players, personal computers, and refrigerators. Many different microcontrollers are available on the market. In this book we shall be looking at programming and system design for the PIC (programmable interface controller) series of microcontrollers manufactured by Microchip Technology Inc.

9.1. MICROCONTROLLER SYSTEMS

A microcontroller is a single-chip computer. Micro suggests that the device is small, and controller suggests that it is used in control applications. Another term for microcontroller is embedded controller, since most of the microcontrollers are built into (or embedded in) the devices they control.

A microprocessor differs from a microcontroller in a number of ways. The main distinction is that a microprocessor requires several other components for its operation, such as program memory and data memory, input-output devices, and an external clock circuit. A microcontroller, on the other hand, has all the support chips incorporated inside its single chip. All microcontrollers operate on a set of instructions (or the user program) stored in their memory. A microcontroller fetches the instructions from its program memory one by one, decodes these instructions, and then carries out the required operations.

Microcontrollers have traditionally been programmed using the assembly language of the target device. Although the assembly language is fast, it has several disadvantages. An assembly program consists of mnemonics, which makes learning and maintaining a program written using the assembly language difficult. Also, microcontrollers manufactured by different firms have different assembly languages, so the user must learn a new language with every new microcontroller he or she uses.

Microcontrollers can also be programmed using a high-level language, such as BASIC, PASCAL, or C. High-level languages are much easier to learn than assembly languages. They also facilitate the development of large and complex programs. In this book we shall be learning the programming of PIC microcontrollers using the popular C language known as mikroC, developed by mikroElektronika. In theory, a single chip is sufficient to have a running microcontroller system. In practical applications, however, additional components may be required so the microcomputer can interface with its environment. With the advent of the PIC family of microcontrollers the development time of an electronic project has been reduced to several hours.

9.2 INTEL 8085 MICROPROCESSOR

The **Intel 8085** is an 8-bit microprocessor introduced by Intel in 1977. It was binary-compatible with the more-famous Intel 8080 but required less supporting hardware, thus allowing simpler and less expensive microcomputer systems to be built.

The "5" in the model number came from the fact that the 8085 requires only a + 5 – volt (V) power supply rather than the + 5V, – 5V and +12V supplies the 8080 needed. Both processors were sometimes used in computers running the CP/M operating system, and the 8085 later saw use as a microcontroller, by virtue of its low component count.

The 8085 had a long life as a controller. Once designed into such products as the DECtape controller and the VT100 video terminal in the late 1970s, it continued to serve for new production throughout the life span of those products (generally longer than the product life of desktop computers).

The 8085 is a conventional von Neumann design based on the Intel 8080. Unlike the 8080 it does not multiplex state signals onto the data bus, but the 8-bit data bus was instead multiplexed with the lower part of the 16-bit address bus to limit the number of pins to 40. Pin #40 is used for the power supply (+ 5 V) and pin #20 for ground. Pin #39 is used as the hold pin. Pins #15 to #8 are generally used for address buses. The processor was designed using nMOS circuitry and the later "H" versions were implemented in Intel's enhanced nMOS process called HMOS, originally developed for fast static RAM products. Only a 5 Volt supply is needed, like competing processors and unlike the 8080. The 8085 uses approximately 6,500 transistors.[1]

The 8085 incorporates the functions of the 8224 (clock generator) and the 8228 (system controller), increasing the level of integration. A downside compared to similar contemporary designs (such as the Z80) was the fact that the buses required demultiplexing; however, address latches in the Intel 8155, 8355, and 8755 memory chips allowed a direct interface, so an 8085 along with these chips was almost a complete system.

The 8085 has extensions to support new interrupts, with three maskable interrupts (RST 7.5, RST 6.5 and RST 5.5), one non-maskable interrupt (TRAP), and one externally serviced interrupt (INTR). The RST n.5 interrupts refer to actual pins on the processor, a feature which permitted simple systems to avoid the cost of a separate interrupt controller.

Like the 8080, the 8085 can accommodate slower memories through externally generated wait states (pin 35, READY), and has provisions for Direct Memory Access (DMA) using HOLD and HLDA signals (pins 39 and 38). An improvement over the 8080 was that the 8085 can itself drive a piezoelectric crystal directly connected to it, and a built in clock generator generates the internal high amplitude two-phase clock signals at half the crystal frequency (a 6.14 MHz crystal would yield a 3.07 MHz clock, for instance).

9.3 PROGRAMMING MODEL

The 8085 is a binary compatible follow up on the 8080, using the same basic instruction set as the 8008 (developed by Computer Terminal Corporation). Only a few minor instructions were new to the 8085 above the 8080 set.

9.4 REGISTERS

The processor has seven 8-bit registers named A, B, C, D, E, H, and L, where A is the 8-bit accumulator and the other six can be used as independent byte-registers or as three 16-bit register pairs, BC, DE, and HL, depending on the particular instruction. Some instructions use HL as a (limited) 16-bit accumulator. As in the 8080, the contents of the memory address pointed to by HL could be accessed as pseudoregister M. It also has a 16-bit stack pointer to memory (replacing the 8008's internal stack), and a 16-bit program counter. HL pair is called the primary data pointers.

9.5 COMMANDS/INSTRUCTIONS

As in many other 8-bit processors, all instructions are encoded in a single byte (including register-numbers, but excluding immediate data), for simplicity. Some of them are followed by one or two bytes of data, which could be an immediate operand, a memory address, or a port number. Like larger processors, it has CALL and RET instructions for multi-level procedure calls and returns (which can be conditionally executed, like jumps) and instructions to save and restore any 16-bit register-pair on the machine stack. There are also eight one-byte call instructions (RST) for subroutines located at the fixed addresses 00h, 08h, 10h,...,38h. These were intended to be supplied by external hardware in order to invoke a corresponding interrupt-service routine, but are also often employed as fast system calls. The most sophisticated command was XTHL, which is used for exchanging the register pair HL with the value stored at the address indicated by the stack pointer.

9.6 8-BIT INSTRUCTIONS

Most 8-bit operations work on the 8-bit accumulator (the A register). For two operand 8-bit operations, the other operand can be either an immediate value, another 8-bit register, or a memory cell addressed by the 16-bit register pair HL. Direct copying is supported between any two 8-bit registers and between any 8-bit register and a HL-addressed memory cell. Due to the regular encoding of the MOV-instruction (using a quarter of available opcode space) there are redundant codes to copy a register into itself (MOV B,B, for instance), which are of little use, except for delays. However, what would have been a copy from the HL-addressed cell into itself (*i.e.,* MOV M,M) instead encodes the HLT instruction, halting execution until an external reset or interrupt occurred.

9.7 16-BIT OPERATIONS

Although the 8085 is an 8-bit processor, it also has some 16-bit operations. Any of the three 16-bit register pairs (BC, DE, HL) or SP could be loaded with an immediate 16-bit value (using LXI), incremented or decremented (using INX and DCX), or added to HL (using DAD). LHLD loaded HL from directly-addressed memory and SHLD stored HL likewise. The XCHG operation exchanges the values of HL and DE. Adding HL to itself performs a 16-bit arithmetical left shift with one instruction. The only 16-bit instruction that affects any flag was DAD (adding HL to BC, DE, HL or SP), which updates the carry flag to facilitate 24-bit or larger additions and left shifts (for a floating point mantissa for instance). Adding the stack pointer to HL is useful for indexing variables in (recursive) stack frames. A stack frame can be allocated using DAD SP and SPHL, and a branch to a computed pointer can be done with PCHL. These abilities make it feasible to compile languages such as PL/M, Pascal, or C with 16-bit variables and produce 8085 machine code.

Subtraction and bitwise logical operations on 16-bits is done in 8-bit steps. Operations that have to be implemented by program code (subroutine libraries) included comparisons of signed integers as well as multiply and divide.

9.8 INPUT/OUTPUT SCHEME

The 8085 supported up to 256 input/output (Input/Output) ports, accessed via dedicated I/O instructions—taking port addresses as operands. This input/output mapping scheme was regarded as an advantage, as it freed up the processor's limited address space. Many CPU architectures instead use a common address space without the need for dedicated input/output instructions, although a drawback in such designs may be that special hardware must be used to insert wait states as peripherals are often slower than memory. However, in some simple 8080 computers, input/output was indeed addressed as if they were memory cells, "memory mapped", leaving the input/output commands unused. input/output addressing could also sometimes employ the fact that the processor would output the same 8-bit port address to both the lower and the higher address byte (i.e. IN 05h would put the address 0505h on the 16-bit address bus). Similar input/output-port schemes are used in the 8080-compatible Zilog Z80 as well as the closely related x86 families of microprocessors.

9.9 DEVELOPMENT SYSTEM

Intel produced a series of development systems for the 8080 and 8085, known as the MDS-80 microprocessor system. The original development system had an 8080 processor. Later 8085 and 8086 support was added including ICE (in-circuit emulators). It was a large and heavy desktop box, about a 20" cube (in the Intel corporate blue color) which included a CPU, monitor, and a single 8 inch floppy disk drive. Later an external box was available with two more floppy drives. It ran the ISIS operating system and could also operate an emulator pod and an external EPROM programmer. This unit used the Multibus card cage which was intended just for the development system. A surprising number of spare card cages and processors were being sold, leading to the development of the Multibus as a separate product.

The later iPDS was a portable unit, about 8" x 16" x 20", with a handle. It had a small green screen, a keyboard built into the top, a 5¼ inch floppy disk drive, and ran the ISIS-II operating system. It could also accept a second 8085 processor, allowing a limited form of multi-processor operation where both processors ran simultaneously and independently. The screen and keyboard could be switched between them, allowing programs to be assembled on one processor (large programs took awhile) while files were edited in the other. It had a bubble memory option and various programming modules, including EPROM and Intel 8048 and 8051 programming modules which were plugged into the side, replacing stand-alone device programmers. In addition to an 8080/8085 assembler, Intel produced a number of compilers including PL/M-80 and Pascal languages, and a set of tools for linking and statically locating programs to enable them to be burnt into EPROMs and used in embedded systems. The hardware support changes were announced and supported, but the software upgrades were not supported by the assembler, user manual or any other means. At times it was claimed they were not tested when that was false.[citation needed]

9.10 APPLICATIONS

For the extensive use of 8085 in various applications, the microprocessor is provided with an instruction set which consists of various instructions such as MOV, ADD, SUB, JMP etc. These instructions are written in the form of a program which is used to perform various operations such as branching, addition, subtraction, bitwise logical and bit shift operations. More complex operations and other arithmetic operations must be implemented in software. For example, multiplication is implemented using a multiplication algorithm.

The 8085 processor has found marginal use in small scale computers up to the 21st century. The TRS-80 Model 100 line uses a 80C85. The CMOS version 80C85 of the NMOS/HMOS 8085 processor has several manufacturers, and some versions (e.g. Tundra Semiconductor Corporation's CA80C85B) have additional functionality, e.g. extra machine code instructions. One niche application for the rad-hard version of the 8085 has been in on-board instrument data processors for several NASA and ESA space physics missions in the 1990s and early 2000s, including CRRES, Polar, FAST, Cluster, HESSI, the Sojourner Mar Rover, and THEMIS. The Swiss company SAIA used the 8085 and the 8085-2 as the CPUs of their PCA1 line of programmable logic controllers during the 1980s.

9.11 MCS-85 FAMILY

The 8085 CPU was only one part of a much larger family of chips developed by Intel, for building a complete system. Although the 8085 CPU itself was not a great success, many of these support chips (or their descendents) later found their use in combination with the 8086 microprocessor. For example, the original IBM PC based on the Intel 8088 processor used many of these chips, and the prominence of IBM PC compatibles is why many of these chips are still in use today, although not as the chips themselves, but with their equivalent functionality embedded into larger VLSI chips, namely the "Southbridge" chips of modern PCs.

- 8007-Ram controller
- 8085-CPU
- 8155-RAM + 3 input/output Ports + Timer

- 8156-RAM + 3 input/output Ports + Timer
- 8185-SRAM
- 8202-Dynamic RAM Controller
- 8203-Dynamic RAM Controller
- 8205-1 Of 8 Binary Decoder
- 8206-Error Detection and Correction Unit
- 8207-DRAM Controller
- 8210-TTL To MOS Shifter and High Voltage Clock Driver
- 8212-8 Bit input/output Port
- 8216-4 Bit Parallel Bidirectional Bus Driver
- 8218/8219-Bus Controller
- 8222-Dynamic RAM Refresh Controller
- 8226-4 Bit Parallel Bidirectional Bus Driver
- 8231-Arithmetic Processing Unit
- 8232-Floating Point Processor
- 8237-DMA Controller
- 8251-Communication Controller
- 8253-Programmable Interval Timer
- 8254-Programmable Interval Timer
- 8255-Programmable Peripheral Interface
- 8256-Multifunction Support Controller
- 8257-DMA Controller
- 8259-Programmable Interrupt Controller
- 8271-Programmable Floppy Disk Controller
- 8272-Single/Double Density Floppy Disk Controller
- 8273-Programmable HDLC/SDLC Protocol Controller
- 8274-Multi-Protocol Serial Controller
- 8275-CRT Controller
- 8276-Small System CRT Controller
- 8278-Programmable Key Board Interface
- 8279 -Key Board/Display Controller
- 8282-8-bit Non-Inverting Latch with Output Buffer
- 8283-8-bit Inverting Latch with Output Buffer
- 8291-GPIB Talker/Listener
- 8292-GPIB Controller
- 8293-GPIB Transceiver
- 8294-Data Encryption/Decryption Unit + 1 input/output Port
- 8295-Dot Matrix Printer Controller
- 8296-GPIB Transceiver
- 8297-GPIB Transceiver
- 8355-16,384-bit (2048 × 8) ROM with input/output
- 8604-4096-bit (512 × 8) PROM
- 8702-2K-bit (265 × 8) PROM
- 8755-EPROM + 2 input/output Ports

QUESTION WITH ANSWERS

Q.1. What are the various registers in 8085?

Ans. Accumulator register, Temporary register, Instruction register, Stack Pointer, Program Counter are the various registers in 8085 .

Q.2. In 8085 name the 16 bit registers?

Ans. Stack pointer and Program counter all have 16 bits.

Q.3. What are the various flags used in 8085?

Ans. Sign flag, Zero flag, Auxillary flag, Parity flag, Carry flag.

Q.4. What is Stack Pointer?

Ans. Stack pointer is a special purpose 16-bit register in the Microprocessor, which holds the address of the top of the stack.

Q.5. What is Program counter?

Ans. Program counter holds the address of either the first byte of the next instruction to be fetched for execution or the address of the next byte of a multi byte instruction, which has not been completely fetched. In both the cases it gets incremented automatically one by one as the instruction bytes get fetched. Also Program register keeps the address of the next instruction.

Q.6. Which Stack is used in 8085?

Ans. LIFO (Last In First Out) stack is used in 8085.In this type of Stack the last stored information can be retrieved first.

Q.7. What happens when HLT instruction is executed in processor?

Ans. The Micro Processor enters into Halt-State and the buses are tri-stated.

Q.8. What is meant by a bus?

Ans. A bus is a group of conducting lines that carriers data, address, & control signals.

Q.9. What is Tri-state logic?

Ans. Three Logic Levels are used and they are High, Low, High impedance state. The high and low are normal logic levels & high impedance state is electrical open circuit conditions. Tri-state logic has a third line called enable line.

Q.10. Give an example of one address microprocessor?

Ans. 8085 is a one address microprocessor.

Q.11. In what way interrupts are classified in 8085?

Ans. In 8085 the interrupts are classified as Hardware and Software interrupts.

Q.12. What are Hardware interrupts?

Ans. TRAP, RST7.5, RST6.5, RST5.5, INTR.

Q.13. What are Software interrupts?

Ans. RST0, RST1, RST2, RST3, RST4, RST5, RST6, RST7.

Q.14. Which interrupt has the highest priority?

Ans. TRAP has the highest priority

Q.15. Name 5 different addressing modes?

Ans. Immediate, Direct, Register, Register indirect, Implied addressing modes.

Q.16. How many interrupts are there in 8085?

Ans. There are 12 interrupts in 8085.

Q.17. What is clock frequency for 8085?

Ans. 3 MHz is the maximum clock frequency for 8085.

Q.18. What is the RST for the TRAP?

Ans. RST 4.5 is called as TRAP.

Q.19. In 8085 which is called as High order/Low order Register?

Ans. Flag is called as Low order register & Accumulator is called as High order Register.

Q.20. What are input & output devices?

Ans. Keyboards, Floppy disk are the examples of input devices. Printer, LED / LCD display, CRT Monitor are the examples of output devices.

Q.21. Can an RC circuit be used as clock source for 8085?

Ans. Yes, it can be used, if an accurate clock frequency is not required. Also, the component cost is low compared to LC or Crystal.

Q.22. Why crystal is a preferred clock source?

Ans. Because of high stability, large Q (Quality Factor) & the frequency that doesn't drift with aging. Crystal is used as a clock source most of the times.

Q.23. Which interrupt is not level-sensitive in 8085?

Ans. RST 7.5 is a raising edge-triggering interrupt.

Q.24. What does Quality factor mean?

Ans. The Quality factor is also defined, as Q. So it is a number, which reflects the lossness of a circuit. Higher the Q, the lower are the losses.

Q.25. What are level-triggering interrupt?

Ans. RST 6.5 & RST 5.5 are level-triggering interrupts.

Q.26. What are tri-state devices and why they are essential in a bus oriented system?

Ans. Tri state logic devices have three states (0, 1 and high impedance). When the enable (may be active high or active low) line is activated, the device works. The disabled enable line makes the device at high impedance state and it is disconnected from the circuit. For example see the tri stated inverter in the figure shown.

In microcomputer system the peripherals are connected in parallel between address bus and data bus. Because of tri stated interfacing devices, peripherals do not load the system buses. Processor communicates with one peripheral or device at a time by enabling the tri state line of the interfacing peripheral or device. Tri state logic is critical to proper functioning of the microcomputer.

Q.27. Why are program counter and stack pointer 16-bit registers?

Ans. Because SP points to the beginning of stack memory (LXI SP 8000H) which is 16-bits. Also PC points to the memory locations (16-bits) of the instructions to be executed to maintain the proper sequence of execution of program.

Q.28. What does it mean by embedded system?

Ans. A specialized computer system that is part of a larger system or machine. Typically, an embedded system is housed on a single microprocessor board with the programs stored in ROM. Virtually all appliances that have a digital interface like watches, microwaves, VCRs, cars etc utilize embedded systems. Some embedded systems include an operating system, but many are so specialized that the entire logic can be implemented as a single program.

Q.29. What are the different addressing modes in 8085?

Ans. **Register :** Data is provided through the registers. Or operand is only register(s). Example: MOV Rd, Rs.

Register indirect : Operand M or register pair. Example: MOV A,M; LDAX B; STAX D; MVI M,32H (exception for immediate addressing mode).

Direct : Operand 8-bit port address or 16-bit memory address. Example: IN 84H, OUT 84H, all CALLs.

Immediate : Instruction having the letter I. Or immediate data to the destination provided. Also all jump instructions as the meaning is jump immediately. Example MVI M, 2H; ADI 47H; LXIH 2050 (exception for direct addressing mode).

Implicit : No operand. Example: XCHG.

Opcode	Operand	Bytes	M-cycles	T-states
MOV	Rd, Rs	1	1	4
	M, Rs		2	7
	Rd, M		2	7
MVI	reg, data	2	2	7
	M, data	2	3	10

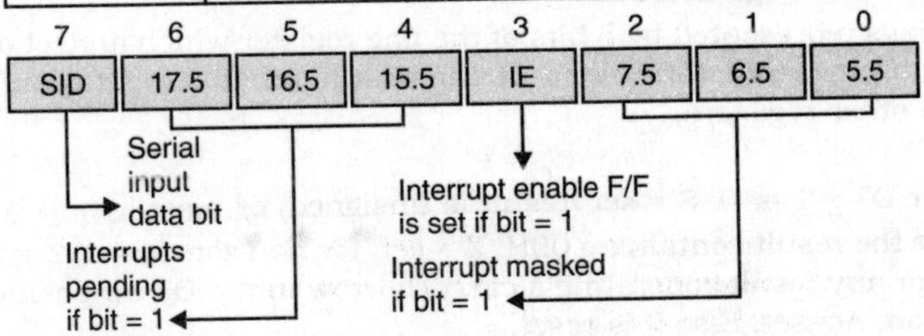

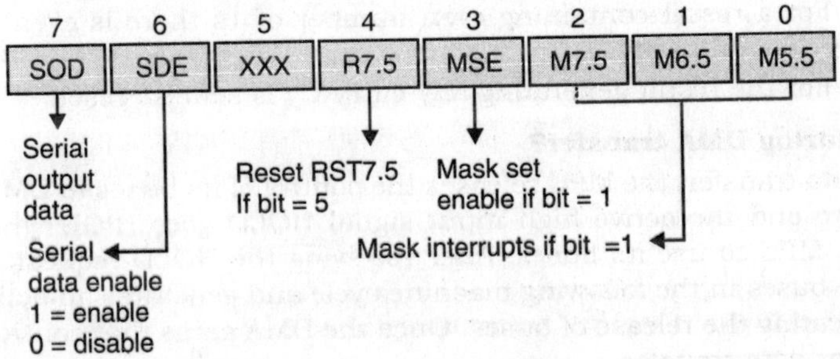

Q.30. What is the difference between MOV and MVI?

Ans. Rd = Destination register, Rs = Source register, M = Memory location pointed out by HL register pair, reg = Register, data = 8-bit data.

Q.31. What are the functions of RIM, SIM, IN?

Ans. Read Interrupt Mask (RIM)

RIM is a multipurpose instruction used to read the status of interrupts 7.5, 6.5, 5.5 and to read serial input data bit. RIM loads 8-bit data in the accumulator with the following interpretation :

Actually RIM does the following three tasks :

Read the interrupt mask (bit 2, 1, 0).

Identify pending interrupts (bit 6, 5, 4).

Receive serial input data bit (bit 7).

Set Interrupt Mask : SIM is a multipurpose interrupt used to implement the 8085 interrupts (RST 7.5, 6.5, 5.5) and serial data output. SIM interprets the accumulator content as follows :

Actually, SIM does the following three tasks :

Mask the interrupts (bit 2, 1, 0).

Reset RST 7.5 (bit 4). This is mainly used to overwrite RST 7.5 without serving it.

To implement serial I/O (bit 7, 6). If bit 6 = 1 is used to enable serial I/O and bit 7 is used to transmit serial output data bit.

```
  1 ↙ Carry
 10011000
 10001001+        7 6 5 4 3 2 1 0
 ---------       |S|Z| |Ac| |P| |Cy|
 100100001
```

Input Data to Accumulator from a Port with 8-bit Address (IN).

The contents of the input port designated in the operand are read and loaded into the accumulator. The operand is an 8-bit address. During execution, this port address is duplicated in the lower order and higher order address buses. Any one of the sets of address lines can be decoded to enable the input port.

Q.32. What are the different flags in 8085?

Ans. The 8085 has 5 flags represented by 5 bits of the flag register which are set or reset after an operation according to data conditions of the result (of that operation) stored in the accumulator and other registers.

They are :

Sign flag (S) : For D7 = 1 or 0, S = set (result is unsigned) or reset (result is signed).

Zero flag (Z) : For the result containing 00H, Z = set, for non zero result Z = reset. Auxiliary carry flag (Ac):- For any result generating a carry/borrow in the D3 bit position and passing it to D4 bit position, Ac=set. Else it is reset.

Parity flag (P) : For a result containing even number of 1s there is even parity and odd number of 1s there is odd parity.

Carry flag (Cy) : For the result generating any carry Cy is set else reset.

Q.33. What happens during DMA transfer?

Ans. To make a fast data transfer, the MPU releases the control of its buses to DMA. DMA acts as an external device and the active high input signal HOLD goes HIGH when the DMA is requesting to the MPU to use its buses. After receiving the HOLD request from DMA, the MPU releases the buses in the following machine cycle and generates an active high output signal HLDA indicating the release of buses. Once the DMA gains that control, it acts in the role of the MPU for data transfer.

Q.34. What do you mean by wait state? What is its need?

Ans. A wait state is a delay experienced byµP when accessing external memory or another device that is slow to respond. the vice versa also cone into scenario.

Now, to be able to access slow memory theµP must be able to delay the transfer until the memory access is complete. One way is to increase theµP clock period by reducing the clock frequency. SomeµPs provide a special control input called READY to allow the memory to set its own memory cycle time. If after sending an address out, theµP dies not receive a READY input from memory, it enters a wait state for as long as the READY line is in 0 state. When the memory access is completed the READY goes high to indicate that the memory is ready for specified transfer.

In the channel. So, only one radio station can broadcast at a time which is wastage of channel space. One way to solve this problem is to use modulation.

Narrow-banding : Suppose a baseband signal from a broadcast system is radiated directly with the frequency range 50Hz to 10Kz *i.e.* with a band-edge ratio (ratio of highest to lowest wavelength) of 200. If an antenna is made for 50Hz, it will be too long for 10 KHz and vice versa. So, we may require a antenna suitable for band-edge ratio 200 which is practically impossible. But frequency translation makes a wideband signal narrowband. Here band width is referred to as the ratio of the edge frequencies of the band. This is narrowbanding.

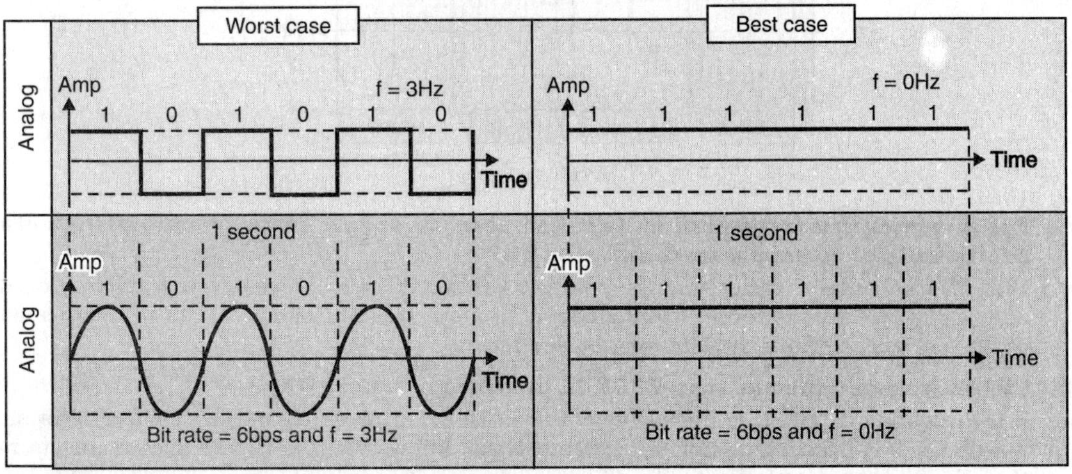

Q.35. Why we use vestigial side band (VSB-C3F) transmission for picture?

Ans. Bandwidth of signals used for picture reception is very large, so standard AM cannot be used. If SSB was used, it would result in half the bandwidth. But, SSB receivers being complex, they are not used. So vestigial transmission is used in which a vestige or trace of the unwanted sideband is transmitted along with the wanted sideband and carrier. Actually in VSB, instead of rejecting one side band completely like SSB, a gradual cut off of one side band is accepted.

Q.36. When transmitting digital signals is it necessary to transmit some harmonics in addition to fundamental frequency?

Ans. Consider a 6bps (taken very slow for easy understanding) digital signal transmission. The best case and worst case digital and analog signals are shown. In worst case and best case, to simulate the digital signal the, necessary analog signal should be of 3Hz and 0Hz respectively. As the bit rate increases, the necessity of high frequency analog signal increases and the possibility of sending a single frequency (fundamental) signal decreases. So, for high bit rate, to improve the shape of the signal to make it recognizable by the receiver, we need to add some odd harmonics.

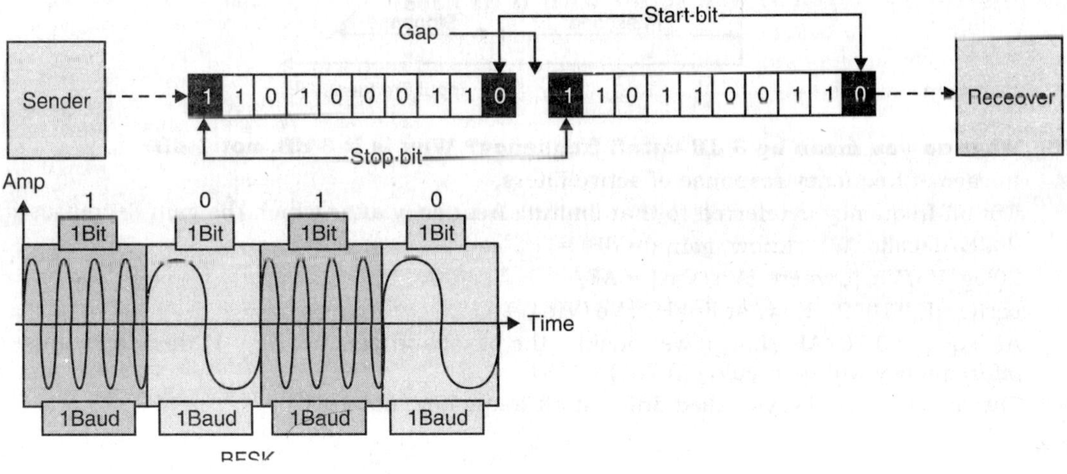

BFSK

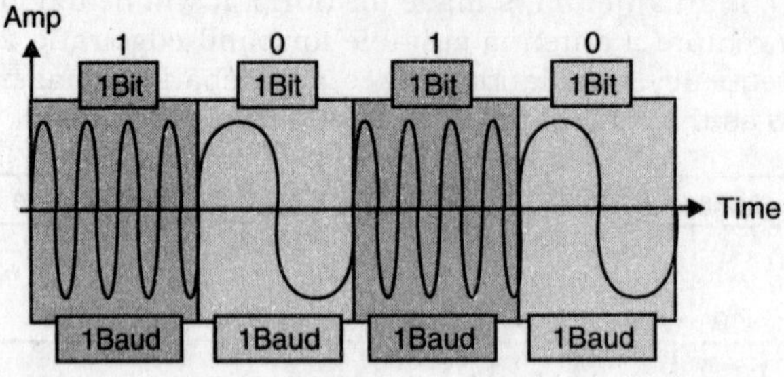

BPSK

Q.37. For asynchronous transmission, is it necessary to supply some synchronizing pulses additionally or to supply start and stop bit?

Ans. Yes, it is necessary. Otherwise the receiver will not be able to detect the starting or the ending of a byte sent by receiver. And in Asynchronous transmission it is not the responsibility of the receiver to group the bits sent by sender..

Q.38. BPFSK is more efficient than BFSK in presence of noise. Why?

Ans. If the question is "BPSK is more efficient than BFSK in presence of noise. Why?" (Not sure whether it is a printing mistake or not but if one knows what is BPFSK please inform me) then the answer is as follows :

Actually, BPSK is not susceptible to the noise degradation that affects ASK or bandwidth limitations of BFSK (BFSK needs more bandwidth than BPSK) . This means that smaller variations in the BPSK signal due to noise can be detected reliably.

Q.39. What is meant by pre-emphasis and de-emphasis?

Ans. **Pre-emphasis :** Improving the signal to noise ratio by increasing the magnitude of higher frequency signals with respect to lower frequency signals.

De-emphasis : Improving the signal to noise ratio by decreasing the magnitude of higher frequency signals with respect to lower frequency signals.

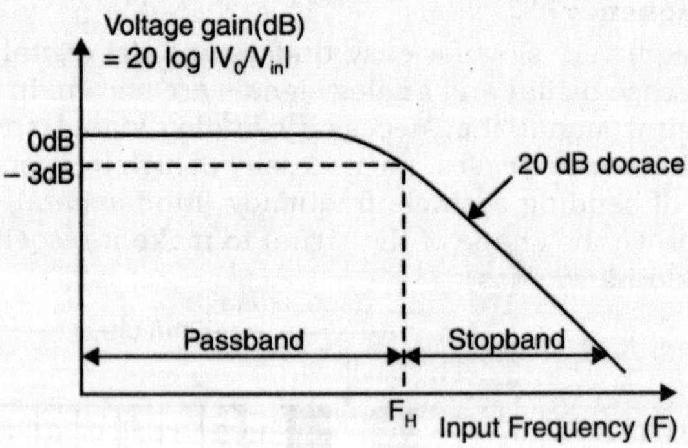

Q.40. What do you mean by 3 dB cutoff frequency? Why is it 3 dB, not 1 dB?

Ans. In case of frequency response of active filters,

Cut off frequency is referred to that limiting frequency after which the gain decreases @ 20dB/decade. We - know, gain (in dB) =
20log|Vo/Vin|, where |Vo/Vin| = AF/
sqrt(1+(F/FH)^2). Now, at F=FH, |Vo/Vin| =
AF/sqrt2 = 0.707AF. Now, if we consider the passband gain i.e. AF = 1, then the gain at cut off frequency will be = 20log (0.707) = 3dB.

That is why it is always called 3dB cut off frequency, not 1dB.

In addition, 3db implies 70% (0.707) of the power, i.e. we r interested to consider the bandwidth range from peak to 70% because up to 70% it's reliable. Hence 3db is called the half power frequency.

Q.41. What do you mean by ASCII, EBCDIC?

Ans. **ASCII :** American Standard Code for Information Interchange. A seven or eight bit code used to represent alphanumeric characters. It is the standard code used for communications between data processing systems and associated equipment.

EBCDIC : Extended Binary Coded Decimal Interchange Code. It is an 8-bit character encoding (code page) used on IBM mainframe operating systems such as z/OS, OS/390, VM and VSE, as well as IBM minicomputer operating systems such as OS/400 and i5/OS (see also Binary Coded Decimal). It is also employed on various non-IBM platforms such as Fujitsu-Siemens' BS2000/OSD, HP MPE/iX, and Unisys MCP. It descended from punched cards and the corresponding six bit binary-coded decimal code that most of IBM's computer peripherals of the late 1950s and early 1960s used.

SUMMARY

1. As in many other 8-bit processors, all instructions are encoded in a single byte (including register-numbers, but excluding immediate data), for simplicity.
2. Some of them are followed by one or two bytes of data, which could be an immediate operand, a memory address, or a port number. Like larger processors, it has CALL and RET instructions for multi-level procedure calls and returns (which can be conditionally executed, like jumps) and instructions to save and restore any 16-bit register-pair on the machine stack.
3. There are also eight one-byte call instructions (RST) for subroutines located at the fixed addresses 00h, 08h, 10h,...,38h. These were intended to be supplied by external hardware in order to invoke a corresponding interrupt-service routine, but are also often employed as fast system calls. The most sophisticated
4. For the extensive use of 8085 in various applications, the microprocessor is provided with an instruction set which consists of various instructions such as MOV, ADD, SUB, JMP etc.
5. These instructions are written in the form of a program which is used to perform various operations such as branching, addition, subtraction, bitwise logical and bit shift operations.
6. More complex operations and other arithmetic operations must be implemented in software. For example, multiplication is implemented using a multiplication algorithm.
7. The 8085 processor has found marginal use in small scale computers up to the 21st century. The TRS-80 Model 100 line uses a 80C85.
8. The CMOS version 80C85 of the NMOS/HMOS 8085 processor has several manufacturers, and some versions.
9. Tundra Semiconductor Corporation's CA80C85B) have additional functionality, eg. extra machine code instructions.
10. The 8085 CPU was only one part of a much larger family of chips developed by Intel, for building a complete system.

EXERCISE

Part-A

1. Differentiate between memory mapped I/O and I/O mapped I/O.
2. Classify the different groups of 8085 instruction set with example.
3. Differentiate between unidirectional buffer and bi-directional buffer.

4. What is the need for ALE signal in 8085 microprocessor?
5. Give the operation of the foll instructions : (*a*) DAA (*b*) DEC.
6. State the functions for ALE and TRAP pins in 8085.
7. If the frequency of the crystal connected to 8085 is 6MHz, calculate the time to fetch and executed NOP instruction.
8. What is a MPU?
9. What do you mean by multiplexing the bus?
10. List out the two parts of an instruction.
11. What is a program counter?
12. What is an instruction?
13. What is PSW?
14. Define - Interrupt.
15. What are the addressing modes for 8085 microprocessor?
16. Where the READY signal used?
17. Define stack.
18. Specify how a program counter is useful in program execution.
19. How the data and address lines are demultiplexed?
20. Show the bit positions of various flags in 8085 flag register?
21. List the various machine cycles of 8085.
22. What is the instruction format of 8085.
23. What are the similarity and difference between subtract and compare instructions?
24. List the type of signals that have to be applied to initiate hardware interrupts in 8085.
25. Write a subroutine to clear the flag register and accumulator?
26. Draw a simple circuit to decode three controls signals RD, WR, IO/M and to produce separate read/write control signal for memory and I/O device?
27. List out the similarities between CALL_RET and PUSH_POP instructions.
28. List four interrupts of 8085.
29. Give the clock out frequency and state time T of 8085 when the crystal frequencies are : (*a*) 5 MHz, (*b*) 6.144 MHz
30. How many times the given loop will be executed? What will be the contents of HL pair when the program control reaches to HLT instruction?

    MVI A, 00h
    LX1H, 5003H
    Loop: DCX H
    DCR A
    JNZ loop
    HLT
31. Define : (*a*) Instruction Cycle, (*b*) M/c cycle, (*c*) T-state.
32. Explain the execution of the instruction CMA M in 8085.
34. If the program counter is always one count ahead of the memory locations from which the machine code is being fetched, how does the microprocessor change the sequence of program execution with a "Jump" execution?
35. Differentiate between hardware interrupts and software interrupts of 8085.
36. What is DAD and what are the flags, affected by this instructions?
37. What is the function performed by SIM Instruction?
38. What is meant by processor cycle?
39. What are the different memory mapping schemes? Give any one advantage and disadvantage for each.

40. Write an 8085 assembly language program, which checks to see if the number is even or odd returns a '0' in B register else returns a '1'.

Part-B

1. (*i*) Draw the pin configuration of 8085 and explain the functional usage of each pin in details.
   (*ii*) Write a 8085 program to generate a delay of 0.6 sec for the given crystal frequency of 5 MHz.
2. (*i*) Explain hardware interrupts in 8085.
   (*ii*) Draw the timing diagram for MVI A, 30H instruction.
   (*iii*) What do you mean by polling in 8085.
3. Write a 8085 assembly language program to generate a delay of 0.1 sec for the given crystal frequency of 2 MHz.
4. (*i*) With suitable examples, explain the addressing modes of 8085.
5. (*i*) Draw the timing diagram of I/O read cycle in 8085 and explain it.
   (*ii*) Interface a 8-bit DIP switch with 8085 such that the address assigned to it is FFH.
6. Explain the 8085 architecture.
7. Write a 8085 program to convert a binary number to decimal.
8. (*i*) Write a 8085 program for multiplying two 8 bit numbers.
   (*ii*) Write an assembly language program to evaluate the expression (A+(B*C))/D, where A, B, C and D are single byte integer.
9. (*i*) Draw the timing diagram of opcode and memory write machine cycles of 8085 for following instructions: MOV V, M; MVI D, data; IN OL; OUT 01; LDA 8000; RST7
   (*ii*) What is the data transfer instruction? Give example.
10. Write an assembly language program based on 8085 microprocessor to implement a look-up table translation to generate the codes for seven-segment display with our digits. (Assume INPUT PORT O D 6 H, and OUTPUT PORT OD 4H)
11. (*i*) Write an 8085 assembly language program to perform 32 bit binary addition.
    (*ii*) What will be the value in ACC, for the given 8085 program below :

```
MVI C, 7F
MVI B, 3E
MOV A,B
RLC
RLC
ANI 7F
HLT.
```

12. (*i*) Write an 8085 assembly language program to fill a block of 1000H bytes starting at address 9000H with characters 2BH.
    (*ii*) Indicate the logic levels on each control or status pin as logic '0' or logic '1' for 8085 microprocessor with the given data below ;

Machine cycle	RD	WR	IO/M	S_0	S_1
Memory write					
Memory read					
I/O read					
I/O write					
O/P code fetch					

14. With example explain different groups of instruction sets.

15. (*i*) Analysis the program and find the content of A register after execution.

```
MVI A, 07
RLC
MOV B,A
RLC
RLC
RLC
ORA B
```

(*ii*) Explain 8085 interrupts in detail.

16. (*i*) Draw and explain the timing diagram for STA 4500 instruction.

(*ii*) Write as assembly language program to find the average of N numbers.

17. Interface 3 numbers of 8KB EPROM and 5 numbers of 8KB RAM to microprocessor 8085 to have a total memory capacity of 64 KB.

18. (*i*) Write an assemble language program to add two 2-digit BCD numbers.

19. Write notes on status flag.

20. Write an assembly language program to find the biggest and smallest in an array of given numbers.

21. Write a program to count continuously in hexadecimal from FFB to 00H in a system with a 0.5 micro sec clock period. Use register C to set up a one milli second delay between each count and display the number at one of the output ports.

22. (*i*) Draw a flowchart and 8085 assembly language program for adding the two hexadecimal numbers 9BH and A7H in register D and E respectively. If the sum is greater than FFH, display 01H at output PORT0: otherwise, display the sum.

(*ii*) Draw the timing diagram for IN and OUT instructions.

23. Explain with the help of suitable diagram how the INTR pin can be used to interrupt the 8085 and how it responds to the signal.

24. (*i*) Explain the logic instructions of 8085 microprocessor with examples.

(*ii*) Write and ALP using 8085 instruction set to add two n-byte numbers stored at memory locations starting at 'X' and 'Y' respectively. Store the result and memory location starting from 'Z'. Draw the flow chart.

25. (*i*) List all the control signals of the Timing and Control Unit. Explain the use of each of these signals.

(*ii*) Write an ALP to evaluate the expression C = A2 + B2.

26. (*i*) What are the features of FIFO and LIFO memory structures?

27. Draw and explain the timing diagram for the execution of the instruction LDA 2080H.

28. (*i*) How do you classify the 8085 interrupts what are they? How is the interrupt branch address obtain in each case?

(*ii*) Draw the timing diagram of DCX instruction.

□□□

Glossary

Glossary

ACCESS TIME: The length of time required for a binary WORD in the memory section of a computer to be read by the CENTRAL PROCESSING UNIT (CPU), or the time to read data from a peripheral data storage area.

ACCUMULATOR: An interface REGISTER (memory) in the ARITHMETIC LOGIC UNIT (ALU), that stores interim arithmetic information for future processing. The accumulator is interface between the ALU and other sections of the computer.

ADDER: The digital logic circuits in the ALU section of a computer which implements the adding process (sum and carry) of two or more binary numbers.

ADDRESS: The location of digital information in the MEMORY UNIT of a computer, or a digital code that designates this location.

ALGORITHM: The sequence of operations that defines a solution to a problem in logic.

ALPHANUMERICS: The term that defines the letters of the alphabet (A to Z) and the ten numerals (0 to 9). The term is generally used to mean any text data.

AMERICAN STANDARD CODE FOR INFORMATION INTERCHANGE (ASCII): A 7-bit binary code, providing 128 different binary combinations for standard American keyboards. ASCII is used to encode all 26 letters of the alphabet (upper and lower case), all ten decimal digits (0 to 9), punctuation marks, standard graphics, and special control codes into machine language. Although ASCII has 128 different codes, only 7 bits are needed for each different code. ASCII characters are generally stored inside 8-bit bytes, providing room for the 128 ASCII codes plus another 128 codes, totaling 256 characters. This 8-bit code is referred to as EXTENDED ASCII.

ANALOG COMPUTER: A non-digital system that manipulates linear, or continually changing data.

AND GATE: See LOGIC GATES

ARITHMETIC LOGIC UNIT (ALU) ; The section of a computer that contains a large amount of logic circuitry and performs the four basic arithmetic functions (addition, subtraction, multiplication, and division). Larger units contain circuitry for higher mathematical functions, such as: quadratic equations, etc.

ASCII DATA (ASCII FILE): Data, or the file containing the data, that is constructed with letters of the alphabet, numerals, punctuation marks, and the standard ASCII formatting commands, such as: "carriage return" or "line feed". These terms refer to text information.

ASSEMBLER: A software program that converts (translates) each symbolic instruction written in ASSEMBLY LANGUAGE into the MACHINE LANGUAGE (binary code) of a computer.

ASSEMBLY LANGUAGE: A programming language (source code) that consists of a group of coded letters or labels, called mnemonics. A mnemonic is a memory assist to help recall data. Each mnemonic represents a single instruction that is translated into the binary code of machine language. Mnemonics are easier to use than machine language instructions. For example, the mnemonic "MUL" tells the computer to "MULTIPLY".

ASYNCHRONOUS: A term that describes a non-clocked, or free-running digital signal that triggers successive computer instructions; the completion of one instruction triggers the next. The speed of operation depends only on the speed of the signal generated through the circuit or network. In contrast with SYNCHRONOUS operation, the computer clock controls the speed of the signals in the system.

BACKPLANE: A part of a computer into which the system's PC boards are plugged to provide a common voltage supply, reference, and SYSTEM BUS for all computer sections. A backplane is often called a MOTHERBOARD.

BASIC: The acronym for Beginner's All-purpose Symbolic Instruction Code. BASIC is an easy to use programming language, originally intended for educational purposes; it is available for personal computers in varying degrees of complexity.

BAUD RATE: A measure of speed (up to 2400 bits per second) of transferring information between two or more sections of a computer system, or between two or more computers.

BAUDOT CODE - A 5-bit digital code having 32 possible combinations of binary "0s" and "1s"; for Teletype (Telex) communications systems. BENCHMARK - The specifications for measuring the characteristics of a computer system, or parts of the system, under clearly-defined conditions.

BINARY CODED DECIMAL (BCD): A coding system in which each decimal system numeral (0 to 9) is represented by a 4-digit (4-bit) binary code.

BINARY SYSTEM (BASE 2): A numbering system consisting of only two digits (0 and 1), as contrasted with a DECIMAL SYSTEM that uses ten digits (0 to 9). In electronics, "binary" and "two-state" are synonymous.

BINARY DIGIT (BIT): The term "BIT" is the contraction of Binary DigiT and is part of a binary WORD that consists of combinations of "0s" and/or "1s". There are only two numerals in binary arithmetic (base 2) and is the basis for binary code (Machine Language), the language of the computer. A "bit" has the same significance in binary arithmetic that a decimal digit has in the more familiar decimal (base 10) arithmetic system.

BISTABLE MULTIVIBRATOR: A digital circuit having two stable electrical states. The output signal of this circuit will remain in either state (0 or 1) indefinitely until an external input signal causes the circuit to switch to its other output state. This circuit is generally called a FLIP-FLOP.

BIT: See BINARY DIGIT.

BLACK BOX: A term that refers to an electronic circuit or system within an enclosure without necessarily providing details of its internal elements. The black box concept often offers a useful approach in the design of a system or in the interconnect between two circuits or systems.

BOOLEAN ALGEBRA: Named after George Boole, a 19th century English Mathematician, who first formulated theorems that included a mathematical analysis of the laws of human logic. It uses algebraic-like notation to describe the interaction of variables having only two states - "true" and "false". In electronics, the states are often referred to as "1" and "0" or, "high" and "low". The terms "high" and "low" refer to the voltage levels of the input and output signals of a logic gate (logic circuit). Boolean algebra is used in the design of logic circuits in computers and similar digital systems. The logic gates called AND, OR, and NOT are the three basic logic gates in Boolean algebra. NAND, NOR, XOR and others are combinations of the three basic logic gate operations.

BOOTSTRAP (BOOT): A software program for initiating the operation of a computer. The function of the program is to set up the input and output (input/output) devices and load the OPERATING SYSTEM from a disk, cassette, or built-in READ ONLY MEMORY (ROM).

BREAKPOINT: Location of a place in a program where program execution can be stopped to permit a visual test, printing, or other performance analyses.

BUBBLE MEMORY: A high-density memory medium upon which a magnetic film is grown on a gadolinium-gallium garnet substrate. A small permanent magnet is mounted inside its package, perpendicular to the surface of the substrate. When an external magnetic field is created with an external coil, magnetic "bubble" domains are formed on the internal magnetic film which represent patterns of "1s". The absence of magnetic bubbles will represent patterns of "0s".

BUFFER: 1. A digital logic circuit inserted between other digital circuits to reduce circuit interaction and/or to provide amplification of a digital signal. 2. An intermediate storage circuit used to compensate for a difference in baud rate or, to compensate for different times of occurrence of different events or instructions. 3. A circuit for converting input or output voltages for signal level compatibility when transmitting data from one device to another.

BUNDLED: A term describing a computer system that includes all necessary hardware and software that will allow the system to operate as advertised.

BUS (BUSS): A series of transmission lines connecting the various elements of a computer for distribution of data, control signals, addresses, and/or voltage supply(s) within a computer. See SYSTEM BUS.

BUS CONTROLLER: A circuit that generates commands and control signals for sequencing and timing of the data transmitted on a bus.

BYTE: A set (group) of consecutive binary digits (bits) that forms a unit (word) or sub-unit of digital information in a computer. Depending on the size of the computer, a digital word may contain one or more bytes. Through common usage, a byte is generally accepted as containing eight bits.

CACHE: A high-speed buffer memory that is similar to a SCRATCH PAD MEMORY, but has a larger capacity; it is located in the CENTRAL PROCESSING UNIT. It is filled from the main memory at low speed with instructions and programs and is operated at a higher speed. A cache can also serve as a storage section between the CPU and a hard disk to speed up access of data on the hard disk.

CD-ROM (COMPACT DISK READ-ONLY MEMORY): An aluminized disk, 4.72" in diameter that provides an optical storage medium for digital data up to 650 Megabytes. A laser beam permanently burns data into its surface which cannot be over-written, altered, or erased, providing read-only memory capability. Larger size disks are generally called "laser disks".

CD-R or WORM (COMPACT DISK, WRITE ONCE, READ MANY MEMORY): A blank disk that can be programmed once, permanently storing up to 650 Megabytes of digital data. Once programmed, it has essentially the same features as a CD-ROM.

CD-E or CD R/W (COMPACT DISK, ERASABLE, READ/WRITE MEMORY): Capable of storing up to 650 Megabytes of digital data. It uses a magneto-optical technology that combines the storage capacity and reliability of optical disks and the erasability of magnetic media.

CENTRAL PROCESSING UNIT (CPU): The section(s) in which all the processing circuits of a computer are located. It incorporates a CONTROL UNIT and an ALU.

CHARACTER GENERATOR: A circuit that accepts digital data and forms the corresponding letters and numerals for a monitor or printer.

CLOCK: A digital pulse generator that controls the timing of a computer and, to a great extent, determines the speed (number of instructions per second) capability of the computer. Generally, it is located in the CPU.

CMOS LOGIC: A CMOS technology monolithic IC "logic family" characterized by low power dissipation per gate, high chip density, and relatively high propagation delay per gate compared with bipolar IC logic families.

COMPILER: A software program that converts (translates) a complete software program written in high-level language SOURCE CODE (such as PASCAL or FORTRAN) into machine language. The entire source code is edited, compiled, and run at one time as compared to an INTERPRETER that is run one line at a time.

COMPLEMENT: Reversal of bit values: "1s" become "0s" and "0s" become "1s".

CONCURRENCY: The independent execution of two or more sequences of events that are either occurring, or appearing to occur simultaneously.

CONSOLE: The term referring to a combination of a DISPLAY and KEYBOARD.

CONTROL BUS: A set of transmission lines whose function is to carry synchronization signals and control data as part of the SYSTEM BUS.

CONTROL UNIT (CU): Part of the CPU containing the CLOCK, PROGRAM COUNTER and INSTRUCTION REGISTER. The CONTROL UNIT also generates control signals and manages the CONTROL BUS.

COUNTER: A circuit whose output(s) change state in a specified sequence on receiving appropriate input signals. The circuit can provide a required output pulse after receiving a specified number of input pulses.

CRASH: The term that describes a situation when a part of, or the complete computer, stops working because of a hardware and/or software malfunction. A head-crash in a disk system refers to the accidental impact of the read/write head on the surface of the disk.

DAISY-WHEEL PRINTER: An impact printer that uses a print element shaped like a flat disk or large thimble to form the alphanumerics and punctuation marks that are part of its print element. Unlike dot-matrix, ink-jet, and laser printers, daisy-wheel printers can not print graphics. See LETTER-QUALITY PRINTER.

DATA BASE: A collection of structured information for creating a specific format out of some, or all, of the collected data.

DATA-BASE MANAGEMENT SYSTEM (DBMS): A software program that controls and supervises the updating, editing, and execution of items from multiple files in a data base environment.

DEBUG: The process of detecting, locating, and correcting a problem in a software program or hardware.

DECREMENT: The reduction of the numerical contents of a counter. A decrement of 1 is usually assumed, unless otherwise specified. It is the complementary operation of INCREMENT.

DEVICE: In a computer system, this term refers to a unit of processing equipment external to the CPU. It is sometimes synonymous with the term PERIPHERAL.

DIODE-TRANSISTOR LOGIC (DTL): One of the first bipolar monolithic IC families of logic gates. A diode in an IC logic gate performs the required logic with a transistor amplifying and inverting the output. The DTL family has been made obsolete by the Transistor-Transistor Logic (TTL) family.

DIRECT ADDRESS: A memory accessing mode in which the contents of the accessed location is called the OPERAND.

DIRECT MEMORY ACCESS (DMA): A method of transferring blocks of data directly between an external device and the computer system memory without the need for intervention by the CPU. This method significantly speeds up the data transfer rate, improving system efficiency.

DISC or DISK (DISKETTE): A thin, flexible, plastic, circular memory medium, coated on both sides with metal oxide and enclosed in a protective jacket. Data is stored magnetically in binary digital form on its surface(s). See FLOPPY DISK.

DISK DRIVE: The mechanical/electronic section that can accept and operate a compatible floppy or hard disk. It may include several motors (for disk rotation and reading/writing head positioning), position sensors, and control circuits.

DISK EMULATOR: See RAM Disk.

DISPLAY: A peripheral device serving as a computer readout, such as a cathode-ray tube (CRT), flat-panel (generally for laptop computers), or other readout devices. The screen can be amber, green, or full-color display. Also called a MONITOR.

DISK OPERATING SYSTEM (DOS): A software program on a disk which coordinates the operation, transfer of data, supervision, and control of a computer. This software program must first be booted into the working memory of the computer from the disk before it can operate.

DISK STORAGE: A method of storing software programs and data on a rotating circular disk (either a floppy or hard disk) coated with magnetic material, such as iron oxide. Data is written (stored) and read (retrieved) by movable read/write heads positioned over data tracks on the surface of the disk. Addressable portions of the disk can be selected for read or write operations.

DOCUMENTATION: Information that explains how to use computer hardware or software. It is usually provided as a manual or stored on a disk.

DOT-MATRIX PRINTER: An impact printer with a computer-driven, multi-pin print element (print-head). It create images by imprinting a series of tiny dots on a paper to print a wide variety of character styles and/or finely detailed graphics. Generally, these printers are extremely fast and are used for draft-quality documents and precise graphics.

DOTS PER INCH (DPI): The measurement of density on dot-matrix printers or other dot-matrix devices. As the DPI increases, image clarity increases.

DOWNLOAD: The transfer of files or data from a source of data to a remote computer. See UPLOAD.

DOWN TIME: A period of time during which a computer is not functioning. It is inoperable because of temporary or permanent failure of hardware or software, or when routine hardware or software maintenance procedures are indicated.

DYNAMIC RANDOM ACCESS MEMORY (DRAM): A type of semiconductor memory in which the presence or absence of a capacitive charge in each element of the memory represents the state of the bit (1 or 0). This charge must be periodically recharged (refreshed) to maintain the desired binary state of the element.

EDITOR: A program for preparing and/or modifying a SOURCE PROGRAM or other file by addition, deletion, or change.

EMITTER-COUPLED LOGIC (ECL): A bipolar monolithic IC logic gate family characterized by very high speed operation and relatively high power dissipation compared with other monolithic IC logic families.

EMULATION: The process of imitation (simulation) of one computer system by another. The imitating program, or device (emulator), accepts the same data, executes the same programs, and achieves the same results.

EPROM (ERASABLE PROGRAMMABLE READ-ONLY MEMORY): A general term for a non-volatile, semiconductor memory that can be programmed, erased, and reprogrammed many times without damage to the device. More specific types of EPROMs include: EAPROM (ELECTRICALLY ALTERABLE PROM), EEPROM (ELECTRICALLY ERASABLE PROM) and UV-EPROM (ULTRAVIOLET ERASABLE PROM). Either memory is suitable for prototype and computer development, or to change memory data when new conditions dictate the change.

EVENT: An occurrence during the execution of a task, such as the completion of an input/output operation.

EXECUTE: The command to run a specified instruction or software program.

EXTENDED BINARY CODED DECIMAL INTERCHANGE CODE (EBCDIC): An 8-bit code developed by IBM for their mainframe computers, providing 256 bit-pattern equivalents of standard keyboard symbols.

FAN-IN: The maximum number of output terminals from other logic gates that can be connected to an input terminal of a specified logic gate.

FAN-OUT: The maximum number of input terminals that can be connected to the output terminal of a specified logic gate.

FETCH: The command to obtain an instruction from a stored program.

FILE: A collection of related data treated as a single unit. In a computer, a file can exist on a disk, magnetic tape, or as an accumulation of information in memory. FIRMWARE - A combination of a software program in hardware, such as a READ-ONLY MEMORY (ROM), or a disk that has files or software programs written on its surface.

FIRST IN, FIRST OUT (FIFO): The term refers to the sequence of entering and then retrieving data from a data storage section of a computer. The first data entered is the first data retrieved.

FLAG: An indicator of a specific condition that informs a section of a program that this condition has already occurred and is identified by the presence or absence of the flag. A flag can be implemented in software and/or hardware.

FLASH MEMORY (FLEXIBLE ARCHITECTURE for SHARED MEMORY): Non-volatile semiconductor memory with access time approximately that of EEPROM or EAPROM and density similar to that of an EPROM. FLASH memory is generally limited to about 100,000 erase cycles. Usually packaged in PCMCIA for plugging into a laptop computer card slot.

FLIP-FLOP CIRCUIT: A logic circuit having two stable output states. It has the ability to change from one state to the other when an input pulse is applied in a specified manner. It is also called a BISTABLE MULTIVIBRATOR.

FLOATING-POINT ARITHMETIC: A method in which the decimal point location of a number in an arithmetic operation is determined by the number's exponent value in base 10. All exponents are equalized prior to the operation to set a decimal point in its proper location in the final computation. Floating-point arithmetic extends a computer's mathematical capability beyond the limit imposed by a fixed word length and contributes to easier programming.

FLOPPY DISK (DISKETTE): A relatively inexpensive, flexible, plastic disk for storing digital data on both sides of its magnetic surface. The amount of data depends on the diameter of the disk and the technology for recording the data. See DISK STORAGE.

FLOW CHART: A symbolic representation of the processing steps performed by a software program or a graphic sequence of logic operations implemented in hardware. A flow chart helps to visualize the procedure(s) necessary to design a software program or a final hardware system.

FONT: A style of print typeface having a specific form and size.

FORMAT: An orderly, structured arrangement of data elements (bits, bytes, and/or fields) that is necessary to produce a larger entity, such as: a list, record, table, file, or dictionary. Also, it is the term that describes the preparation of a magnetic disk to allow it to accept digital data.

FORTRAN: A science-oriented high-level software language, derived from the contraction of the terms FORmula TRANslator.

FREQUENCY SHIFT KEYING (FSK): A technique of digitally keying (modulating) an audio tone (carrier wave) so that a commercial telephone line can carry digital data.

FULL DUPLEX: A data transmission mode that provides simultaneous and independent transmission and reception. A conventional telephone communication is an example of this technique. See HALF DUPLEX.

GATE: See LOGIC GATE.

GATE ARRAY: A group of standard logic gates that can be interconnected into a complete circuit or system. Also called LOGIC ARRAY.

GENERAL PURPOSE INTERFACE BUS (GPIB): A BUS specification standard (IEEE 488) for controlling peripheral devices.

GIGO (GARBAGE IN, GARBAGE OUT): The term describing the output of a computer whose operation or accuracy is faulty.

GRAPHICS: Schematic drawings, pictures, line drawings, and/or diagrams generated by data entered into a computer via a keyboard or a data base.

HALF-DUPLEX: A communications mode that allows transmission and reception of digital data between computers, but not simultaneously.

HANDSHAKING: A communications synchronizing technique carried out before and after any transfer of digital data. It consists of a sequence of signals for non-clocked (asynchronous) systems in which a reply is needed to complete a data transfer operation.

HARD COPY: A printed copy of a file, message, or graphic of the visual display on the screen of a computer monitor.

HARD-DISK DRIVE: A sealed unit containing high-density, high-speed, rigid metal disks, and recording heads to store digital data. It reads and writes data faster than floppy disks.

HARDWARE: The physical equipment of a computer system consisting of mechanical and electrical/electronic components.

HEXADECIMAL: The base 16 number system using 16 symbols (0 to 9 and A to F) to represent 16 decimal numerals (0 to 15).

HIGH-LEVEL LANGUAGE (HLL): A programming language (source code) consisting of a unique group of symbols and command statements representing a series of machine operations. A COMPILER or INTERPRETER translates (converts) a HLL into MACHINE LANGUAGE. BASIC, FORTRAN, PASCAL, ALGOL, and ADA are some examples of high-level languages.

IEEE 488: See GENERAL PURPOSE INTERFACE BUS

IMPACT PRINTER: A computer-driven mechanical imprinting device where the characters are formed by the printer-head key striking a ribbon to imprint the character's image onto a paper.

INCREMENT: The increase in the numerical contents of a counter. An increment of one is usually assumed, unless otherwise specified. See DECREMENT.

INK-JET PRINTER: A printer that forms characters by electrostatically aiming and depositing a tiny patterned drop of ink onto the paper to be printed.

INPUT/OUTPUT (input/output) SECTION: The section that interfaces between the computer's SYSTEM BUS and the peripherals feeding data into and taking data out of the computer. Depending on the number of peripherals in a system, the input/output sections can have a single PORT or multiple ports.

INSTRUCTION: A software statement that specifies a machine operation. Also called a COMMAND.

INSTRUCTION SET (INSTRUCTION REPERTOIRE): A description of the total operational capabilities of a computer provided by the computer or the CPU (MPU) manufacturer. It consists of a listing of binary words for each executable commands. It is sometimes called the computer's MICROCODE.

INTERPRETER: A high-level language translator that converts individual high-level computer language program instructions (source code) into machine instructions. It translates and executes each statement line-by-line during the running of the program.

INTERRUPT: The suspension of normal program execution to perform a higher priority service routine as requested by a peripheral device. After completion of the service routine operation, the interrupted program routine is resumed at the point where it was interrupted.

JUMP: An instruction that causes the computer to fetch the next instruction to be executed from a location other than the next sequential location in memory.

KEYBOARD: A peripheral device consisting of alphanumerics, punctuation marks, and other special function keys that are mechanically arranged to allow the entry of data, commands, and other information into the system.

LAN: See LOCAL-AREA NETWORK.

LANDSCAPE: A printer feature, generally controlled by software, which rotates the output image by 90° to print across the length rather than the width of the paper.

LASER (LIGHT AMPLIFICATION by STIMULATED EMISSION of RADIATION): A system that generates high-intensity, highly-focused light for many purposes, including printers, high-density memory media, and a light generator for fiber optic transmission systems.

LASER DISK: See CD-ROMLASER PRINTER: A computer-driven photocopier that creates an original image of the text or graphics from the output of the computer. A computer-controlled laser beam "paints" the desired image inside the photocopier and then prints the image on a sheet of paper.

LAST-IN, FIRST-OUT (LIFO): A method of storing and retrieving data in a stack, table, or list.

LIGHT PEN: A light-sensitive stylus for forming graphics by touching coordinates on a display screen, thereby seeming to draw directly on the screen.

LIBRARY: A collection of standard software instructions, programs, routines, and subroutines in a computer's memory.

LINKAGE: Instructions that connect one program to another, providing continuity of executions between the programs LOCAL AREA NETWORK (LAN) – A combined hardware/ software technique for interconnecting company related multiple computers or computer terminals through a high-speed networking system.

LOGIC GATE: A digital circuit resulting in an output whose state (0 or 1) depends on the specific combination of the states of input signals. Definitions of the more commonly used logic gates are listed below:

AND: All inputs must be in a "1" state to produce a "1" state output.

NAND (NOT AND): All inputs must be in a "1" state to produce a "0" state output.

OR (NOT OR): Any one input, or more, in a "1" state will produce a "0" state output.

NOT (INVERTER): A logic gate having only one input and one output. If the input is in a "1" state, the output is in a "0" state and vice–versa.

OR: Any one input, or more, in a "1" state will produce a "1" state at the output.

XOR (EXCLUSIVE OR): If any of the inputs are in a "1" state, but not if two or more inputs are "1", the output is in a "1" state.

LOOK AHEAD: 1. A feature of a CPU which allows the masking of an interrupt request until the current sequential instruction has been completed. 2. A feature of an adder circuit in the ALU section which allows the circuit to look ahead to see that all the generated arithmetic carrys are available for addition.

LOOPING: The repetition of program instructions until a conditional exit situation is encountered.

MACHINE LANGUAGE: Sets of numeric binary code instructions in a computer which execute its operations. All other programming languages (SOURCE PROGRAMS) must be translated into machine language (OBJECT PROGRAM) before entering the CPU.

MACRO: A combination of commands, instructions, or keystrokes which may be stored in a computer's memory to be executed as a single command by a single keystroke or a simultaneous combination of keystrokes.

MACROASSEMBLER: An assembly language translator that converts macro expressions into several machine language instructions. Although macros simplify program coding and speed up execution of a program, a code for each macro must also be generated.

MAINFRAME COMPUTER: Second largest of the computer family, in capability and, generally, in size, having a WORD-WIDTH of 32 bits and higher. The largest computer is the SUPER COMPUTER with a WORD WIDTH of 64 to 128 bits.

MEMORY: A data storage structure in a computer that accepts binary information for storage in electrical, mechanical, or magnetic form, and retains the information for as long as needed. Selected data can be written, moved, displayed, copied, or erased. The variety of memory media includes: paper, magnetic, bubble, optical laser, and semiconductor memory. Each one has a different method of operation and a specific criterion for being selected.

MICROCOMPUTER: A microprocessor-based computer, consisting of an MPU, internal semiconductor memory, input and output sections, and a system bus, all on one, or several monolithic IC chips inserted into one or several PC boards. The addition of a power supply and connecting cables, appropriate peripherals (keyboard, monitor, printer, disk drives, etc.), an operating system and other software programs can provide a complete microcomputer system. The microcomputer is generally the smallest of the computer family, however, the improvement in performance capability of newer microcomputer systems can make the microcomputer as powerful as larger systems.

MICROPROCESSOR UNIT (MPU): The Central Processor Unit (CPU) implemented in monolithic IC technology, usually, but not necessarily, on one VLSI chip. In many cases, the SYSTEM BUS is also included on the MPU chip.

MINICOMPUTER: Considered to be more capable than a microcomputer but less powerful than a mainframe. Generally, the WORD-WIDTH of the minicomputer is between 12 to 32 bits. MNEMONIC - A symbolic label or code reminder that assists the user in remembering a specific operation or command.

MODEL: A representation of a process or system that can be controlled to demonstrate the effects that various actions will have on the process or system.

MODEM: An acronym for MOdulator/DEModulator that refers to specific equipment that provides a means of communication between two computer systems over conventional telephone lines. Each remote computer requires its own MODEM and a compatible communications software program for proper interfacing.

MONITOR: The visual readout device of a computer system. A monitor can be in several forms:, a Cathode Ray Tube (CRT), a Aiquid Crystal Display (LCD), or a flat-panel, full-color display. See DISPLAY.

MOUSE: A manually operated input device for moving or entering positional information and other data or commands by accessing (pointing to) images on a monitor.

MULTITASKING: The technique of using several applications programs (tasks) in a computer system or on several terminals in a network at the same time. Multitasking can simultaneously work with several programs or interrelated tasks that share memories, codes, buffers, and files.

MULTIUSER: The term describing the capability of a computer system to be operated at more than one terminal at the same time.

NEGATIVE LOGIC: This term refers to logic in which the negative voltage represents the "1" state and the zero voltage represents the "0" state.

NESTING: Embedding commands or data in levels of other data so that specific routines or instructions can be executed or accessed continuously in loops, without returning to the main program.

NIBBLE: A sequence of four adjacent bits, or a half-byte. A hexadecimal or BCD coded digit can be represented by a nibble.

NODE: The endpoint of a network branch or the junction of two or more branches.

NON-VOLATILE MEMORY: A memory where stored data remains undisturbed by the removal of electrical power.

OBJECT CODE: Machine language code produced by a translator program, such as an assembler, interpreter, or compiler. Instructions in object code can be executed by a Central Processing Unit (CPU).

OPERATING SYSTEM (OS): A structured software program (set of programmed routines) that manages, controls, coordinates, and sequences the hardware and software resources of a computer system. See BOOTSTRAP. The operating system is sometimes called SYSTEMS SOFTWARE.

OPERATION CODE (OP-CODE): Part of a computer instruction word that designates the function performed by a specific instruction. For example, op-codes for arithmetic instructions include: "ADD", "SUB", "MUL" and "DIV".

OVERFLOW: An error condition occurring in a computer when a mathematical operation produces a result having a magnitude that exceeds the capacity of the computer's arithmetic register.

PASCAL: A high-level programming language that is structured to encourage efficient programming habits (documentation) and is used extensively in educational institutions and engineering environments.

PARALLEL OPERATION: A method of data transmission in which all bits of a digital word are handled simultaneously with each bit on a separate line. Although faster and simpler to install and operate than SERIAL OPERATION, this method requires more transmission lines (real estate).

PARITY: A method of verifying the accuracy of binary data after it has been transferred to or from a storage area.

PARTITIONING: The logical grouping of electronic functions within a given set of hardware components.

PCMCIA (PERSONAL COMPUTER MEMORY CARD INTERNATIONAL ASSOCIATION): A package in a plastic card form containing semiconductor memory, particularly FLASH. These cards are plugged into the card slot in laptop computers.

PERIPHERAL: A term designating the various kinds of machines and devices that work in conjunction with a computer but are not necessarily part of the computer structure. Typically, peripherals refer to: printers, keyboards, monitors, scanners, CD ROM drives, and plotters. A hard drive, floppy disk drive, and a MODEM are considered to be peripheral devices even though they may be physically located inside a computer.

PLUG-COMPATIBLE: A term that describes the ability of peripherals to be interchanged without modification.

POLLING: A process in which a number of peripheral devices, remote stations, or nodes in a computer network are interrogated, one at a time, to determine if service is required.

POP: The instruction that removes a word from the top of a stack.

PORT: An input/output channel (either parallel or serial), terminated at a connector on the computer. It interconnects the computer's input and/or output terminals to an appropriate source and/or destination.

PORTRAIT: A term that designates the position of conventional printing across the width of a page. POSITIVE LOGIC – This logic represents the reverse of NEGATIVE LOGIC. It is the more commonly used form of logic. A positive voltage represents a "1" state and a negative (or zero) voltage represents a "0" state.

PRINT SPOOLER: A device for temporarily storing data to be printed when the printer is functioning. It provides uninterrupted data entry and editing while the printer is active and while other data awaits transmission to the printer.

PROGRAM: A complete sequence of computer software instructions necessary to provide an application, solve a specific problem, perform an action, or respond to external stimuli in a prescribed manner. As a verb, it means to develop a program.

PROGRAM COUNTER (PC): A special-purpose register in the CPU which contains the address of the next instruction to be fetched and executed.

PROGRAMMABLE LOGIC ARRAY (PLA): An unprogrammed, general-purpose logic structure in monolithic IC form consisting of an array of similar, and/or compatible logic gates. Also called PROGRAMMABLE ARRAY LOGIC (PAL).

PROGRAMMABLE READ-ONLY MEMORY (PROM): A blank read-only memory (ROM) that is programmed with external programming equipment after manufacture. Once programmed, it is not re-programmable and is considered to be a ROM.

PROPAGATION DELAY: The time required for the output of a logic gate to respond to a combination of input pulses.

PUSH: The instruction used to deposit a word on top of a stack.

PUSH-DOWN STACK: A dedicated temporary storage register in a computer, sometimes part of a system memory, structured so that data (words) in the stack are retrieved in reverse order of entry. See LIFO.

RAM DISK (DISK EMULATOR): A portion of a hard drive configured to emulate a RAM. It accesses information quickly, but its data must be saved in a non-volatile memory for future use. Otherwise, the information is lost when power is removed.

RANDOM ACCESS: A technique of accessing (reading) a word of data from a memory structure by the CPU. Since a word in the memory can be accessed directly, the time required is independent of its location (address) in the memory structure. It is sometimes called a "direct access" method.

RANDOM ACCESS MEMORY (RAM): A volatile, semiconductor storage structure that accesses temporary data with a random or direct accessing method. It is more accurately referred to as "erasable read/write" memory. Data in this memory can be read by the CPU, edited, altered, erased, or new information written over existing data by computer commands. Its data must be saved for future access by writing it into a non-volatile memory. See NON-VOLATILE MEMORY and VOLATILE MEMORY.

READ-ONLY MEMORY (ROM): A semiconductor memory whose data cannot be erased, or overwritten; it can only be accessed (read) for use by the CPU. The data in a ROM is of a permanent nature and is programmed by the ROM manufacturer. In many cases, its programmed information identifies the dedicated function of a computer. A ROM can also be in the physical form of a module (tape or disk) that is plugged into a computer to change its operation from one program to another.

REAL-TIME DATA: Time-dependent data processed by a computer whose output data is capable of controlling other time-related events, such as traffic control. Real time is the actual time it takes for an event to occur.

REGISTER: A temporary storage unit for quick, direct accessibility of a small amount of data for processing. Most computers include a set of internal registers that can be accessed more quickly than the system's main memory.

RS-232-C: Identifies an interconnection standard for serial operation. It specifies the configuration and type of connectors in the computer's serial I/O port(s) and peripheral serial port(s).

RESIDENT SOFTWARE: The program(s) residing in the main memory of a computer system. For convenience of operation, several software programs can reside in RAM after the computer is turned on and booted, and they can be accessed quickly from within another program.

SCRATCH PAD MEMORY: A group of internal registers used for the temporary storage of data being collected and sorted for immediate processing. It is analogous to a pad of paper for quickly jotting down notes. SECTOR - A section of a recording track on a magnetic disk.

SEMICONDUCTOR MEMORY: Data storage devices formed with semiconductor components (generally monolithic ICs). See RAM, ROM, PROM, EPROM, EAPROM, EEPROM, and FLASH.

SERIAL OPERATION: A method of data transmission where the data is handled in sequence, one bit at a time.

SERVICE ROUTINE: A set of instructions for performing a programmed operation, typically, in response to an interrupt command.

SHIFT: A computer operation consisting of moving a group of adjacent data bits either to the left or to the right by a prescribed number of positions. The move is done in a SHIFT REGISTER for a carry-over operation.

SIMPLEX: A data transmission mode that provides transmission in one direction only.

SIMULATION: The imitation of a logical operation of one computer by another to measure and evaluate the operation of the computer being designed. Simulation is primarily intended to provide an analysis of program logic, independent of hardware environment, and is extremely useful for debugging a new software program prior to committing it to ROM.

SOFTWARE: Programs, languages, procedures, and documentation for a computer system. Software includes: operating systems (system software), language translators (assemblers, interpreters and compilers), subroutine libraries, application programs, and the information in instruction manuals.

SOURCE CODE (SOURCE PROGRAM): A set of computer instructions in hard-copy or stored form. When written in a language other than machine language, the source code requires translation by an assembler (or macroassembler), interpreter, or compiler into object code.

STACK: A dynamic, sequential data list usually contained in the computer system's main memory. It has special provisions to access data from either end. Storage and retrieval of data from the stack is performed automatically by the CPU.

STATIC MEMORY: A type of semiconductor read/write memory (RAM) that does not require periodic refresh cycles. As long as electrical power is ON, the data in a static memory is maintained.

STORAGE: This term is interchangeable with the term MEMORY.

SUBROUTINE: A short program segment that performs a specific function and is available for general use by other programs and routines.

SUPERCOMPUTER: The largest mainframe computer featuring exceptionally high speed operation while manipulating huge amounts of information.

SUPPORT CHIPS: Computer-related circuits other than the CPU. Examples are: main memory (RAM and ROM), input/output ports, and system bus.

STATE: The logic input or output condition of a binary digital circuit - the state is either a "0" or "1".

STORAGE CAPACITY: The amount of data that can be retained in a memory unit. It is expressed either by a number of bits or bytes (8-bit words).

STREAMER/STREAMING TAPES: A small tape recorder, usually in cassette form, is used to store data from a hard disk for the purpose of backing up this data.

SYNCHRONOUS COMMUNICATION: A method of transferring binary data, in serial form, between computers or between a computer and its peripherals. Transmission of data is at a rate set by the computer's clock with synchronization bits located at the beginning of each message or block of data.

SYNTAX: A set of grammatical rules defining valid use of specific commands or instructions in a computer language.

SYSTEMS SOFTWARE: A general term for software that supervises, sequences, and coordinates programs. Systems software may include programs, such as: operating systems, assemblers, interpreters, compilers, software debugging programs, text editors, utilities and peripheral drivers.

TEXT EDITOR: See EDITOR.

THIRD-PARTY SOFTWARE: Software developed by a software company rather than by a computer manufacturer or user.

THROUGHPUT: The number of instructions executed per second, measured in millions of instructions per second (MIPS) or billions of instructions per second (BIPS). TOGGLE - To switch back and forth between two states or conditions of operation, as in a toggle switch.

TOP-DOWN HIERARCHICAL DESIGN: A hardware and/or software design approach that starts at the most general level of a machine or software program. It proceeds, step-by-step, to lower levels, adding detail as the design progresses.

TRACK: A ring on the surface of a magnetic disk. TRACTOR-FEED - A pin-fed device for advancing continuous form paper through a computer printer. TRANSISTOR-TRANSISTOR LOGIC (T2L) - A logic gate family that provides higher-speed and higher-power than the obsolete DTL logic family. The first transistor in the circuit performs the required logic. Another transistor amplifies and inverts the output. Improved pin-compatible versions of this logic family are called TTL-Schottky (T2L-S) and Low Power TTL-S (LPT2L-S).

TRANSLATOR: See ASSEMBLER, MACROASSEMBLER, INTERPRETER, and COMPILER.

TRI-STATE LOGIC: The term that designates the possible conditions of a specific logic gate output: "0", "1" or "undefined".

TRUNCATE: The dropping of digits or characters from one end of a data item causing loss of accuracy or information.

TRUTH TABLE: A tabulation of all possible combinations of states at the inputs of a logic gate which will result in a specific logic state at the output of the gate.

TURNKEY SYSTEM: A complete computer system ready to operate without any hardware or software modification or addition.

UNFORMATTED (UNINITIALIZED) DISK: A blank magnetic disk with no track/sector identification recorded on it that allows users to implement their own track/sector identifications.

USART: UNIVERSAL SYNCHRONOUS/ASYNCHRONOUS RECEIVER TRANSMITTER - A circuit that converts serial-to-parallel or parallel-to-serial operation for communication between two computers or between a computer and its peripherals.

USER-FRIENDLY PROGRAM: A software program that has been designed to easily direct the user through the operation or application of a program. A menu-driven program is considered to be "user-friendly".

UTILITY: A software program designed to perform a computer system's routine housekeeping functions, like copying, deleting files, and/or providing techniques to simplify the execution of a program.

VOCABULARY: A list of operating codes or instructions available to the software programmer for writing a program in a specific language.

VOLATILE MEMORY: A memory whose contents are irretrievably lost when power is removed. If data in RAM must be saved after power shut-down, back-up in nonvolatile memory (magnetic disk, tape, or CD-R) is essential.

WAIT STATE: An internal condition of delay in processing time executed by the CPU when a synchronizing control signal is not present. Wait states synchronize the timing of a CPU with the relatively slower access time of the computer's main memory.

WINCHESTER DRIVE: See HARD-DISK DRIVEWINDOWING - The ability of a program to divide a display screen into smaller sub-units that permit portions of different sections of a program, or different programs, to be displayed on the screen, edited, and copied independently.

WORD PROCESSING (WP): The term refers to a program, that allows creating, editing, formatting, displaying, printing, and storage of text with great flexibility and ease. Different WP programs provide different, and sometimes, more desirable capabilities than others.

WORD: The set of binary bits handled by a computer as a primary unit of data. The width (number of bits) of a computer word depends on the hardware design. Wider words imply higher levels of precision, higher speed, and more intricate instructions. Typically, each location in memory contains one word.

WORM (WRITE-ONCE, READ-MANY): A high-density optical disk memory available in a variety of formats from 5.25" to 14". The WORM can be programmed once, permanently saving a user's data. It then becomes an optical disk read-only memory having essentially the same features as a CD-ROM. Also called CD-R (CD-RECORDABLE).

WORKSTATION: The work area and/or equipment used for computer operations, including Computer-Aided Design (CAD). The equipment generally consists of a monitor, keyboard, printer and/or plotter, and other output devices.

WRITE: The process of storing data into a memory.

X-Y PLOTTER: A computer-driven printing mechanism that draws coordinate points in graph form.

ZERO-WAIT STATE: A condition that results when no delays are inserted into the system software to deliberately detain operation of the CPU. This can be done with high-speed memory and proper design of computer architecture. Zero-wait state is desirable for the high-speed operation of a computer.

ZOOM: The process of proportionately enlarging or reducing an image displayed on a computer monitor.

Appendices

Appendices

FUNDAMENTALS TIPS

A.1. COMPUTER ORGANIZATION

CO is how operational attributes are linked together and contribute to realize the architectural specification. Computer architecture is the architectural attributes like physical address memory, CPU and how they should be made to coordinate with each other keeping the future demands and goal in mind.

Computer architecture comes before *CO*. It is like building the design and architecture of house takes maximum time and organization is the building house by bricks or by latest technique keeping the basic layout and architecture of house in mind.

Architecture describes what the computer does and organization describes how it does.

Architecture of organization are independent. You can change the organization of computer without changing its architecture.

A.2. ARITHMETIC ALGORITHMS

The solution of any problem that is stated by a finite number of well-defined procedural steps is called an algorithm.

(1) Addition and subtraction of signed-magnitude numbers:

		Subtract Magnitudes		
Operation	***Add magnitudes***	***When A > B***	***When A < B***	***When A = B***
$(+A) + (+B)$	$+(A + B)$			
$(+A) + (-B)$		$+(A - B)$	$-(B - A)$	$+(A - B)$
$(-A) + (+B)$		$-(A - B)$	$+(B - A)$	$+(A - B)$
$(-A) + (-B)$	$-(A + B)$			
$(+A) - (+B)$		$+(A - B)$	$-(B - A)$	$+(A - B)$
$(+A) - (-B)$	$+(A + B)$			
$(-A) - (+B)$	$-(A + B)$			
$(-A) - (-B)$		$-(A - B)$	$+(B - A)$	$+(A - B)$

Hardware for signed magnitude addition and subtraction:

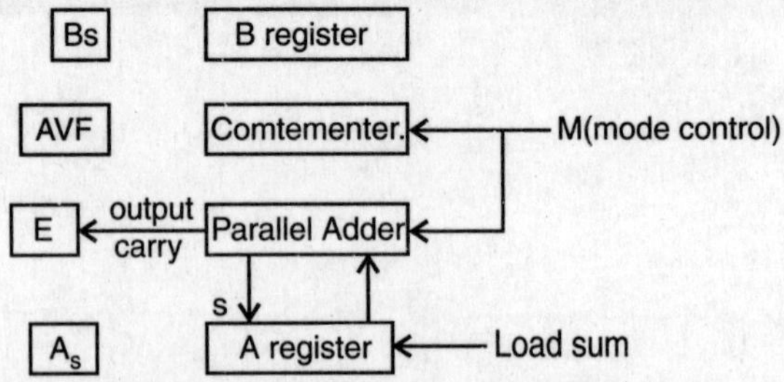

Fig. A.1.

Flowchart for add and subtract operations:

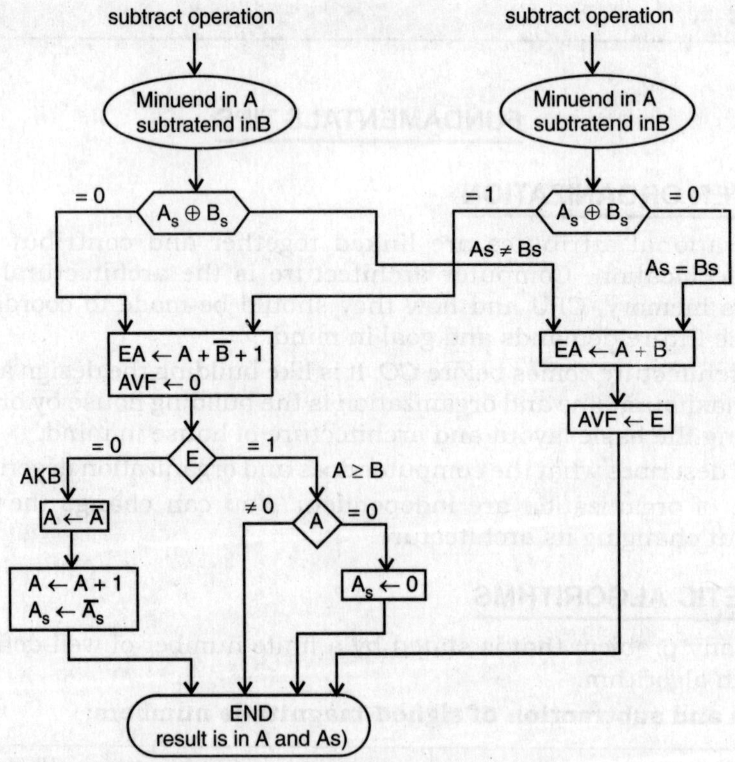

Fig. A.2.

A.3. ADDITION AND SUBTRACTION WITH SIGNED-2'S COMPLEMENT DATA

Hardware for signed-2's complement addition and subtraction.

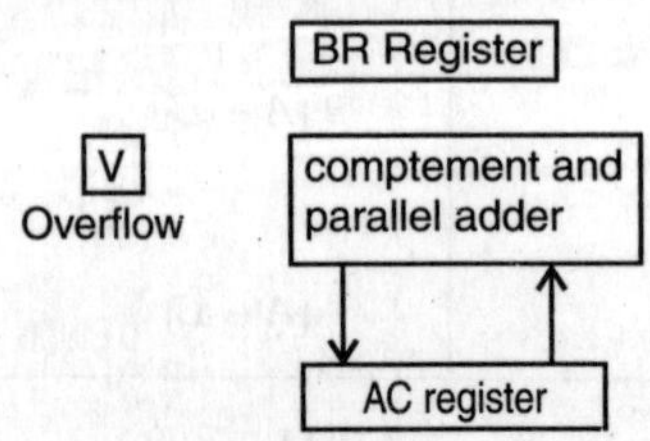

Fig. A.3.

Algorithm for adding and subtracting number is signed-2's complement representation.

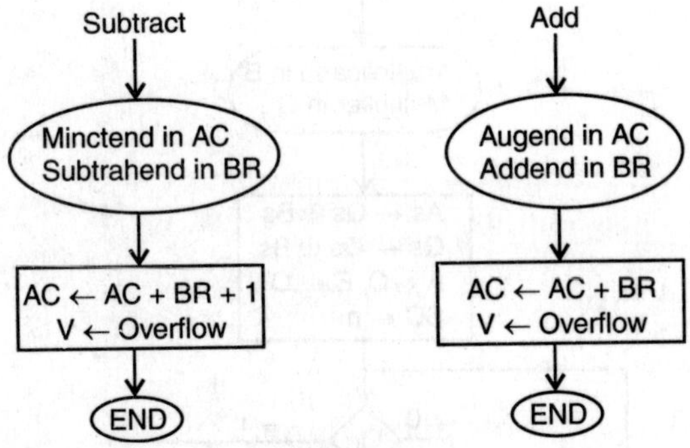

Fig. A.4.

A.4. MULTIPLICATION ALGORITHMS

15	Multiplicand
13	Multiplier
45 → P_1	(First partial product)
15 × → $P2$	(Second partial product)
195 → P	(Final product)

101	= equal bits
110	= equal bits
000	→ (First partial product)
1010	→ (Second partial product)
10100	→ (Third product)
11110	→ (Final product)

Hardware for multiply operation:

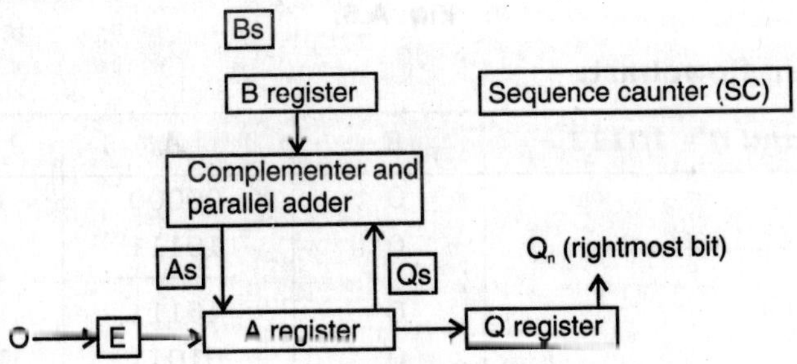

Fig. A.5.

Hardware Algorithm: Flowchart for multiply operation

η = no. of bits in multiplier

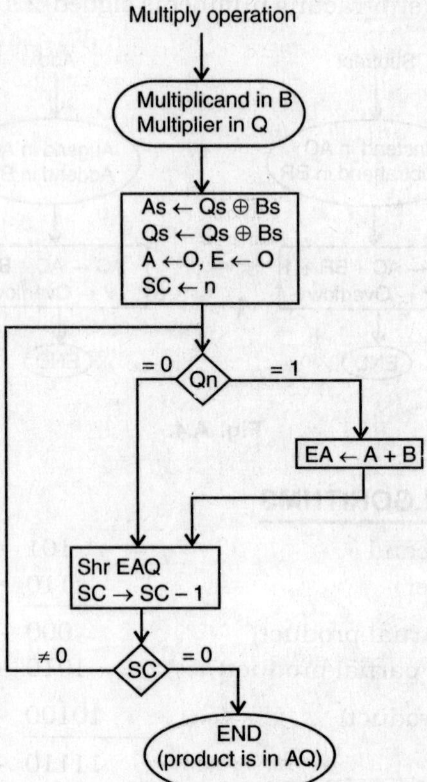

Fig. A.6.

Example of given flowchart:

Multiplicand B = 10111	***E***	***A***	***Q***	***SC***
Multiplier in *Q*	0	00000	10011	101
Q_n = 1; add *B* + *A*	0	10111		
First partial product	0	10111		
Shift right EAQ	0	01011	11001	100
Q_n = 1; add *B* in *A*		10111		
Second partial product	1	00010		
Shift right *EAQ*	0	10001	01100	011
Q_n = 0, shift right *EAQ*	0	01000	10110	010
Q_n = 0, shr *EAQ*	0	00100	01010	001
Q_n = 1, add *B* in *A*		10111		
Fifth partial product	0	11011		
Shr *EAQ*	0	01101	10101	000
Find product in *AQ* = 0110110101				

A.5. BOOTH MULTIPLICATION

Booth algorithm for multiplication of signed-2's complement numbers

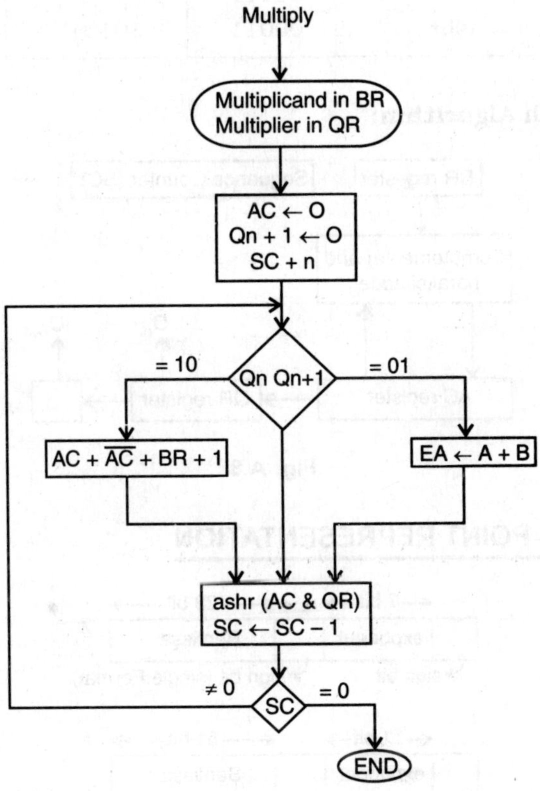

Fig. A.7.

Example:

$(-9) \times (-13) = +117$

Multiplicand $9 = 1001$

$\Rightarrow$ $1001\ 0110 + 1 = 0111$

$-9 = 10111 = BR$

$-13 = 10011 = QR$

$BR = 10111$

Q_n	Q_{n+1}	$\overline{BR} + 1 = 01001$	AC	QR	Q_{n+1}	SC
		Initial	00000	10011	0	101
1	0		01001			
			01001			
		ashr	00100	11001	1	100
1	1	ashr	00010	01100	1	011
			10111			
0	1	Add BR in AC	11001			
		ashr	11100	10110	0	010

0	0	ashr	11110	01011	0	001
			01001			
1	0		00111			
		ashr	00011	10101	1	000

Ans. is 0001110101.

Hardware for Booth Algorithm:

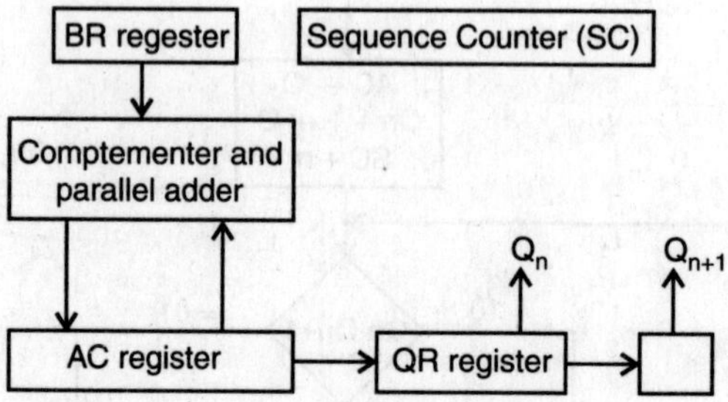

Fig. A.8.

A.6. IEEE FLOATING POINT REPRESENTATION

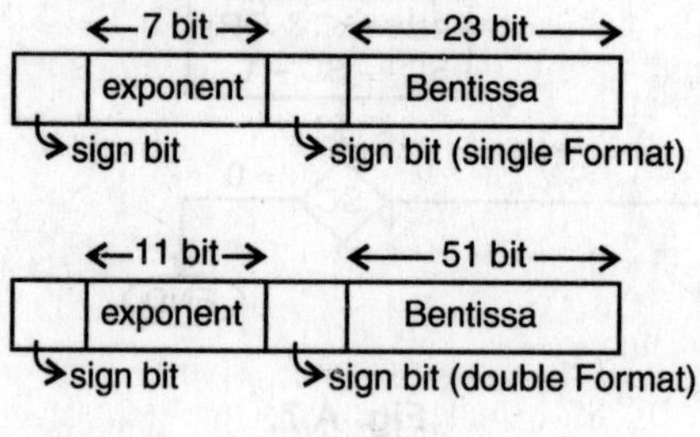

Fig. A.9.

All points are represented in $m \times r^e$.

IEEE: Initiate of electronics and electrical engineering.

A.7. DESIGN OF FAST ADDERS

Faster algorithms and their implementation:

(1) Fast Adder: In the parallel binary adder the carry generated by the *i*th adder is fed as carry input to the $i + 1$ adder. In this adder the output (C_{out} $S_3S_2S_1S_0$) is available only after the carry is propagated through each of the adders that means from *LSB* to *MSB* adder to intermediate adders.

Hence, the addition process can be considered to be completed only after the above carry propagation delay through adders, which is proportional to the number of stages in bit.

One of the methods of making this faster look ahead carry addition which eliminates the ripple carry delay. This method is based on the carry generating and the carry propagating functions of the full adder.

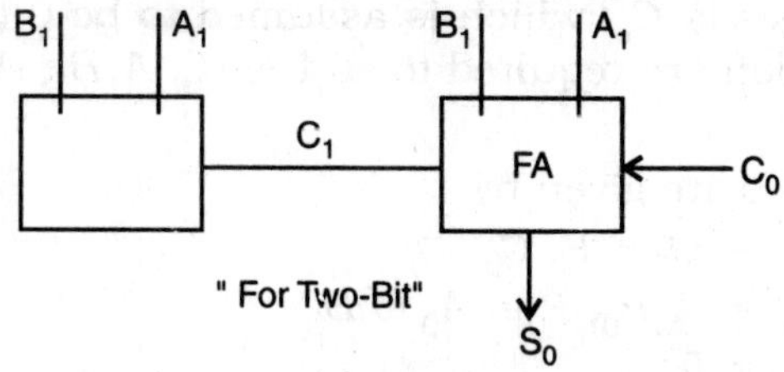

" For Two-Bit"

Fig. A.10.

4-Bit Carry Look Ahead Adder: The carry look ahead adder is based on the principle of looking at the lower order bits of the Augend and Addend. If high order carry is generated. This adder reduces the carry delay by reducing the number of gates or number of stages through which carry signal must propagate. To explain its operation consider the truth table of full adder.

ROW	*A*	*B*	C_{in}	*S*	C_{out}	
0	0	0	0	0	0	No carry generation
1	0	0	1	1	0	*i.e.*, $C_{out} = 0$
2	0	1	0	1	0	
3	0	1	1	0	1	Carry propagation
4	1	0	0	1	0	*i.e.*, $C_{out} = C_{in}$
5	1	0	1	0	1	
6	1	1	0	0	1	Carry generation
7	1	1	1	1	1	*i.e.*, $C_{out} = 1$

In row 0 and 1 the carry output C_{out} is always zero and independent of carry input (C_{in}) while in row six and seven the C_{out} is always 1 independent of C_{in}. These are known as carry generate combination.

In row 2, 3, 4 and 5 the carry output is equal to the carry input, *i.e.* $C_{out} = 1$ only when $C_{in} = 1$. These are carry propagate combinations.

Let G_i represents the unity carry ($C_{out} = 1$). Generate condition and P_i represent the propagate condition of *i*th stage of parallel adder.

From truth table, G_i is obtained by summing up the combinations corresponding to sixth and seventh rows as follows:

$$G_i = A_iB_i\,C_{in} + A_iB_i\overline{C_{in}} = A_iB_i\,(C_{in} + \overline{C_{in}})$$

$$G_i = A_iB_i \qquad ...(1)$$

Similarly, the carry propagation condition (P_i) occurs when either $A_i = 1$ and $B_i = 0$ or vice-versa. Now P_i is given by

$$P_i = A_i\overline{B_i} + \overline{A_i} + B_i$$

$$P_i = A_i \oplus B_i$$

Consider the addition of 2, 4 bit binary no's $A\,(A_3A_2A_1A_0)$ and $B\,(B_3B_2B_1B_0)$. The unity carry O/P of the $(i-1)$ stage as follows:

$$C_i\,(C_{out}) = G_i + P_i\,C_{i-1}$$

$$C_i = 0, \text{ then } C_{i-1} = \text{initially carry}$$

If $\quad C_i = 1, \text{ then } \overline{C_{i-1}} = 0$

Where C_i for *LSB* stages is C_{in} which is assumed to be 0 (zero). In a four bit binary adder four stages of addition are required to add A_0B_0, A_1B_1, A_2B_2, A_3B_3.

Therefore,

for $i = 0, 1, 2, 3$ and C_i's are given by

$$C_0 = G_0 + P_0 C_{in}$$

where $G_0 = A_0B_0,\ P_0 = A_0 \oplus B_0$

and $C_{in} = 0$

$$C_0 = A_0B_0$$

$$C_1 = G_1 + P_1C_0 = G_1 + P_1(G_0 + P_0C_0)$$

$$G = G_1 + P_1G_0 + P_0P_1G_n$$

where $G_1 = A_1B_i$

and $P_1 = A_1 \oplus B_1$

$$G_2 = G_2 + P_2C_1 = G_2 + P_2(G_1 + P_1G_0 + P_0P_1C_{in})$$

$$G_2 = G_2 + P_2G_1 + P_2P_1G_0 + P_0P_1P_2\ C_{in}$$

where $G_2 = A_2B_2$

and $P_2 = A_2 \oplus B_2$

Similarly, $C_3 = G_3 + P_3C_2 = G_3 + P_3(G_2 + P_2G_1 + P_2P_1G_0 + P_0P_1P_2\ C_{in})$

$$C_3 = G_3 + P_3G_2 + P_2P_3G_1 + P_1P_2P_3G_0 + P_0P_1P_2P_3\ C_{in}$$

where $G_3 = A_3B_3$

and $P_3 = A_3 \oplus B_3$

Sum of *A* and *B* is given by,

$$S = C_3\ S_3S_2S_1S_0$$

where, $S_i = A_i \oplus B_i \oplus C_{i-1}$

For $i = 0, 1, 2, 3$

i.e., $S_0 = A_0 \oplus B_0 \oplus C_{in}$

$$S_1 = A_1 \oplus B_1 \oplus C_0$$

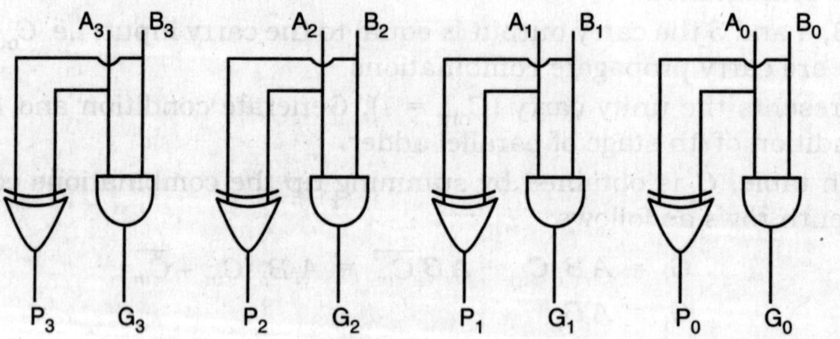

Fig. A.11.

Since, $S_0 = A_0 \oplus B_0 \oplus C_{in}$

$\Rightarrow$ $S_0 = P_0 \oplus C_{in}$

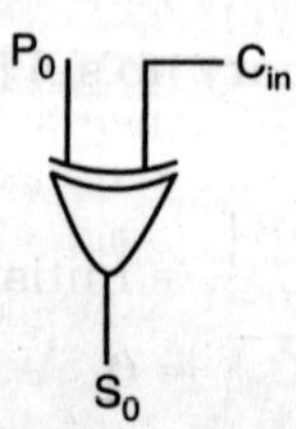

$$S_1 = A_1 \oplus B_1 \oplus C_0$$

$$\Rightarrow \quad S_1 = P_1 \oplus C_0$$

$$\Rightarrow \quad S_1 = P_1 \oplus (G_0 + P_0 C_{in})$$

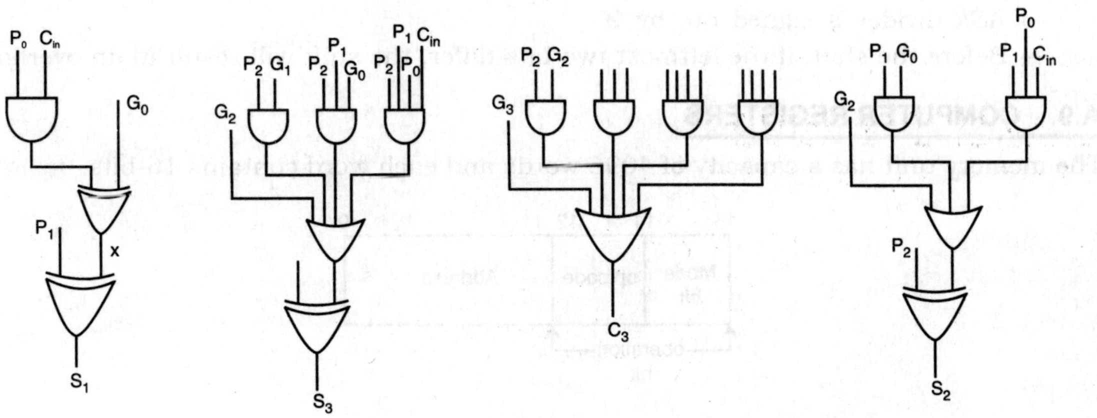

Fig. A.12.

$$S_0 = A_0 \oplus B_0 \oplus C_{in}$$

$$\Rightarrow \quad S_0 = P_0 \oplus C_{in}$$

$$S_1 = A_1 \oplus B_1 \oplus C_0$$

$$\Rightarrow \quad S_1 = P_1 \oplus C_0$$

$$\Rightarrow \quad S_1 = P_1 \oplus (G_0 + P_0\, C_{in})$$

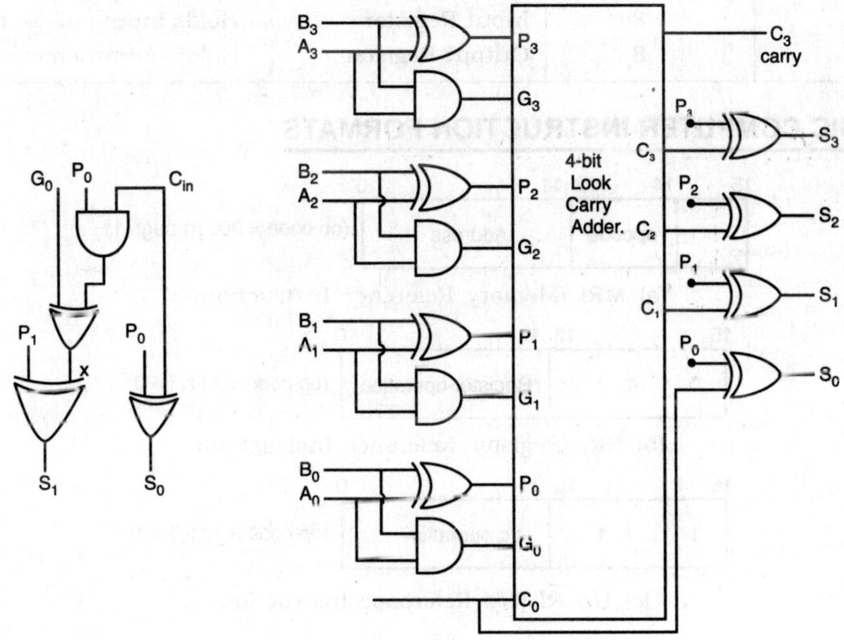

Fig. A.13.

A.8. ARITHMETIC SHIFT

→ An arithmetic shift is meant for signed binary number (integer).
→ *ASL* multiplies a signed no. by 2.
→ *ASR* divides a signed no. by 2.
→ Before the shift, if the leftmost two bits differ, the shift will result in an overflow.

A.9. COMPUTER REGISTERS

The memory unit has a capacity of 4096 words and each word contains 16-bits.

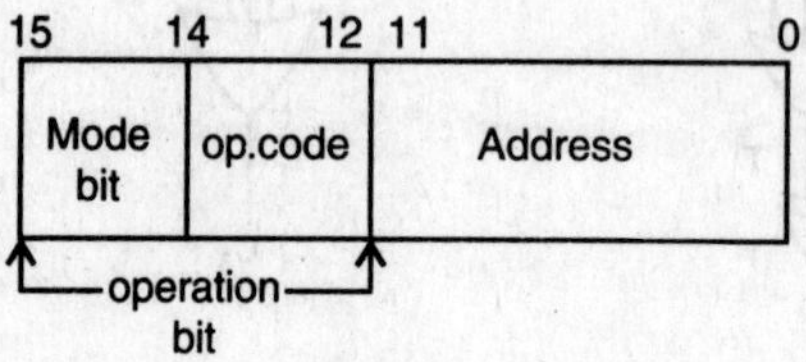

A.10. LIST OF REGISTERS FOR THE BASIC COMPUTERS

Register Symbol	*Number of Bis*	*Register Name*	*Function*
DR	16	Data Register	Holds memory operand.
AR	12	Address Register	Holds address for memory.
AC	16	Accumulator	Processor register.
IR	16	Instruction Register	Holds instruction code.
PC	12	Program Counter	Holds address of instruction.
TR	16	Temporary Register	Holds temporary data.
INPR	8	Input Register	Holds input character.
OUTR	8	Output Register	Holds output character.

A.11. BASIC COMPUTER INSTRUCTION FORMATS

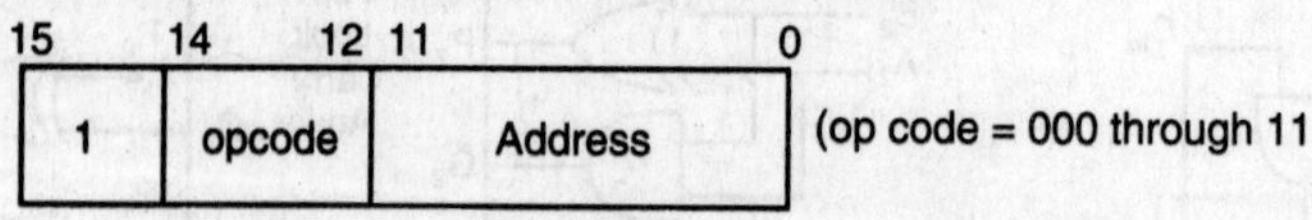

(a) MRI (Memory Reference Instruction)

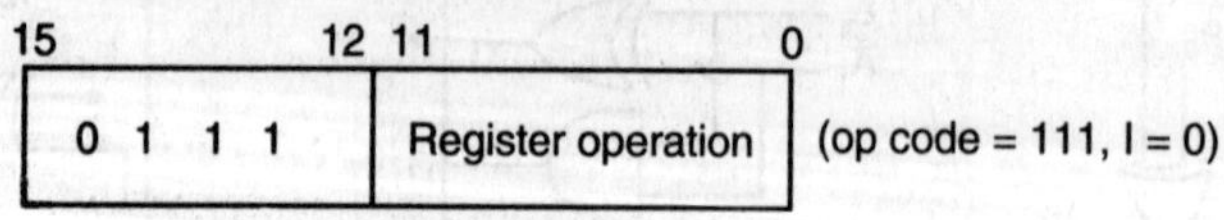

(b) RRI (Register Reference Instruction)

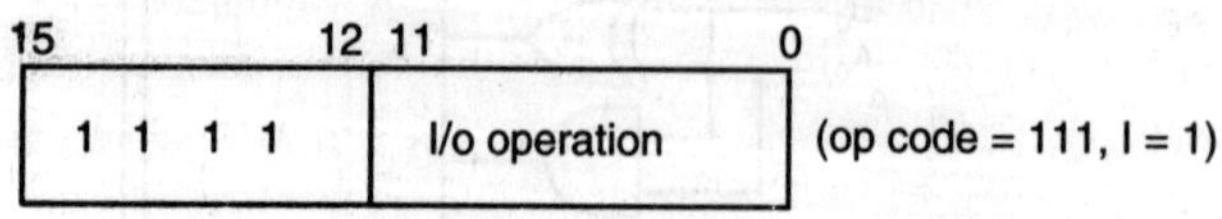

(c) I/o RI (I/o Reference Instruction)

A.12. BASIC COMPUTER REGISTERS AND MEMORY

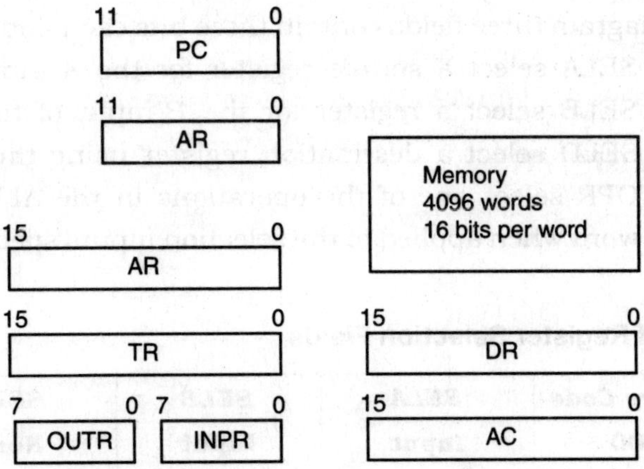

A.13. GENERAL REGISTER ORGANIZATION

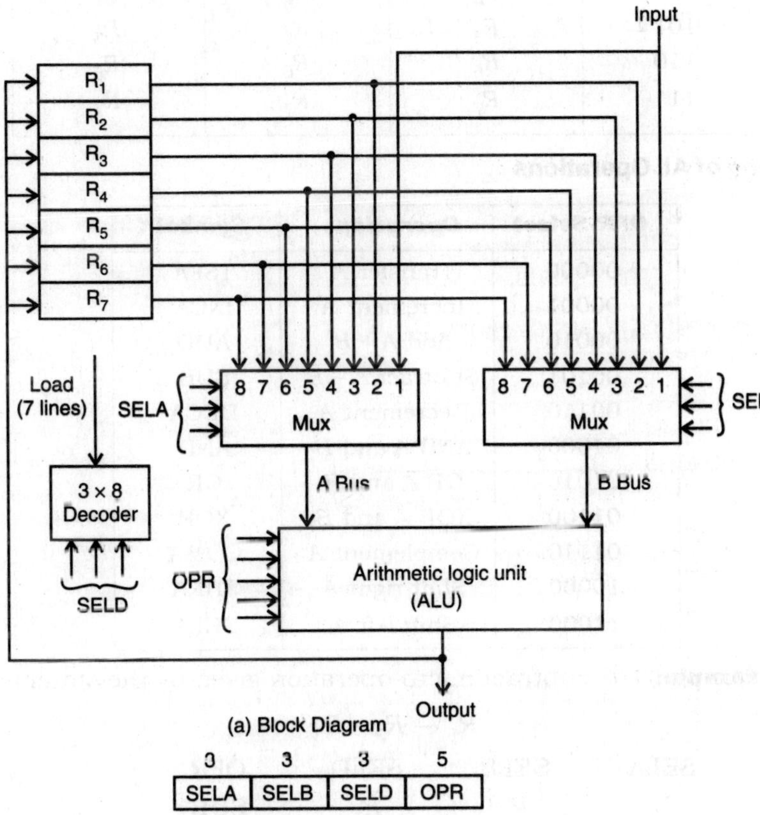

(a) Block Diagram

(b) Control word

Fig. A.14.

A.14. CONTROL WORD

There are 14 binary selection inputs and the unit, and their combined value specifies a control word. In left diagram three fields contain three bits each and one field has five bits.

— The 3 bits of SELA select a source register for the A input of the ALU.

— The 3 bits of SELB select a register for the B input of the ALU.

— The 3 bits of SELD select a destination register using the decoder.

— The 5 bits of OPR select one of the operations in the ALU.

The 14-bit control word when applied to the selection inputs specify a particular micro-operation.

(A) Table: Encoding of Register Selection Fields

Binary Code	*SELA*	*SELB*	*SELD*
000	**Input**	**Input**	**None**
001	R_1	R_1	R_1
010	R_2	R_2	R_2
011	R_3	R_3	R_3
100	R_4	R_4	R_4
101	R_5	R_5	R_5
110	R_6	R_6	R_6
111	R_7	R_7	R_7

B. Table: Encoding of All Operations

OPR Select	*Operations*	*Symbol*
00000	Transfer A	TSFA
00001	Increment A	INCA
00010	Add $A + B$	ADD
00101	Subtract $A - B$	SUB
00110	Decrement A	DECA
01000	AND A and B	AND
01010	OR A and B	OR
01100	XOR A and B	XOR
01110	Complement A	COMA
10000	Shift right A	SHRA
11000	Shift left A	SHLA

Example: The subtract micro-operation given by the statement:

$$R_1 \leftarrow R_2 - R_3$$

Field:	SELA	SELB	SELD	OPR
Symbol:	R_2	R_3	R_1	SUB
Control word:	010	011	001	00101

Note: A memory unit that stores control words is called control memory.

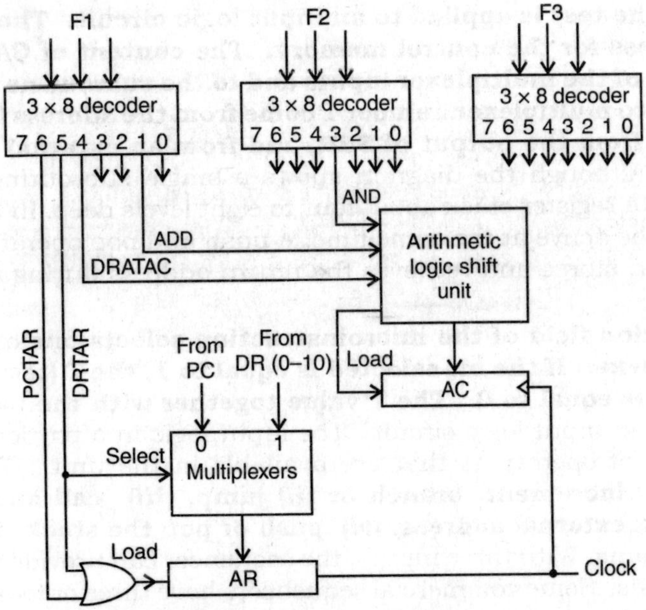

Fig. A.15. Decoding of micro-operation fields

The other outputs of the decoders that are associated with an AC operation must also be connected to the arithmetic logic shift unit in a similar fashion.

A.15. MICROPROGRAM SEQUENCER

The basic components of a micro programmed control unit are the control memory and the circuits that select the next address. The address selection part is called a microprogram sequencer. A microprogram sequencer can be constructed with digital functions to suit a particular application. However, just as there are large ROM units available in integrated circuit packages, so are general-purpose sequencers suits for the construction of microprogram control units. To guarantee a wide range of acceptability, an integrated circuit sequencer must provide an internal organization that can be adapted to a wise range of applications.

The purpose of a microprogram sequencer is to present an address to the control memory so that a microinstruction may be read and executed. The next-address logic of the sequencer determines the specific address source to be loaded into the control address register. The choice of the address source is guided by the next-address information bits that the sequencer receives from the present microinstruction. Commercial sequencers include within the unit an internal register tack used for temporary storage of addresses during microprogram looping and subroutine calls. Come sequencers provide an output register which can function as the address register for the control memory.

The illustrate the internal structure of a typical microprogram sequencer we will show a particular unit that is suitable for use in the microprogram computer example developed in the preceding section. The block diagram of the sequencer is shown in Fig. 7.8. The control memory is included in the diagram to show the interaction between the sequencer and the memory attached to it. **There are two multiplexers in the circuit. The first multiplexer selects an address from one of four sources and routes in into a control address register CAR. The second multiplexer tests the value of a selected status bit**

and the result of the test is applied to an input logic circuit. The output from CAR provides the address for the control memory. The content of CAR is incremented and applied to one of the multiplexer inputs and to the subroutine register SBR. The other three inputs to multiplexer number 1 come from the address field of the present microinstruction, from the output of SBR, and from an external source that maps the instruction. Although the diagram shows a single subroutine register, atypical sequencer will have a register stack about four to eight levels deep. In this way, a number of subroutines can be active at the same time. A push and pop operation, in conjunction with a stack pointer, stores and retrieves the return address during the call and return microinstructions.

The CD (condition field of the microinstruction selects one of the status **bits in the second multiplexer. If the bit selected is equal to 1, the T (test) variable is equal to 1; otherwise, if is equal to 0. The T value together with the** two bits from the BR (branch) field go to an input logic circuit. The input logic in a particular sequencer will determine the type of operations that are available in the unit. **Typical sequencer operations are: (*i*) increment, branch or (*ii*) jump, (*iii*) call and (*iv*) return from subroutine, load an external address, (*vi*) push or pop the stack, and other address sequencing operations.** With three inputs, the sequencer can provide up to eight address sequencing operations. Some commercial sequencers have three or four inputs in addition to the T input and thus provide a wide range of operations.

The input logic circuit in Fig. A.16., **has three inputs, IO, I1 and T, and three outputs, S_0, S_1, and L. Variable S_0 and S_1 select one of the source addresses for CAR. Variable L enables the load input in SBR. The binary values of the two selection variables determine the path in the multiplexer. For example, with S_1S_0 = 10, multiplexer input number 2 is selected and establishes a transfer path from SBR to CAR. Note that each of the four inputs as well as the output of MUX 1 contains a 7-vit address.**

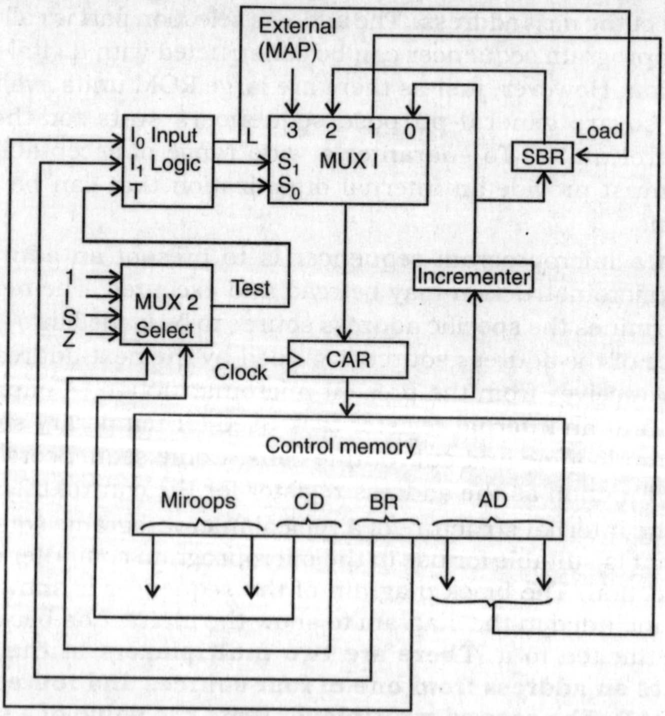

Fig. A.16. Microprogram sequencer for a control memory

The truth table for the input logic circuit is shown in Table 7.4. Inputs I_1 and I_0 are identical to the bit values in the BR field. The function listed in each entry was defined in Table 7.1. The bit values for S_1 and S_0 are determined from the stated function and the path in the multiplexer that establishes the required transfer. The subroutine register is loaded with the incremented value of CAR during a call microinstruction (BR = 01) provided that the status bit condition is satisfied (T = 1). The truth table can be used to obtain the simplified Boolean functions for the input logic circuit:

$$S_1 = I_1$$
$$S_0 = I_1 I_0 + I'_1 T$$
$$L = I'_1 I_0 T$$

Table A.1. Input Logic Truth Table for Microprogram Sequencer

BR Field	Input I_1 I_0 T	MUX 1 S_1 S_0	Load SBR L
0 0	0 0 0	0 0	0
0 0	0 0 1	0 1	0
0 1	0 1 0	0 0	0
0 1	0 1 1	0 1	1
1 0	1 0 ×	1 0	0
1 1	1 1 ×	1 1	0

The circuit can be constructed with three AND gates, an OR gate, and inverter.

Note that the incrementer circuit in the sequencer of Fig. A.16. is not a counter constructed with flip-flops but rather a combinational circuit constructed with gates. A combinational circuit incrementer can be designed by cascading a series of half-adder circuits (see Fig. A.16). The output carry from one stage must be applied to the input of the next stage. One input in the first least significant stage must be equal to 1 to provide the increment-by-one operation.

A.16. PERFECTING MICROINSTRUCTION

The disadvantage of microprogram control is that it leads to lower operating speed because of the time it takes to fetch the microinstructions from control memory. This disadvantage can be remove and faster operation can be achieved if the next microinstruction is perfected while other microinstruction is being executed.

Thus the execution time can be overlapped with the patched time but there are same problems. When we prefect next microinstruction some time status plug and the result of currently executed microinstruction are required to determine the address of next microinstruction.

A.17. WIDE BRANCH ADDRESSING

As the no of branches increases generating address for each branch become difficult. To handle this type of situation the best and simple way is used:

Programmable Logic Array (PLA) to generate the required branch address and this way of generating the branch address is called as wide branch addressing. In this method the of code of instruction is translated into the starting address of the corresponding micro routine. The opecode bits of the IR is connected to the inputs to the PLA hence PLA acts as a and outputs the address of the desired micro routine.

A.18. STATIC RAM

The static RAM consists essentially of internal flip-flops that store the binary information. The stored information remains valid as long as power is applied to the unit.

A.19. DYNAMIC RAM

The dynamic RAM stores the binary information in the form of electric charges that are applied to the capacitors. The capacitors are provided in-side the chip by MOS transistors. The stored charge on the capacitors tend to discharge with time and the capacitors must be periodically recharge by refreshing dynamic memory. Refreshing is done by cycling through the words every few milliseconds to restore the decaying charge. The dynamic RAM offers reduced Power consumption and larger storage capacity in a single memory chip.

Most of the memory (main) in general purpose computer is made up of RAM Ics, but a portion of the memory may be constructed with ROM chips. Originally, RAM was used to refer to a random access memory, but now it is used to designate a read/write memory to differentiate if from rad only memory. Although ROM is also a random access. ROM is used to store programs that are permanently resident in the computer.

Among the other things, the ROM portion of main memory is needed for storing an initial programs called a bootstrap loader. The boot strap loader is a program whose function is to start the computer software operating when power is turned on. Since RAM is Volatile, its contents are destroyed when power is turned off. The contents of ROM remains unchanged after power is turned off and on again.

RAM and ROM chips are available in a variety of sizes. If the memory needed for computer is larger than the capacity of one chip, it is necessary to combine a number of chips to form the required memory size.

The next address may also a function of external input conditions. While the micro-operations are being executed, the next address is computed in the next address circuit and then transferred into the control address register to read the next microinstruction.

The next address generator is sometimes called a microprogram sequencer, as it determines the address sequence that is read from control memory.

The control data register holds the present microinstruction while the next address is computed and read from memory. The data register is sometimes called a pipeline register. It allows the execution of the micro-operation specified by the control word simultaneously with the generation of next microinstruction.

The main advantage of the microprogrammed control is the fact that once the hardware configuration is established, there should be no need for further hardware or wiring changes.

A.20. MICROPROGRAMMING

Microprogramming generally involves the use of ROM to store effectively the control signals.

When a computer is microprogrammed, the individual operations between and on registers are called micro-operations. For instance, transferring the program's contents in to the memory address register is a micro-operation. Similarly, incrementing the program counter is a micro-operation. Transferring the contents of accumulator in the another register is also a micro-operation. In each of these cases a micro-operation is initiated by raising a single control signals. Thus sequencing micro-operations involves sequencing the appropriate control signals.

The use of micro program control in a processor to decode and generate various control signals with in a CPU represents a marked and signified and improvement in the system design and operation of the control portion of the CPU.

A.21. IEEE STANDARD 754 FLOATING POINT NUMBER

IEEE Standard 754 floating point is the most common representation today for real number on computers, including Intel-based PC's, McIntoshes, and most Unix platforms. This article gives a brief overview of IEEE floating point and its representation. Discussion of arithmetic implementation may be found in the book mentioned at the bottom of this article.

What are Floating Point Numbers ?

There are several ways to represent real numbers on computers. Fixed point places a radix point somewhere in the middle of the digits, and is equivalent to using integers that represent portions of some unit. For example, one might represent1/100 that of a unit; if you have four decimal digits, you could represent 10.82, or 00.01. Another approach is to use retionals, and represent every number as the ratio of two integers.

Floating-point representation the most common solution-basically represents real in scientific notation. Scientific notation represents numbers as a base number and an exponent. For example, 123.456 could be represented as 1.23456×10^2. In hexadecimal, the number 123. abc might be represented as $1.23abc \times 16^2$.

Floating-point solves a number of representation problems. Fixed-point has a fixed window of representation, which limits it from representing very large or very small numbers. Also, fixed-point is prone to a loss of precision when two large number are divided.

Floating-point, on the other hand, employs a sort of "sliding window" of precision appropriate to the scale of the number. This allows it to represent numbers from 1,000,000,000,000 to 0.0000000000000001 with ease.

Storage Layout

IEEE floating point numbers have three basic components; the sign, the exponent, and the mantissa is composed of the *fraction* and an implicit leading digit (explained below). Thc cxponent base (2) is implicit and need not be stored.

The following figure shows the layout for single (32-bit) and double (64-bit) precision floating-point values. The number of bits for each field are shown (bit ranges are in square brackets):

	Sign	Exponent	Fraction	Bias
Single Precision	1[31]	8[30–23]	23[22–00]	127
Double Precision	1[63]	11[62–52]	52[51–00]	1023

The Sign Bit

The sign bit is as simple as it gets. 0 denotes a positive numbcr; 1 denotes a negative number. Flipping the value of this bit flips the sign of the number.

The Exponent

The exponent field needs to represent both positive and negative exponents. To do this, a *bias* is added to the actual exponent in order to get the stored exponent. For IEEE single–precision floats, this value is 127. Thus, an exponent of zero means that 127 is stored in the exponent field. A stored value of 200 indicates an exponent of (200–127), or 73. For reasons discussed later, exponents of –127 (all 0s) and + 128 (all 1s) are reserved for special numbers.

For double precision, the exponent field is 11 bits, and has a bias of 1023.

The Mantissa

The *mantissa*, also known as the *signific and*, represents the precision bits of the number. It is composed of an implicit leading bit and the fraction bits.

To find out the value of the implicit leading bit, consider that any number can be expressed in scientific notation in many different ways. For example, the number five can be represented as any of these:

$$5.00 \times 10^6$$
$$0.50 \times 10^2$$
$$5000 \times 10^{-3}$$

In order to maximize the quantity of representable numbers, floating-point numbers are typically stored in *normalized* form. This basically puts the radix point after the first non-zero digit. In normalized form, five is represented as 5.0×10^0.

A nice little optimization is available to us in base two, since the only possible non-zero digit is 1. Thus, we can just assume a leading digit of 1, and don't need to represent it explicitly. As a result, the mantissa has effectively 24 bits of resolution, by way of 23 fraction bits.

Putting it All Together

So, to sum up:

1. The sign bit is 0 for positive, 1 for negative.
2. The exponent's base is two.
3. The exponent field contains 127 plus the true exponent for single-precision, or 1023 plus the true exponent for double precision.
4. The first bit of the mantissa is typically assumed to be 1, *f*, where *f* is the field of fraction bits.

A.22. RANGES OF FLOATING-POINT NUMBERS

Lets consider single-precision floats for a second. Note that we're taking essentially a 32-bit number and re-jiggering the fields to cover a much broader range. Something has to give, and it's precision. For example, regular 32-bit integers, with all precision centered around zero, can precisely store integers with 32-bits of resolution. Single-precision floating-point, on the other hand, is unable to match this resolution with its 24 bits. It does, however, approximate this value by effectively truncating from the lower end. For example:

	11110000	11001100	10101010	00001111	//	32-bit integer
=	+1.1110000	11001100	10101010	x 2^{32}	//	Single-precision Float
=	11110000	11001100	10101010	00000000	//	Corresponding Value

This approximates the 32-bit value, but doesn't yield an exact representation. On the other hand, basis the ability to represent fractional components (which integers lack completely), the floating-point value can represent numbers around 2^{127}, compared to 32-bit integers maximum value around 2^{32}.

The range of positive floating point numbers can be split into normalized numbers (which preserve the full precision of the mantissa), and *denormalized* numbers (discussed later) which use only a portion of the fraction's precision.

	Denormalized	Normalized	Approximate Decimal
Single Precision	$\pm 2^{-149}$ to $(1-2\_^{-23})\times2^{-126}$	$\pm 2^{-126}$ to $92-2^{-23}) \times 2^{127}$	$\pm \sim 10^{-44.85}$ to $\sim 10^{38.53}$
Double Precision	$\pm2^{-1074}$ to $1^{-52})\times2^{-1022}$	$\pm2^{-1022}$ to $(2-2^{-52})\times2^{1023}$	$\pm \sim 10^{-323.3}$ to $\sim 10^{308.3}$

Since the sign of floating point number is given by a special leading bit, the range for negative numbers is given by the negation of the above values.

There are five distinct numerical ranges that single-precision floating-point numbers are **not** able to represent:

1. Negative numbers less than $-(2-2^{-23}) \times 2^{127}$ (*negative overflow*)
2. Negative numbers greater than -2^{-149} (*negative underflow*)
3. Zero
4. Positive numbers less than 2^{-149} (*positive underflow*)
5. Positive numbers greater than $(2-2^{-23}) \times 2^{127}$ (*positive overflow*)

Overflow means that values have grown too large for the representation, much in the same way that you can overflow integers. Underflow is a less serious problem because is just denotes a loss of precision, which is guaranteed to be closely approximated by zero.

Here's a table of the effective range (excluding infinite values) of IEEE floating-point numbers:

	Binary	Decimal
Single	$\pm (2-2^{-23}) \times 2^{127}$	$\sim \pm 10^{38.53}$
Double	$\pm(2-2^{-52}) \times 2^{1023}$	$\sim \pm 10^{308.25}$

Note that the extreme values occur (regardless of sign) when the exponent is at the maximum value for finite numbers (2^{127} for single–precision, 2^{1023} for double), and the mantissa is filled with 1s (including the normalizing 1 bit).

Special Values

IEEE reserves exponent field values of all 0s and all 1s to denote special values in the floating-point scheme.

Zero

As mentioned above, zero is not directly representable in the straight format, due to the assumption of a leading 1 (wed need to specify a true zero mantissa to yield a value of zero). Zero is a special value denoted with an exponent field of zero and a fraction field of zero. Note that – 0 and + 0 are distinct values, though they both compare as equal.

Denormalized

If the exponent is all 0s, but the fraction is non-zero (else it would be interpreted as zero), then the value is a *denormalized* number, which does not have an assumed leading 1 before the binary point. Thus, this represents a number $(-1)^5 \times 0.f \times 2^{-126}$, where s is the sign bit and f is the fraction. For double precision, denormalized numbers are of the form $(-1)^s \times 0.f \times 2^{-1022}$. from this you can interpret zero as a special type of denormalized number.

Infinity

The values + infinity are denoted with an exponent of all 1s and a fraction of all 0s. the sign bit distinguishes between negative infinity and positive infinity. Being able to denote infinity as a specific value is useful because it allows operations to continue past overflow situations. *Operations with infinite values are well defined in IEEE floating point.*

Not a Number

The value NaN (*Not a Number) is used to represent a value that does not represent a real number. NaN's are represented by a bit pattern with an exponent of all 1s and a non-zero fraction. There are two categories of NaN; QNaN (QuietI NaN)* and SNaN (signalling NaN).

A QNaN is a NaN with the most significant fraction bit set. QNaN's propagate freely through most arithmetic operations. These values pop out of an operation when the result is not mathematically defined.

An SNaN is a NaN with the most significant fraction bit clear. It is used to signal an exception when used in operations. SNaNs can be handy to assign to uninitialized variables to trap premature usage.

Semantically, QNaNs denote *indeterminate* operations, while SNaN's denote *invalid* operations.

Special Operations

Operations on special numbers are well-defined by IEEE. In the simplest case, any operation with a NaN yields a NaN result. Other operations are as follows:

Operation	Result
n ÷ ± Infinity	0
± Infinity × ± Infinity	± Infinity
± nopnzero ÷ 0	± Infinity
Infinity + Infinity	Infinity
± 0 ¸ ± 0	*NaN*
Infinity − Infinity	*NaN*
± Infinity ¸ ± Infinity	*NaN*
± Infinity × 0	*NaN*

To sum up, the following are the corresponding values for a given representation;

Float Values (b = bias)

Sign	Exponent (e)	Fraction (f)	Value
0	00..0	00..00	+ 0
0	00..00	00..01 : 11..11	Positive Denormalized Real $0.f \times 2^{(-b+1)}$
0	00..00 : 11..10	XX..XX	Positive Normalized Real $1.f \times 2^{(e-b)}$
0	11..11	00..00	+ Infinity
0	11..11	00..01 : 0.1..11	SNaN
0	11..11	10..00 : 11..11	QNaN

1	00..00	00..00	– 0
1	00..00	00..01 : 11..00	Negative Denormalized Real $-0.f \times 2^{(-b+1)}$
1	00..01 : 11.10	XX..XX	Negative Normalized Real $-1.f\,2^{(e-b)}$
1	11.11	00..00	– Infinity
1	11..11	00..01 : 01..11	SNaN
1	11..11	10..00 : 11.11	QNaN

A.23. PERIPHERAL DEVICES

The devices which are connected to a computer and controlled by processor are called **Peripheral Devices**.

Example: Keyboards, Mouse, Display unit, Printers, etc.

A.24. INPUT-OUTPUT INTERFACE

"A hardware unit that plays an important role between the CPU and peripheral devices to supervise the input and output transfers is known as interface".

Input-output interface provides a method for transferring information between internal storage and external I/o devices. Peripherals connected to a computer need special communication links for interfacing them with the CPU. The purpose of the communication link is to restive the differences that exist between the central computer and each peripheral. The major differences are:

1. Peripherad are electromechanical and electromagnetic devices and their manner of operation is different from the operation of the CPU and memory, which are electronic devices.
2. The data transfer rate of peripherals in usually slower than the transfer rate of the CPU.
3. Data codes and formats in peripherad. Different from the word format in the CPU and memory
4. The operating modes of peripherals the different from each other and each must be controlled as so into to disturb the operation of other peripherals connected to the CPU.

To resolve these differences computer systems include special hardware components between the CPU and peripherals to supervise and synchronize all input and output transfers. These components are called interface units because they interface between the processor bus and the peripheral device.

There are fear types of communication that an interface may receive. They are:

1. **Control command:** A control command is issued to activate the peripheral and to inform it what to do. The particular control commands issued depends on the peripheral.

Example: A magnetic tape unit may be instructed to back space the tape by one record, or to start the tape. moving in the forward direction.

2. **Status Command:** A status command is used to text various status conditions in the interface and the peripherals.
3. **Data output command:** A data output command causes the interface to respond by transferring dataform the bus into one of its registers.
4. **Data Input Command:** The data input command is the opposite of the data output. In this call the interface receives written of data from the peripheral and places it in its buffer register.

Input-Output Versus Memory Bus

In addition to communicating with input-output the processor must communicate with the memory unit. Like the input-output bus, the memory bus contains data, address, and read write control lines. The computer buses can communicate to memory and input-output in three different ways.

1. Use two separate buses, one for memory and the other for input-output.
2. Use one common bus for both memory and input-output but have separate control lines for each.
3. Use one common bus for memory and input-output with common control lines.

A.25. ISOLATED VERSUS MEMORY MAPPED INPUT-OUTPUT

In isolated input-output configuration the input and output instruction for both interface register and memory are different. When the CPU fetches and decodes the instructions for input-output interfaces, it enables the input-output read or input-output write control line and informs the other components attached to the common bus hot the address in the address bus is for interface register and not for memory words.

Similarly: When the CPU fetches and decodes the instruction from memory, it places the memory address on the address bus and ended the memory areas or memory write control line, which inform the external components that the address is for the memory and input-output addresses are is dated in the method.

Alternative, method is known as memory mapped input-output. In this method the computer use the same address space, for memory and input-output both. In memory mapped input-output there is no specific input or output instructions. The am instruction cache used for manipulating input-output data residing in interfaces and can also be used to manipulate memory words. Computers with memory. Mapped input-output can use memory type instructions to access input-output data and hence, allows the computer to use the same instructions for memory transfers or input output transfers.

Example 1: What is the difference between isolated input-output and memory mapped input-output? What are the advantage and disadvantages of each?

Example 2: What are the three methods of input-output bus structures?

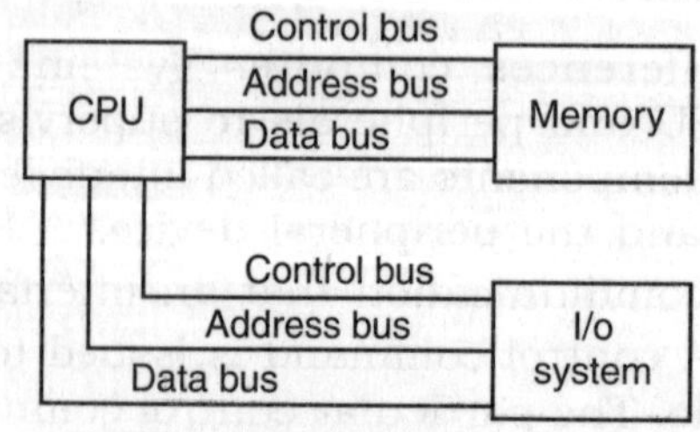

Fig. A.17.

(a) (Isolated input-output) (separate memory and input-output buses)

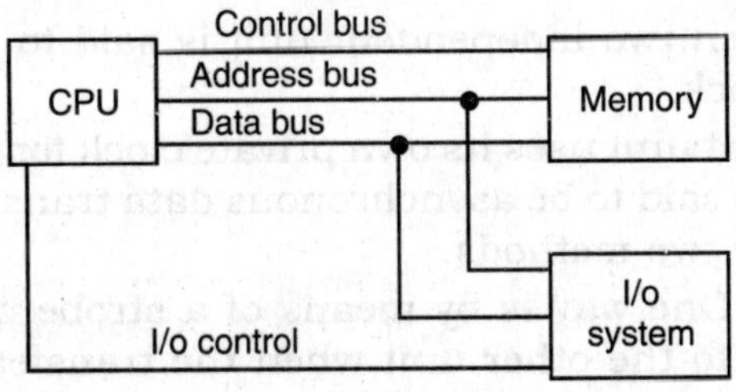

Fig. A.18.

(b) Shared address and data lines.

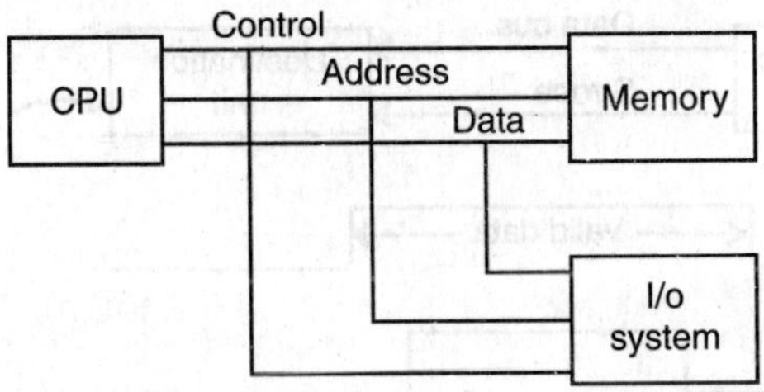

Fig. A.19.

(c) Shared address, data and control lines memory-mapped input-output)
Example of input Interface.

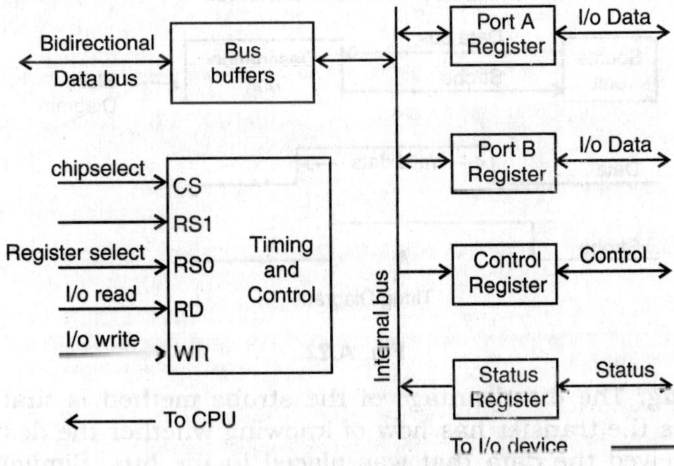

Fig. A.20.

CS	RSI	RSO	Register selected
0	X	X	None: data bus in high impudence
1	0	0	Port A register
1	0	1	Port B register
1	1	0	Control register
1	1	1	Status register

A.26. ASYNCHRONOUS DATA TRANSFER

The data transfer between two independent unit is said to be synchros data transfer if they share a common clock.

But if two independent unit uses its own private clock for data transfers between them when the data transfer is said to be asynchronous data transfers. This asynchronous way of data transfer active by two methods.

1. **Strobe Control:** One way is by means of a strobe pulse supplied by one of the units to indicate to the other unit when the transfer has to occur. This method is known as strobe control. The store control method employs a single control line. The strobe may be activated by either the source or the destination unit.

(a) Source- Initiated strobe for data transfer

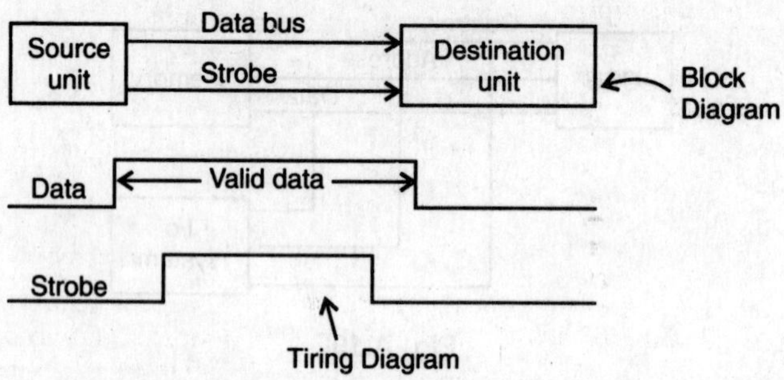

Fig. A.21.

(b) Destination-Initiated strobe for dat transfer

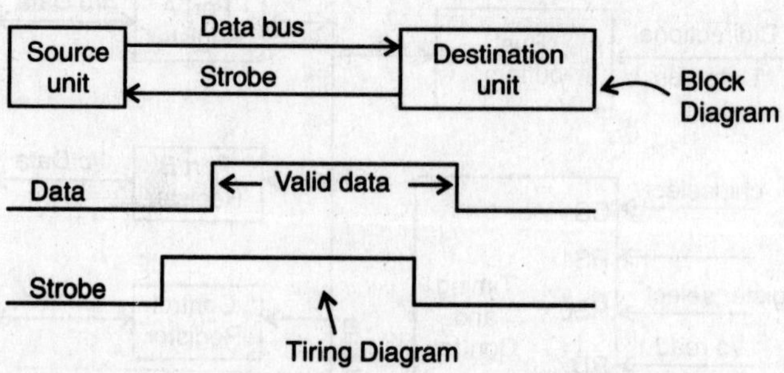

Fig. A.22.

2. **Handshaking:** The disadvantage of the strobe method is that the source unit that initiates the transfer has how of knowing whether the destination unit has actually received the data that was placed in the bus. Similarly, a destination unit that initiates the transfer has no way of knowing whether the source unit has actually placed the data or the bus.

The handshaking added this problem by introducing a second control signal line that provides a reply to the unit that initiates the transfer.

One control line is in the same direction as the data flow in the bus from the source to desecration. it is used by the source unit to in form the distortion unit whether there are valid data or not. The other control line is in the other direction from destination to the source.

(a) **Source:** Initiated Handshaking:

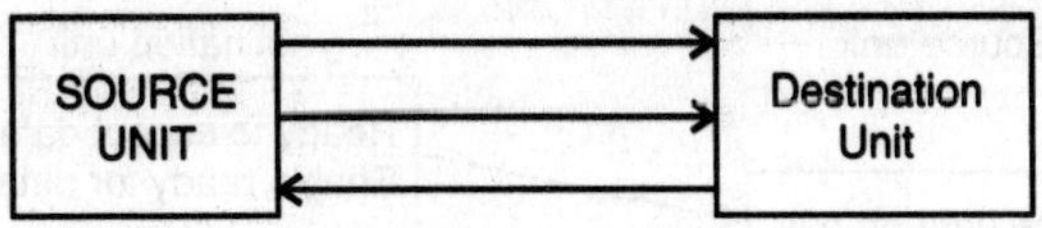

(a) Block Diagram

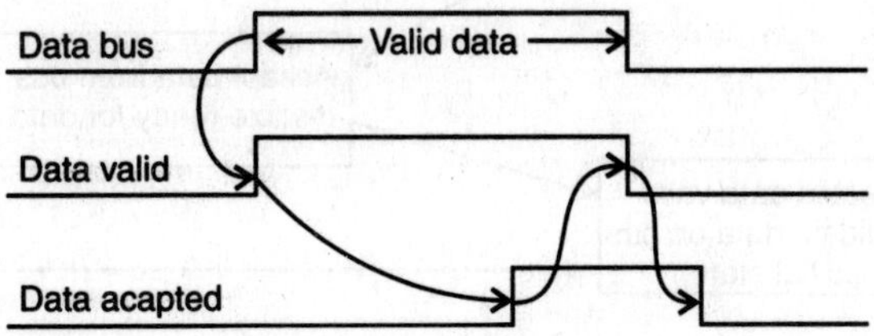

(b) Timing Diagram

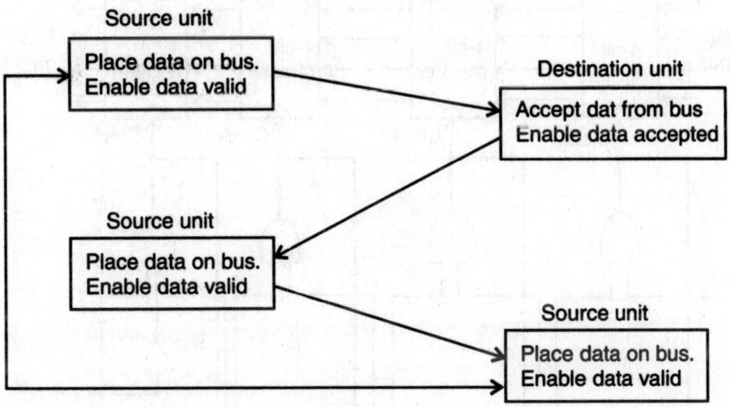

Fig. A.23.

(*b*) Destination Initiated Handshaking

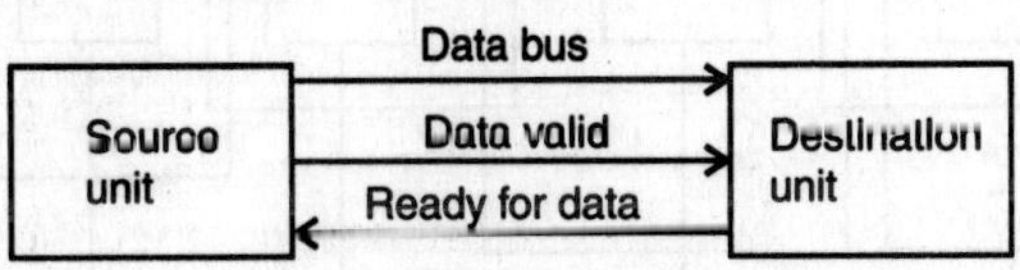

(a) Block Diagram

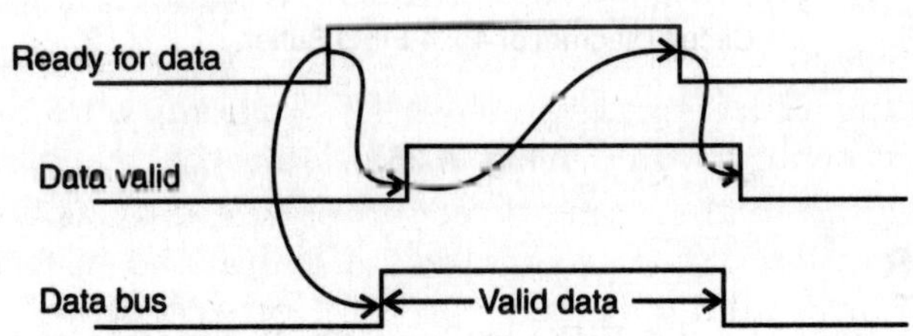

(b) Timing Diagram

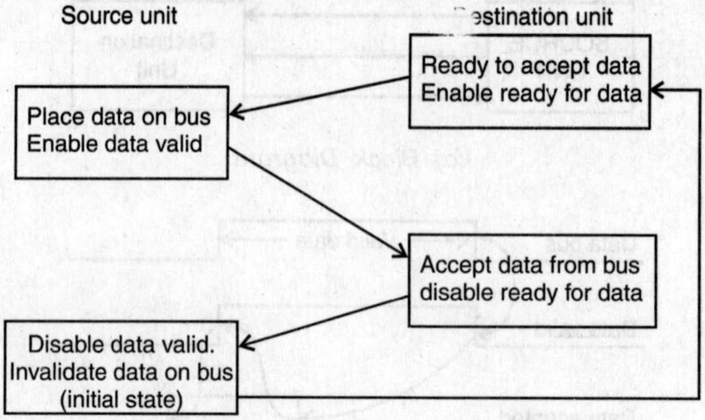

Fig. A.24.

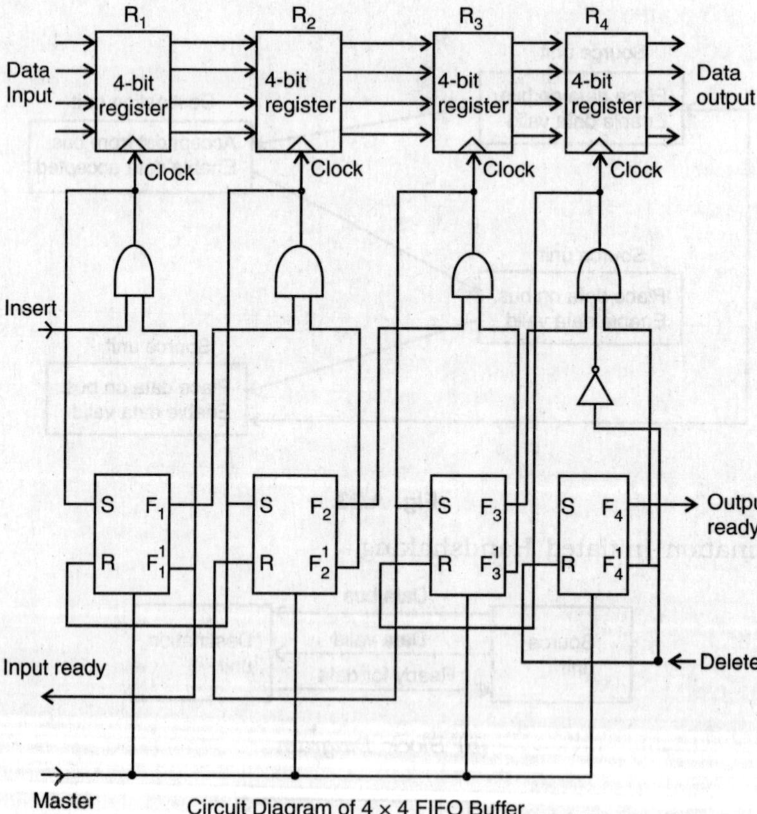

Fig. A.25.

A.27. FIFO BUFFER

The logic diagram of a typical 4 × 4 FIFO buffer side. It consists of four 4-bit register R_I,

$$I = 1, 2, 3, 4, \text{ and a control register with flip-flop}$$

$$F_i, i = 1, 2, 3, 4, \text{ one for each register}$$

The FIFO an store four words of four bits each. The number of bits per word can be increased by increasing the number of bits in each register and the number of words can be increased by increasing the number of register.

A flip-flop F_i in the control register that set to 1 indicates that a 4-bit data word is store in the corresponding register R_1. A 0 to F_i indicate that the corresponding register doesn't contain valid data. The control directs the improvements of data through the registers. Whenever the F_i bit of the control register is set (F_i = 1) the Fi + 1 bit is reset (F'_{i+1} = 1), to accept the data from register R1. The same clock transition sets F_{i+1} to 1 and reset F_i to 0. This causes the control flag to move one position to the right together with the data. Data in the register move down the FIFO toward the output as long as there are empty locations a head of it, This ripple through operation stop when the data reach a register R1 with the next Flip-flop F_{i+1} being set it 01, or bit the last register R4. An overall master cleat is used to initaize all control register flip-flop to 0.

Insertion: Data are inserted into the buffer provided that the input ready signal is enabled. This occurs when the first control flip-flop F1 is reset, indicating that register R1 is empty. Data are loaded from the input lines by enabling the clock, in R1 through the insert control line. The same clock sets F1, which disables the input ready control , indicating that the FIFO is row busy and unable to accept more data. The ripple through process begins provided that R2 and F_1 is cleared. This erables the input ready line, indicating that the input are now available for another data word. If the FIFO is full, F_1 remains set and the input ready line stays in the 0 state.

Detection: The data falling, through the registers stack up at the output end. The output ready control line is enabled when the last control flip-flop F_4 is set, indicating that there are valid data in the output register R_4. The output data form R4 are accepted by a destination unit, which then enables the delete control signal. This resets F_4, causing output ready to disable, indicating that the data on the output are no longer valid. Only after the delete signal goes back to 0 can the data from R_3 move into R_4. If the fIFO is empty, there will be no data in R_3 and F_4 will domain in the reset-states.

A.28. AS SYNCHRONOUS SERIAL TRANSMISSION

A serial asynchronous data transmission technique used special bits that are inserted at both ends of the character code. Each character consists of three parts.

1. a start bit,
2. the character bits,
3. otop bito.

A.29. USART OR UART

USART and Universal synchronous asynchronous receiver transmitter. Intel has developed 8251, a USART which is capable of transmittal data either in synchronous or asynchronous mode.

Example 1: What is the basic advantage of using interrupt initiated data transfer over transfer under program control without an interrupt?

Sol. The problem with programmed input-output is that the CPU has to with a long time for the input-output module of concern to be ready for reception or transmission of data. The CPU, while editing, must repeatedly interrogate the status of the input-output modules. As a result, the level of the system degraded need lessly. An alternative is for the CUP to issue an input-output command to a module and then go on to do some other useful work. The input-output module will then interrupt the CPU to request service when it is ready to exchange data with the CPU.

A.30. DIRECT MEMORY ACCESS (DMA)

"Direct Memory Access is a technique for moving data directly between the main memory and the peripheral devices without the intervention of CPU.

Cycle Stealing: The DMA device take memory bus from the processor and steal one memory cycle required fro read or write.

DMA controller: In DMA the processor realizes the control of the buses to a device called DMA controller.

Bus request (BR): It is the input signal used by the DMA controller to request the CPU to relinquish control if the buses. When this input is active, the CPU terminates the execution of the current instruction and places the address bus, the data bus and control signal into a high impedance state.

Bus grant (BG): It is activated by CPU to inform the external DMA that the buses are in the high impedance state. The DMA that originates the Bus request can now take control of the buses to process the data transfer.

Example 1: Why are read and write control lines in DMA (controller bidirectional? Under what condition and for what purposes are they used as inputs? Under what condition and for what purposes are they used as output?

Solution: When the CPU communicates with the DMA controller the read and write lines are used as inputs from the CPU to the DMA controller when the DMA controller communicate with memory, the read and write lines are used as outputs from the DMA to memory.

1. CPU bus signals for DMA transfer.

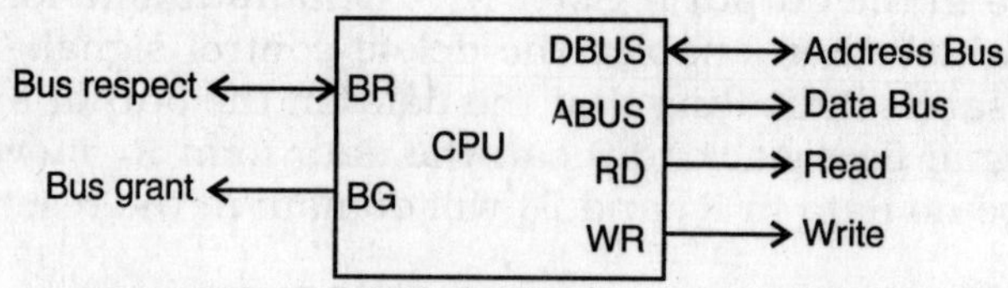

Fig. A.26.

2. Block Diagram of DMA Controller.

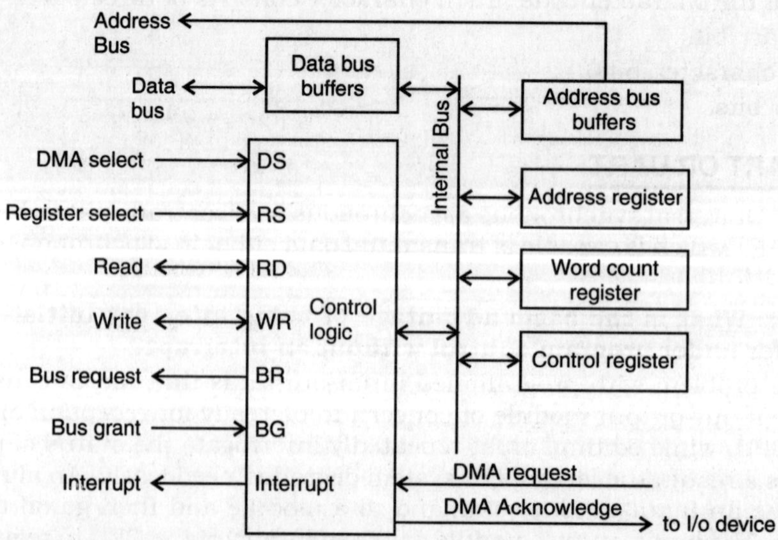

Fig. A.27.

3. DMA transfer in a computer system.

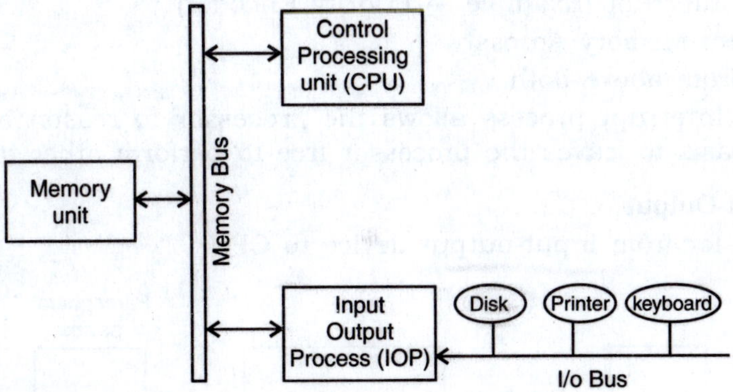

Fig. A.28.

CPU-IOP Communication

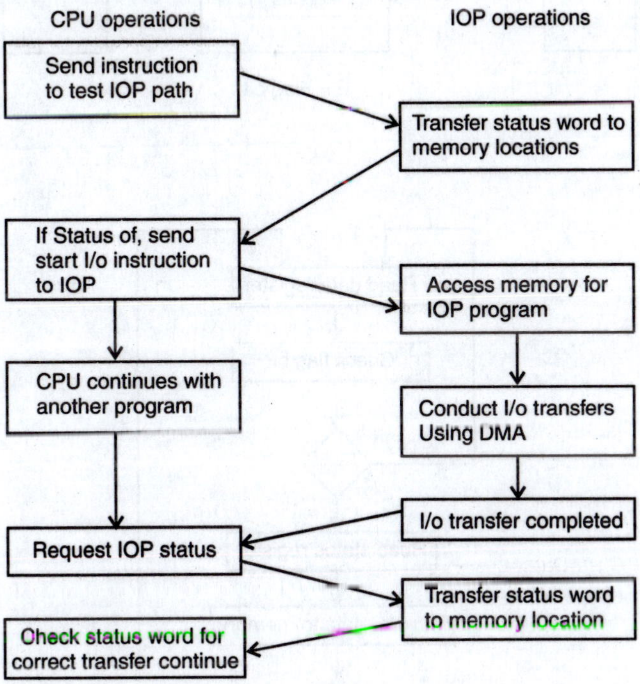

Fig. A.29.

A.31. MODES OF TRANSFER

Data transfer to and from peripherals may be handled in one of the three possible modes.

1. Programmed input-output

   To reduce drawback of (1) we create (2)

2. Interrupt-Imitated input-output

   →Priority Interrupt

→Serial interrupt (Example → Daisy-chin-up)
→parallel interrupt (Example → Priority Encoder)

3. DMA [Direct Memory Access]
   It is fast from above both
   Interrupt: Interrupt process allows the processor to reason or a service on a demand basis to leaves the processor free to perform other task.

Programmed Input-Output

Data transfer from input-output device to CPU

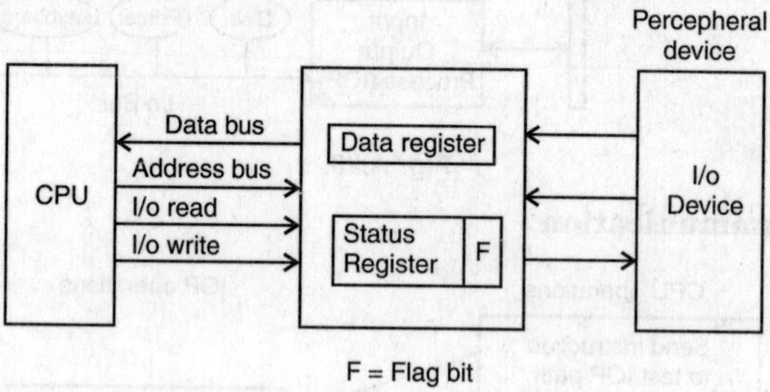

(*a*)

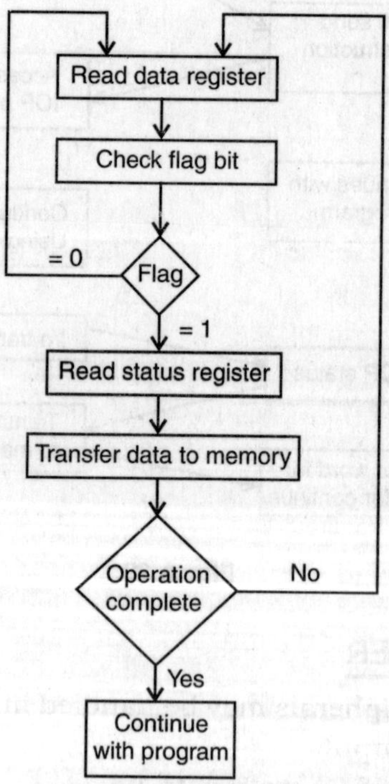

Flowchart for CPU program to input data.

(*b*)

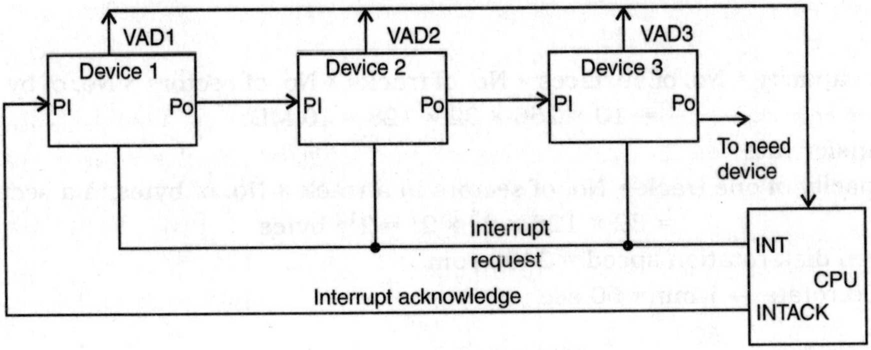

(c)

Fig. A.30.

Daisy-Chain Priority interrupt.

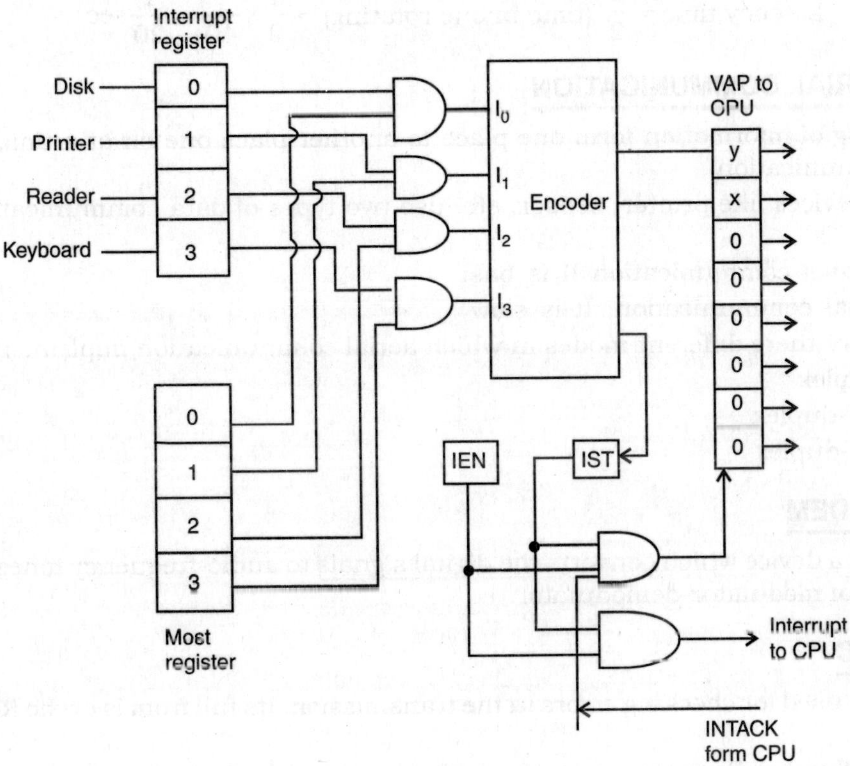

Fig. A.31.

"Priority Encoder or Priority Interrupt Hardware

A certain magnetic disk has following speciates

→

→

→

→

→

(i) Disk capacity = No. of surfaces × No. of tracks × No. of sectors × No. of by in a sector

$$= 10 \times 256 \times 32 \times 128 = 10 \text{ MB.}$$

(ii) Transfer rate

Capacity of one track = No. of sectors in a track × No. of bytes in a sector

$$= 32 \times 128 \times 2^5 \times 2^7 = 2^{12} \text{ bytes.}$$

Given disk rotation speed = 2400 rpm

2400 rotate → 1 min= 60 sec

$$1 \text{ rotation} \rightarrow \frac{60}{2400} = \frac{1}{40} \text{sec}$$

$$\text{in } \frac{1}{40} \text{sec} = 212 \times 40 \text{ by too}$$

Transferred = 160 k Bytes

$$\text{Latency time} = \frac{1}{2} \text{ (time in one rotating)} = \frac{1}{2} \times \frac{1}{40} = \frac{1}{20} \text{sec}$$

A.32. SERIAL COMMUNICATION

Transferring of information form one place to another place one bit at a time (known as serial communication).

Some devices like printer, sensor, etc. use two types of data communication to send data.

1. Parallel communication II is bast
2. Serial communication. It is slow

There are there different modes in which aerial communication implement

1. Simplex
2. Half-duplex
3. Full-duplex

A.33. MODEM

A modem is a device which converts the digital signals to audio-frequency tones. Modem is short form of modulator-demodulator.

A.34. CRC

It is method used for checking errors in the transmission. Its full from is cyclic Redundancy check.

Binary Number Formats

Name	**Size (bits)**	**Example**
Bit	1	1
Nibble	4	0101
Byte	8	0000 0101
Word	16	0000 0000 0000 0101
Double word	32	0000 0000 0000 0000 0000 0000 0000 0101

A.35. ANOTHER METHOD FOR REPRESENTING FLOATING POINT NUMBERS

Floating-Point Numbers

All the numbering systems which we have studied so for fall into category of base r fixed point numbers.

Fixed point number means the fractional point (decimal point) if once finalized will be fixed.

Disadvantage of fixed point no. Vapresertation is that it cannot represent very large number and very small number.

→ The remedy of this disadvantage is Scientific notation.

→ In Scientific Notation, number is in the from of

$$N = M \times 10^{E}$$

Where M and E are fixed points number called Mantissa and Exponent respectively.

Now with the version of Scientific. Number (notation) is called FLOATING NUMBER POINT.

For into take an example → consider 9 no. $(10000 \times 34)_{10}$

This no. can be represented as,

$1000\ 234 \times 10^{-3}$
$1000\ 23.4 \times 10^{-2}$
$1000\ 2.34 \times 10^{-1}$
1000.234×10^{0}
100.01234×10^{1}
10.00234×10^{2}
$1.000\ 234 \times 10^{3}$
0.1000234×10^{4}

So, in representation of $(1000.234)_{10}$, you must absorved that decimal point is not fixed. It can FLDAT from number to number depending upon Power of 10.

→ In

$$N = \pm M \times ME$$

B = Base of plodding point no. system

It is 10 for decimal system and 2 for Binary system

B = 10 B = 2

FORMAT OF FLOATING POINT NUMBER

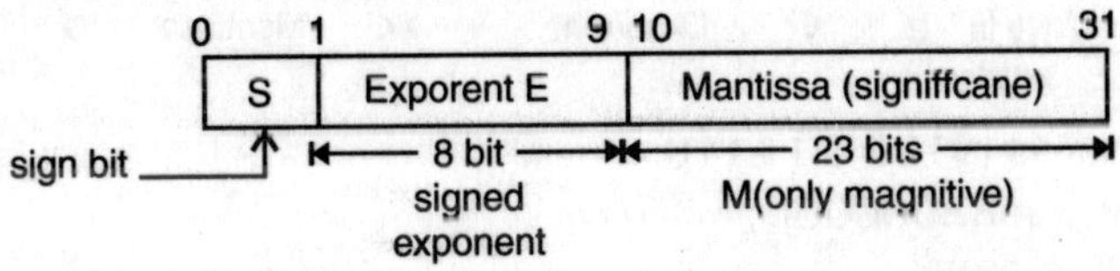

Format of a 32 bit floating point number.

Fight bits are used for exponent E, with remaining 23 bits for mantissa.

E con range from –128 to +127

→ Floating point number are standardized by IEEE. According to it we. represent floating point number is form.

$$\pm 1.M \times 2^{E^{1}-127} \quad \text{(single precision -32 bits)}$$

$\pm 1.M \times 2^{E^1-1023}$ (Double precision –64 bits)

Here M = Mantissa

E = E^1 – 127

$\Rightarrow$ E^1 = E + 127 E = exponent of number

The range for E1 is 0 to 255.

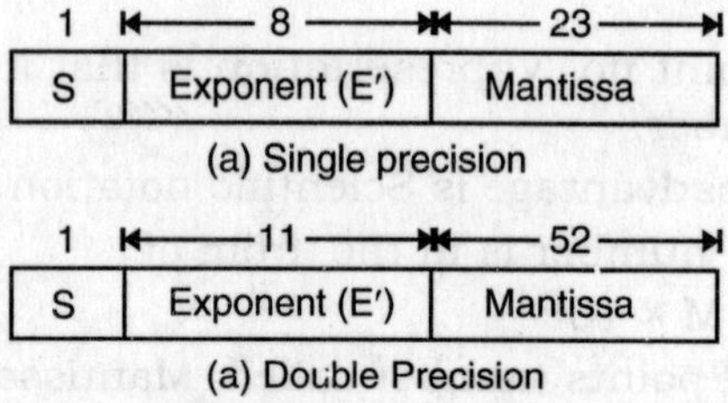

IEEE 754 format

What is 1 M?

Whenever any number is given to you to represent it in floating point no. system, you have to bring down to the format 1.M

Let the no. is 0.0110×2^6

Presently no. has form 0.M where M = 0110

But we want 1.M

So, now benefit no. to left side till you get 1,

$0.0110 \times 2^6 \rightarrow$ Actua ln umber

$0.110 \times 2^5 \rightarrow$ After 1st shift } shift operation

$1.10 \times 2^4 \rightarrow$ after 2nd shift

Now stop here as we have 1.M Now,

M = 10

S = plus (positive)

E′ = E + 127 = 4 + 127 = 131 = (10000011)2

Therefore the floating point no. *i.e.*,

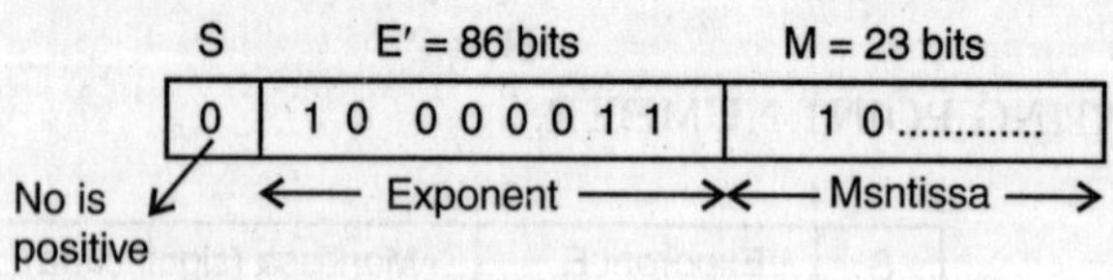

$\therefore$ finally number is $(0100\ 00\ 1110\ \text{-----})_2$

Index this no. *i.e.*, $(43800000)_{16}$

A.36. DE-NORMALISED/NORMALISED FLOATING POINT NUMBER

→ While representing 0.0110×2^6 number in floating point, you have shifted the bits and adjusted exponent, till you get 1.M

The number we get after proffering this operation is called "Naimalised number.

→ De-normalised number and 'RAW' number which does not follow 1.M format.

Range →

• Each floating point format (single precision/double precision) has capability to represent the no. within 'Certain' Range.

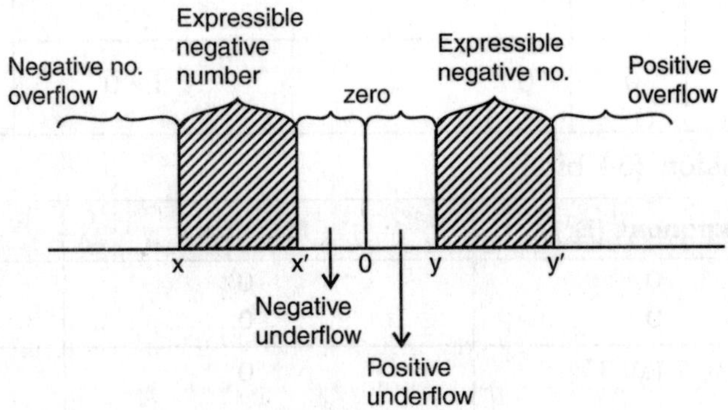

Fig. A.32.

This figure shows the number that can be expressed (+ve or – ve), by shaded part. Below and above the shaded parts, numbers are not representable.

• Fine regions are not included in the groups

1. Negative number, those are under negative overflow.
2. Negative number, those are under negative underflow.
3. Zero.
4. Positive number those are under positive underflow.

→ What is overflow and underflow?

• Overflow → Overflow occur when an arithmetic operation results in magnitude greater than can be expressed with an exponent.

Underflow → Underflow occurs which fractional part is too small. Underflow is less serious problem because the result canbe approximated to zero.

• To represent zero '0' special bit pattern to designated in IEEE format when e = Exponent = 0 and dignificand (Mantissa) = 0, the no. is zero.

• In IEEE format largest exponent field E = $(11111111)_2$. Corresponds to 255 is reserved to denote infinite number that result from some types of overflow. In this standard, certain invalid bit combinations considered to being a special not a number (NAN) category.

Table Interpretation of IEEE 750 Floating-Print Numbers

(a) Single Precision (32 bits)

	Sign	Exponent (E^1)	Mantissa	Value
1. Positive zero	0	0	0	0
2. Negative zero	1	0	0	– 0
3. Plus infinity or min 48 infinity	0 or 1	255 (all 1's)	0	(–1)s ˝
4. NAN	0 or 1	255 (all 1's)	≠ 0	NAN

5. Normalized non-zero number	0 or 1	0 < e < 255	f may be zero	$(-1)^s\ 2^{E'-127}$ (1.<M) or not
6. Denomalized	0	0	f ≠ 0	$(-1)^{s_2-126}$(0.M)

(b) Double precision (64-bits)

	Sign	Exponent (E^1)	Mantissa	Value
1. 2.	0 1	0 0	0 0	0 0
3.	0 or 1	2047 (all 1's)	0	(–1)s ∞
4.	0 or 1	2047 (all 1's)	≠ 0	NAN
5.	0 or 1	0 < e < 2047	f may be zero or not	$(-1)^s.2^{E^1-1023}$ (1.M)
6.	0 or 1	0	f ≠ 0	$(-1)^s\ 2^{-1022}$ (0.M)

Represent following in single precision format.

For 'O', signata = 0, M = 0, and E = –127

$$(-1)^s\ 1.M \times 2^{E^1-127}$$

∴ $(-1)^0\ 1.0 \times 2^0$

∴ E = 0

S	Exponent (E^1)	Mentis
0	0000 0000	0000 ----0

∴ Flatten point no. is 000000000011 (hex)

(ii) –1

Sin bit = 1 → No. is –Ve

(1)10 = 1.0 × 20 → normalized

∴ M = 0

F = E^1 – 127

∴ E^1 = 127 = $(01111111)_2$

S	(E^1)	Mentisa
S	01111111	00------0
1 bit	8 bits	23 bits

∴ Flatting pint nos. is $(10111111100....0)_2$ and (BF800000)16 hex.

A.37. 2D ORGANIZATION

Semiconductor 2D memory (16W, 46/W)

Static memory cells can be organized into a 2D-memory system as shown in fig. The figure illustrates the organization of 156 word, 4 bit per word 200 Memory.

There is an address decoder to decoder the address in 4-bit MAR, which selects one out of 16 word wires. V is the voltage applied to a selected word wire. A number of memory calls equal to the bits/word are connected to each word wire. For i = 1, 2, 3, 4, the bit wires B_{0i} and B_{1i} i of all words are connected together and fed to this sense amplifier S_{Ai} as shown is figure.

To write, 1001 (say), in word 2 the following is done.

- Word address 0010 is stored in MAR. When decoded it selects word W_2 and the voltage line on W_2 is made V.
- The data 1001 is stored in MBR. SA_1 brings voltage on B_{11} to 1, i

SA_2 brings voltage on B_{02} to 0

SA_3 brings voltage on B_{03} to 1

SA_4 brings voltage on B_{14} to 1

Thus, call 1 of word 2 will other a 1, cell 2 of word 2 will store a 0, cell 3 of word 2 will store a 0, cell 4 cell 4 of word 2 will store a 1. To read the contents of award, the address must be needed in MAR and all the memory cells belonging to this word experience voltage V/2. Depending on whether a cell Head a 0 or 1 stored the voltages on the corresponding bit writes do, what is sensed by the respective sense amplifier. If the voltage in B_{0i} dip SA_i will store a 0 in bit 1 of MBR and if the voltage in B1i dip, SA_1 will store 1 in bit in of MBR.

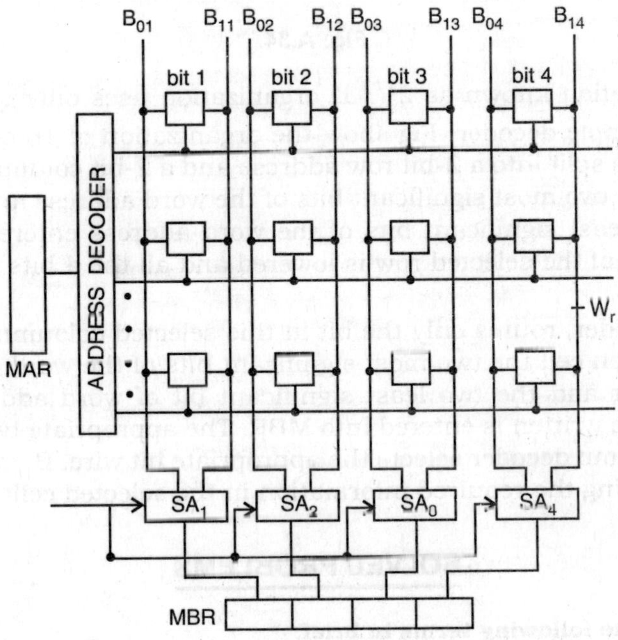

Fig. A.33.

A.38. $2^{1/2}$ D ORGANIZATION

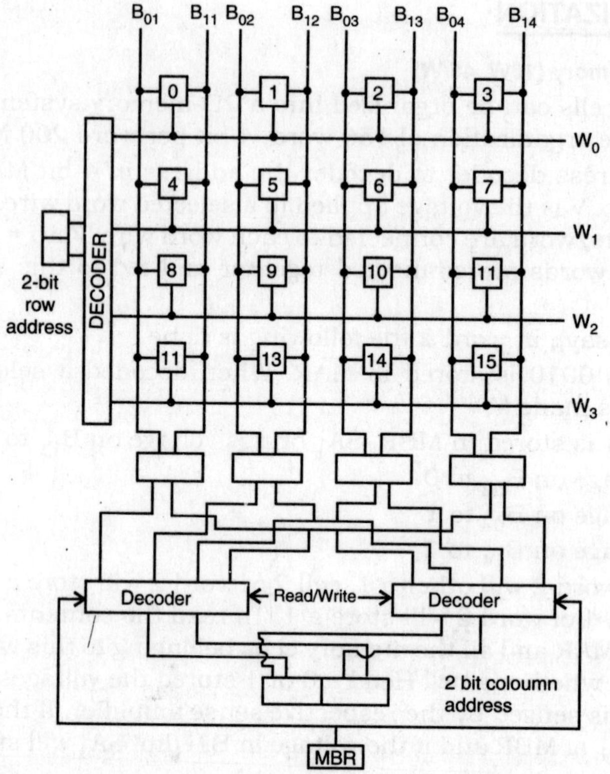

Fig. A.34.

Another organization known as $2^{1/2}$ D organization uses only $\sqrt{W}$ words wires and allows the use of simple decoder. Fig show the organization of 16 cells for 4 wires. The 4-bit word address is split into a 2-bit row address and a 2-bit column address. To read a bit form the memory two most significant bits of the word address are entered bits of the row registered two least significant bits of the word address entered into the coloumn register. The voltage of the selected row is lowered and all the 4 bits cells in this row are read.

The column decoder, routes only the bit in this selected coloumn as data output. To write a bit in the given cell the two most significant bits of the word address are entered into the row register and the two least significant bit of word address in the column register. The bit to be written is entered into MBR. The appropriate two is selected by the decoder and the column decoder selects the appropriate bit wire, B_{1i} or B_{0i} and its voltage is lowered while writing the required information in the selected cells.

SOLVED PROBLEMS

Problem 1. Define the following terms in brief:

(a) Register transfer language

(b) Micro operation

(*c*) **Registers**

(*d*) **Control function**

Solution: (*a*) Register transfer language: The symbolic notation used to describe the micro-operation transfers among registers is called a register transfer language. The term register transfer implies the availability of hardware logic circuits that can perform a stated micro-operation and transfer the result of the operation to the same or another registers the word "language" is borrowed from programmes who apply this term to programming language.

(*b*) **Micro operation:** The operations executed on data stored in registers are called micro-operations. A micro-operation is an elementary operation performed on the information stored in one or more registers. The result of operation may replace the previous binary information of a register or may be transferred to another register *e.g.,* shift, count, clear and load.

(*c*) **Registers:** The register that gold the address of the memory units is called a memory address register and is designated by the name MAR other designations for registers are:

PC (Program counter)

IR (Instruction register)

R_1 (Processor Register)

(*d*) **Register Transfer:** Information transferred from one register to another is designated in symbolic form by means of a replacement operator. The statement $R_2 \leftarrow R_1$, denotes a transfer of the content of transfer R_1 into register R_2. It designates a replacement the contents of R_2 by the content of R_1. By definition, the content of source register R_1 does not change after the transfer.

(*e*) **Control Function:** The control function is a Boolean variable that is equal to 1 or 0. The control function is included in the statement as follows:

P, $\quad R_2 \leftarrow R_1$

The control condition is terminated with a colon, it symbolizes the requirement that the transfer operation be executed by hardware only of P = 1.

Problem 2. Describe three state bus buffer. Draw the diagram of three state buffer.

Solution: A bus system can be constructed with three state gates instead of multiplexers. A three state gate is a digital circuit that exhibits three states. Two of these states are signals equivalent to logic 1 and 0 as in a conventional gate the third state is a high independence state behaves like an open circuit, which means that the output is disconnected and does not have a logic significance. Three state gates may perform any conventional logic such as AND or NAND. However, the one most commonly used in the design of a bus system is the buffer gate.

The graphic symbol of a three state buffer gate is shown below:

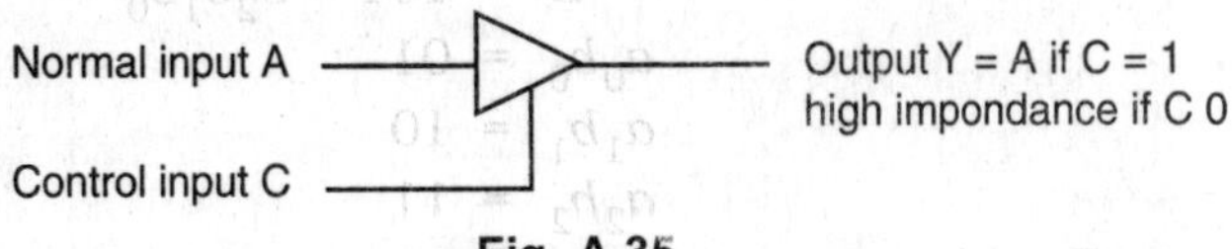

Fig. A.35.

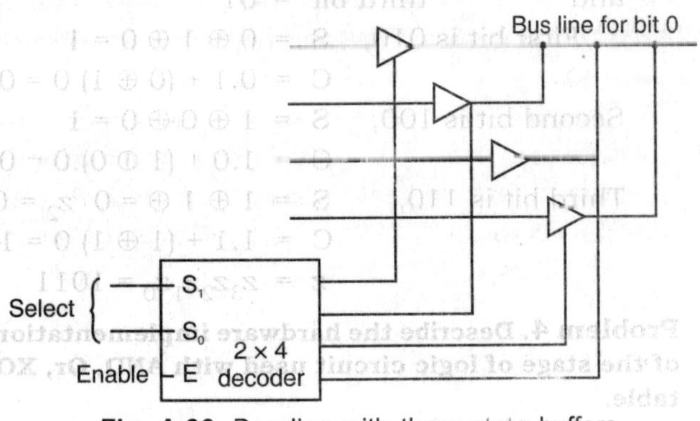

Fig. A.36. Bus line with three state buffers

To construct a common bus for four registers of n-bit using three state buffers, we need n circuit with four buffers. Each group of four buffers receives one significant bit from the four registers. Each common output produces one of the lines for the common bus for a total of n lines only one decoder is necessary to select between the four registers.

Problem 3. Consider the following circuit A = $a_2a_1a_0$ and B = $b_2b_1b_0$ are three bit binary numbers input to the circuit. The output is Z is z_3z_2,z_0. R_0 R_1 and are registers with loading clock shown. The registers are loaded with their input data with the falling edge of the clock pulse. The bits of the input numbers A, B and full address are shown in the circuit.

(a) Find the output 2 for the inputs

A = 110

B = 101

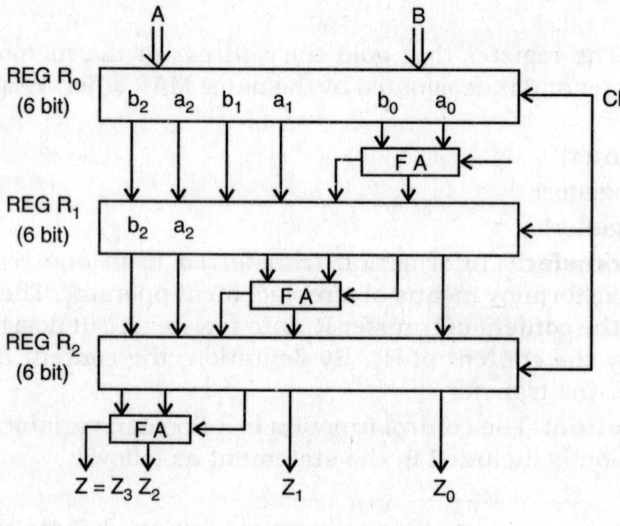

Fig. A.37.

Solution: Given value, A = 110 = $a_2a_1a_0$

B = 101 = $b_2b_1b_0$

a_0b_0 = 01

a_1b_1 = 10

a_2b_2 = 11

For bit a_0b_0 = 01

and third bit = 01

∴ First bit is 010, S = $0 \oplus 1 \oplus 0 = 1$

C = $0.1 + (0 \oplus 1)\,0 = 0 + 0 = 0$

Second bit is 100, S = $1 \oplus 0 \oplus 0 = 1$

C = $1.0 + (1 \oplus 0).0 = 0$ $z_1 = 1$

Third bit is 110, S = $1 \oplus 1 \oplus = 0$ $z_2 = 0$

C = $1.1 + (1 \oplus 1)\,0 = 1$ $z_3 = 1$

z = $z_3z_2z_1z_0$ = 1011

Problem 4. Describe the hardware implementation of logic-micro operation. Draw the diagram of the stage of logic circuit used with AND, Or, XOR and NOT GATE with the help of function table.

Solution: The hardware implementation of logic micro-operations requires that logic gates be inserted for each bit or pair of bits in the register to perform the required logic function. Although there are 16 logic micro-operations, most computers use only four AND, OR XOR and complement.

S_1	S_0	Output	Operation
0	0	$E = A \wedge B$	AND
0	1	$E = A \vee B$	OR
1	0	$E = A \oplus B$	XOR
1	1		
1	1	$E = A$	Complement

Function Table

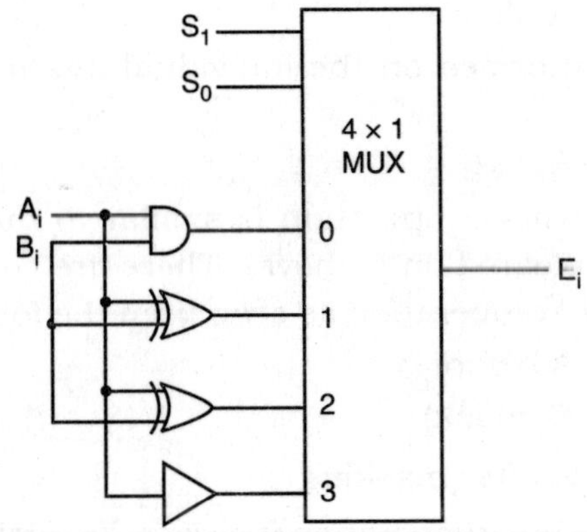

Fig. A.38. Logic diagram of one stage of logic circuit

One stage of circuit that generates the four basic logic micro-operation. It consists of four gates and multiplexer. Each of the four logic operations is generated through a gate that perform the required logic the outputs of gates are applied to the data inputs of the multiplexer. The two selection inputs S_1 and S_0. Choose one of the data inputs of the multiplexer and direct its value to the output. For a logic circuit with *n* bits the diagram mist be repeated *n* times for $i = 0, 1, 2, 3, n$ the selection variables are applied to all stages.

Problem 5. Define the term and describe its operation:

(*a*) Selective set
(*b*) Selective complement
(*c*) Selective clear
(*d*) Mask operation
(*e*) Ensert operation
(*f*) Clear operation

Solution. (*a*) Selective set: The selective set operation sets to 1. The bits in the register A where there are corresponding is in register B. It does not affect that have O's in the B.

e.g., 1010 A before
<u>1100</u> B(logical operand)
1110 A (after)

The two left most bits of B are 1's B are 1's, so the corresponding bits of A are set to 1. The two bits of A with corresponding O's in B remain unchanged. Therefore the OR micro-operation can be used to selectively get bits of a register.

(b) Selective complement: The selective complement operation complements bits in A where there are corresponding 1's in B. It does not affect bit positions that have O's in B.

e.g.,	1010	A before
	1100	B (logic operand)
	0110	

Therefore exclusive–OR operation can be used to selectively complement bits of register.

(c) Selective clear: Selective clear operation clears to O. The bits in A only where there are corresponding 1's in B.

e.g.,	1010	A before
	1100	B (logic operand)
	0010	A after

The boolean operation performed on the individual bits in AB^1. The corresponding logic operation

$$A \leftarrow A \cap B^1$$

(d) Mask operation: The mask operation is similar to the selective clear operation except that the bits of A are cleared only above. There are corresponding O's in B. The mask operation is an AND micro-operation as seen from the following numerical example

1010	A before
1100	B (mask)
1000	A after masking

(e) Insert operation: The insert operation inserts a new value into a group of bits this is done by first masking the bits and then OR in them the required value.

e.g.,	0110	1010	A before
	0000	1111	B (mask)
	0000	1010	After masking
	0000	1010	A before
	1001	0000	B (insert)
	1001	1010	A after insertion

(f) Clear operation: Clear operation compares the words in A and B and Produces an all O's result if the two numbers are equal. This operation is achieved by an exclusive–OR micro-operation.

e.g.,	1010	A
	1010	B
	0000	$A \leftarrow A \oplus B$

Problem 6. Show the bock diagram of the hardware that implements the following register transfer statements

$$yT_1 : R_2 \leftarrow R_1, R_1 \leftarrow R_2$$

Solution:

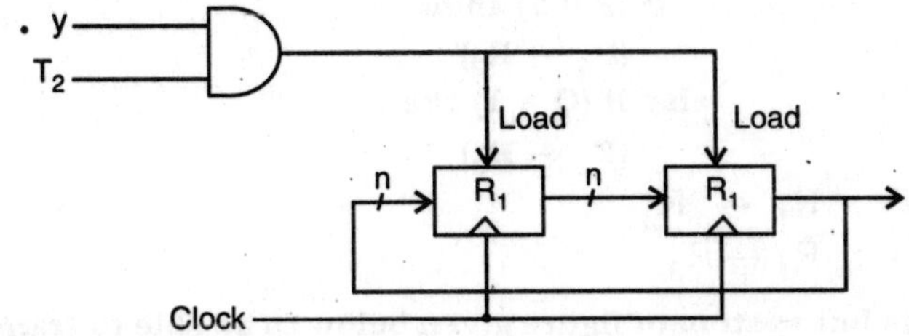

Fig. A.39.

Problem 7. The outputs of four registers R_0, R_1, R_2 and R_3 are connected through 4–to–1 line multiplexers to the inputs of a fifth register, R_5. Each register is eight bits long. The required transfers are dictated by four timing variables to through T_3 as follows:

$$T_0 : R_5 \leftarrow R_0$$
$$T_1 : R_5 \leftarrow R_1$$
$$T_2 : R_5 \leftarrow R_2$$
$$T_3 : R_5 \leftarrow R_3$$

The timing variables are mutually exclusive, which means that only one variable is equal to 1 at any given time, while the other three are equal to 0. Draw a block diagram showing the hardware implementation of register transfers, include the connection S necessary from the four timing variables to the selection inputs of the multiplexers and to load input of register R_5.

Solution:

T_0	T_1	T_2	T_3	S_1	S_0	R_3 load
0	0	0	0	×	×	0
1	0	0	0	0	0	1
0	1	0	0	0	1	1
0	0	1	0	1	0	1
0	0	0	1	1	1	1

$$S_1 = T_2 + T_3$$
$$S_0 = T_1 + T_3$$
$$\text{load} = T_0 + T_1 + T_2 + T_3$$

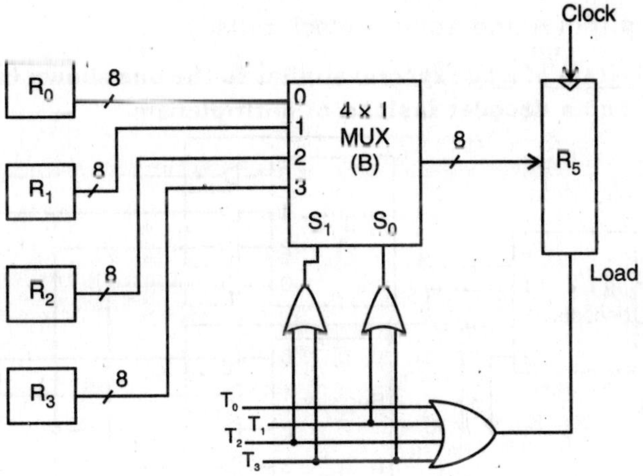

Fig. A.40.

Problem 8. Represent the following conditional, control statement by two register transfer statements with control functions.

If (P = 1) then

$(R_1 \leftarrow R_2)$

else if (Q = 1) then

$(R_1 \leftarrow R_3)$

Solution:

$$P : R_1 \leftarrow R_2$$
$$P^1Q : R_1 \leftarrow R_3$$

Problem 9. What has to the bus system of figure given below to be able to transfer information from any register to any other register ? Specifically. Show the connections that must be included to provide a path from the outputs of register C to the inputs of register A.

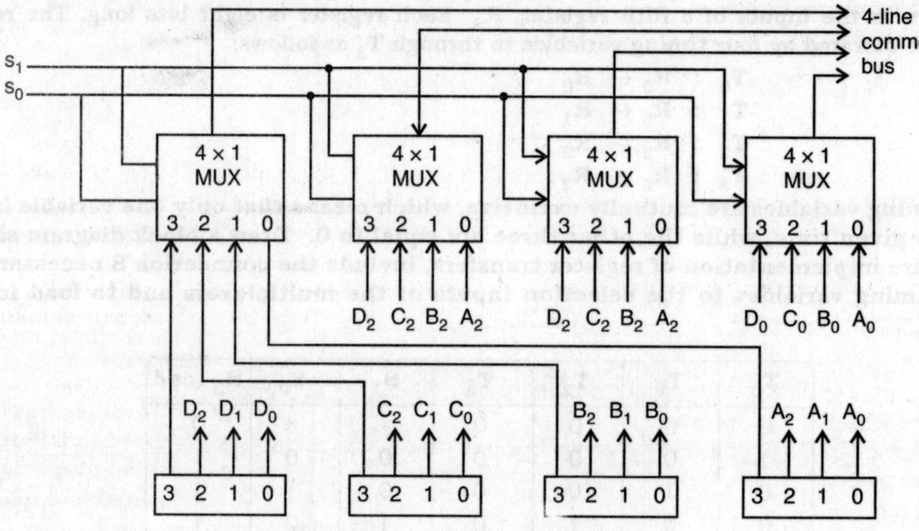

Fig. A.41. Bus system for Four register

Solution: Connect the 4–line common bus to the fair inputs of each register. Provide a "load" control input in each register provide a clock input for each register to transfer from register C to register A.

Apply S_1S_0 = 10 (to select C for the bus)

Enable the load input of A and apply a clock pulse.

Problem 10. Draw a diagram of a bus system similar to the one shown in the figure below but use three-state buffers and a decoder instead of multiplexers.

Solution:

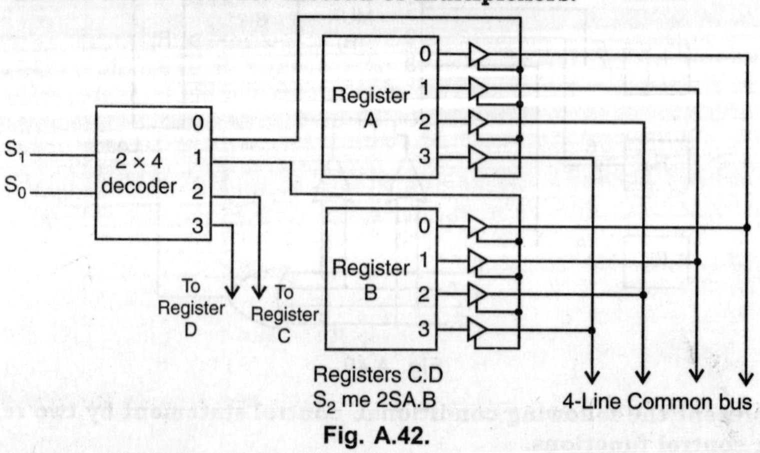

Fig. A.42.

Probelem 11. A digital computer has a common bus system for 16 registers of 32 bits each. The bus is constructed with multiplexers.

(*a*) How many selection inputs are there in each multiplexers ?

(*b*) What size of multiplexers are needed ?

(*c*) How many multiplexers are there in the bus ?

Solution: (*a*) 4 selection lines to select one of 16 register

(*b*) 16 × 1 multiplexers are needed.

(*c*) 32 multiplexers, one for each bit of the registers.

Problem 12. The following transfer statements specify a memory. Explain the memory operation in each case.

(*a*) $R_2 \leftarrow M\ [AR]$

(*b*) $M\ [AR] \leftarrow R_3$

(*c*) $R_5 \leftarrow M\ [R_5]$

Solution: (*a*) Read memory word specified by the address in AR into register R_2

(*b*) Write content of Register R_3 into the memory word specified by the address in AR

(*c*) Read memory word specified by the address in R_5 and Transfer content to R_5 (destroys previous value)

Problem 13. Draw the block diagram for the hardware that implements the following statements:

$$x + yz : AR \leftarrow AR + BR$$

Where AR and BR are two n-bit registers and *x*, *y* and *z* are control variables. Include the logic gates for the control function. (Remember that the symbol + designates an OR operation in a control or Boolean function but that it represents an arithmetic plus in a micro-operation)

Solution:

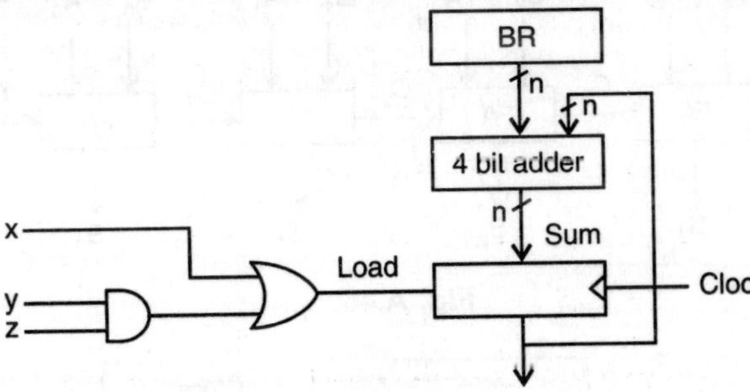

Fig. A.43.

Problem 14. Show the hardware that implements the following statement. Include the logic gates for the control function and a block diagram for the binary counter with a count enable input.

$$xyT_0 + T_1 + y'T_2: AR \leftarrow AR + 1$$

Solution:

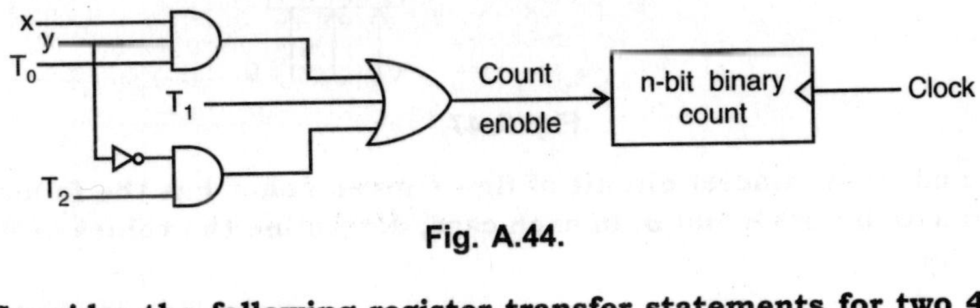

Fig. A.44.

Problem 15. Consider the following register transfer statements for two 4-bit register R_1 and R_2

$$xT : R_1 \leftarrow R_2$$
$$x^1T : R_1 \leftarrow R_2$$

Every time that variable T = 1, either the content of R_2 is added to the content of R_1 if x = 1, or the content of R_2 is transferred to R_1 if x = 0. Draw a diagram showing the hardware implementation of the two statements. Use block diagrams for the two. 4-bit registers, 4-bit adder and a quotable 2-to-1 line multiplexer that selects the inputs to R_1. In the diagram, show how the control variables x and T select the inputs of the multiplexers and the load input of register R_1.

Solution:

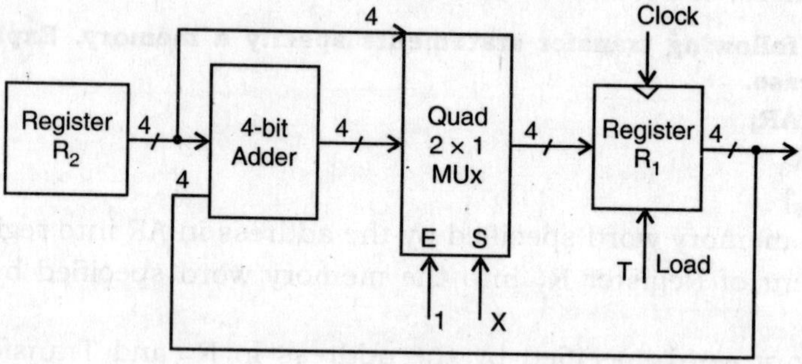

Fig. A.45.

Problem 16. Using a 4-bit counter with parallel load and a 4-bit adder as in figure given below, draw a block diagram that shows how to implement the following statements

$$x : R_1 \leftarrow R_1 + R_2 \quad \text{Add } R_2 \text{ to } R_1$$
$$x^1y : R_1 \leftarrow R_1 + 1 \quad \text{Increment } R_1$$

Where R_1 is a counter with parallel load and R_2 is a 4-bit register.

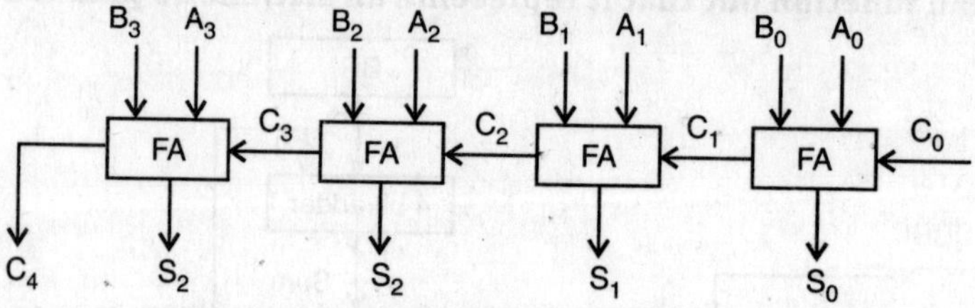

Fig. A.46.

Solution:

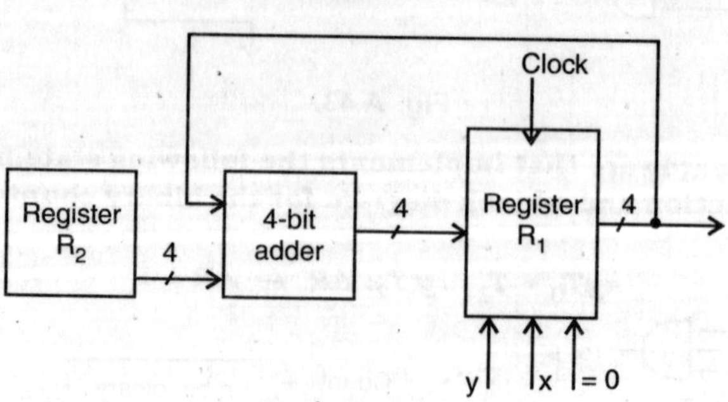

Fig. A.47.

Problem 17. The adder-subtracter circuit of figure given below has the following values for input mode M and data inputs A and B. In each case, determine the values of the outputs: S_3, S_2, S_2, S_1, S_0 and C4

	M	A	B
(a)	0	0111	0110
(b)	0	1000	1001
(c)	1	1100	1000
(d)	1	0101	1010
(e)	1	0000	0001

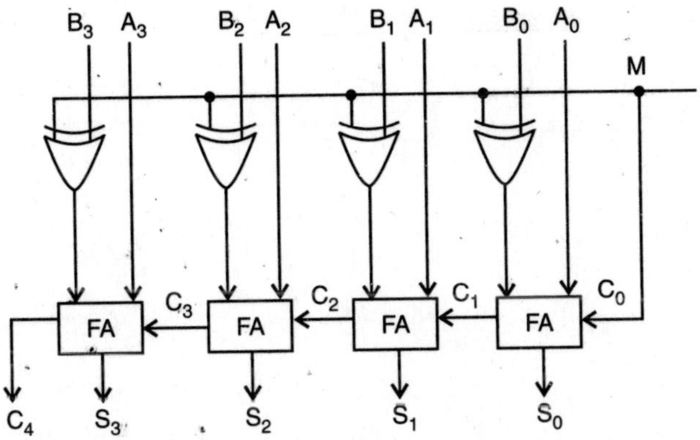

Fig. A.48.

Solution.

M	A	B	Sum	C_4
0	0111	0110	1101	0
0	1000	1001	0001	1
1	1100	1000	0100	1
1	0101	1010	1011	0
1	0000	0001	1111	0

7 + 6 = 13
8 + 9 = 16 + 1
12 – 8 = 4
5 – 10 = – 5 (in 2's complement)
0 – 1 = – 1 (in 2's complement)

Problem 18. Design a 4-bit combinational circuit decrements using four full adder circuits.

Solution: A 's complement of 1 = P + 1111

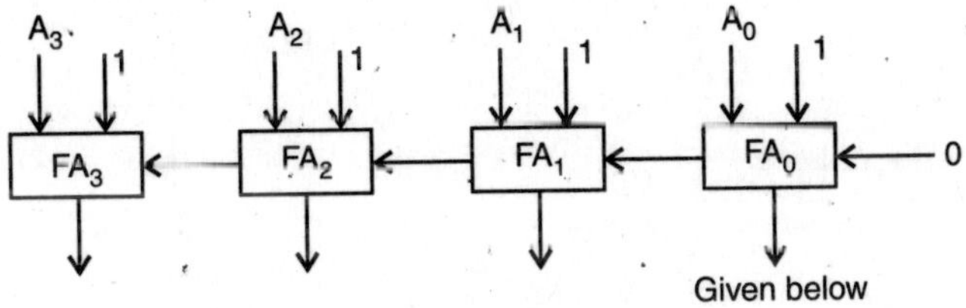

Fig. A.49.

□□□

INDEX

INDEX

□□□

OTHER DETAILS

(i) Who recommended you this book ? (Please tick (✓) in the box near the option relevant to you.

☐ Teacher ☐ Friends ☐ Bookseller

....

....

....

....

(ii) Name of the recommending teacher, his designation and address :

....

....

....

....

(iii) Name and address of the bookseller you purchased the book from :

....

....

....

....

(iv) Name and address of your institution (Please mention the University or Board as the case may be)

....

....

....

....

(v) Your name and complete postal address :

....

....

....

....

(vi) Course you are enrolled in :

....

....

....

....

The best assessment will be awarded each month: The award will be in the form of our publications as decided by the Editorial Board.

Please mail the filled up coupon at your earliest to :

S.K.KATARIA & SONS®
4760-61/23, Ansari Road,
Darya Ganj, New Delhi-110002
☎: 23243489, 23269384
Email : katariabooks@yahoo.com
katson_sanjeev@yahoo.co.in